FUNDAMENTAL INTERACTIONS IN PHYSICS AND ASTROPHYSICS

Studies in the Natural Sciences

A Series from the Center for Theoretical Studies

University of Miami, Coral Gables, Florida

Volume 1 – IMPACT OF BASIC RESEARCH ON TECHNOLOGY
Edited by Behram Kursunoglu and Arnold Perlmutter • 1973

Volume 2 – FUNDAMENTAL INTERACTIONS IN PHYSICS
Edited by Behram Kursunoglu, Arnold Perlmutter, Steven M. Brown, Mou-Shan Chen, T. Patrick Coleman, Werner Eissner, Joseph Hubbard, Chun-Chian Lu, Stephan L. Mintz, and Mario Rasetti • 1973

Volume 3 – FUNDAMENTAL INTERACTIONS IN PHYSICS AND ASTROPHYSICS
Edited by Behram Kursunoglu, Steven Weinberg, Arthur S. Wightman, Geoffrey Iverson, Arnold Perlmutter, and Stephan Mintz • 1973

FUNDAMENTAL INTERACTIONS IN PHYSICS AND ASTROPHYSICS

A Volume Dedicated to P.A.M. Dirac on the Occasion of his Seventieth Birthday

Conference Committee

Behram Kursunoglu (Chairman)
Steven Weinberg
Arthur S. Wightman

Editors

Geoffrey Iverson
Arnold Perlmutter
Stephan Mintz

Center for Theoretical Studies
University of Miami
Coral Gables, Florida

PLENUM PRESS • NEW YORK-LONDON

Library of Congress Cataloging in Publication Data

Coral Gables Conference on Fundamental Interactions at High Energy, 9th, University of Miami, 1972.
Fundamental interactions in physics and astrophysics.

(Studies in the natural sciences, v. 3)
"A volume dedicated to P. A. M. Dirac on the occasion of his seventieth birthday."
Sponsored by the Center for Theoretical Studies, University of Miami.
Includes bibliographical references.
1. Nuclear reactions—Congresses. 2. Dirac, Paul Adrien Maurice, 1902- —Congresses. I. Iverson, Geoffrey J., ed. II. Perlmutter, Arnold, 1928- ed. III. Mintz, Stephan, ed. IV. Dirac, Paul Adrien Maurice, 1902- V. Miami, University of, Coral Gables, Fla. Center for Theoretical Studies. VI. Title. VII. Series.

QC793.9.C68 1972 539.7'6 73-18315
ISBN 0-306-36903-6

Lectures from the 1972 Coral Gables Conference on Fundamental Interactions at High Energy, January 19-21

A Division of Plenum Publishing Corporation
227 West 17th Street, New York, N.Y. 10011

United Kingdom edition published by Plenum Press, London
A Division of Plenum Publishing Company, Ltd.
Davis House (4th Floor), 8 Scrubs Lane, Harlesden, London, NW10 6SE, England

Printed in the United States of America

COUNCIL DEDICATION

During the 1972 meeting of the Scientific Council of the Center on Tuesday, the 18th of January, the Council has passed a motion to dedicate the ninth annual conference on Fundamental Interactions at High Energy to one of the greatest men of science of all time, Professor Paul Adrien Maurice Dirac, on the occasion of his 70th birthday. The Council takes the greatest pride and deepest joy in communicating this tribute to the participants of this conference. Because of this, these proceedings will gain particular significance in bearing this tribute to this great man. Professor Dirac, we are certain that all the participants and our many colleagues who are not here join us in looking forward to your acceptance also of the dedication of our 1982 conference.

SCIENTIFIC COUNCIL MEMBERS

Gerald Edelman
Maurice Goldhaber
Foy Kohler
Robert Mulliken
Lars Onsager
Julian Schwinger
Edward Teller
Behram Kursunoglu

P. A. M. DIRAC

PREFACE

The present volume is a compilation of the talks presented at the 1972 Coral Gables Conference on Fundamental Interactions at High Energy held at the University of Miami by the Center for Theoretical Studies. The volume contains, in addition, contributions by B. Kursunoglu and G. Breit, which were not actually presented, but are included as tributes to Professor P.A.M. Dirac, to whom the Conference is formally dedicated.

Again this year the theme, style and format of each session was in most cases the responsibility of the section leaders who also constituted the Conference Committee. This organization of the conference meant that each section was coherent and essentially self-contained, and as well, allowed for spirited panel discussions to critically summarize, and to indicate new directions for future research.

This volume is divided into four sections on Constructive Field Theory, and Advances in the Theory of Weak and Electromagnetic Interactions, Cosmic Evolution,and New Vistas in the Theory of Fundamental Interactions. Each section represents a thorough, penetrating survey of one of the most active research programs of theoretical physics.

Thanks are due to typists Mrs. Helga S. Billings, Mrs. Jackie Zagursky, Miss Connie Retting, and to Mrs. Norma Gayle Hagan for her industrious supervision of the many programs involved in the conference.

This conference received some support from the Atomic Energy Commission and the National Aeronautics and Space Administration.

The Editors

Miss Sevil Kursunoglu, Professor Behram Kursunoglu, Professor P. A. M. Dirac, Mrs. Behram Kursunoglu, Mr. Maurice Gusman, Mrs. P. A. M. Dirac, and Mrs. Edward Teller (left to right) at the 1972 Coral Gables Conference on Fundamental Interactions of the Center for Theoretical Studies.

CONTENTS

Section One

CONSTRUCTIVE FIELD THEORY

Section Two

ADVANCES IN THE THEORY OF WEAK AND ELECTROMAGNETIC INTERACTIONS

Section Three
COSMIC EVOLUTION

Section Four
NEW VISTAS IN THE THEORY OF FUNDAMENTAL INTERACTIONS

CONSTRUCTIVE FIELD THEORY
INTRODUCTION TO THE PROBLEMS

A.S. Wightman*
Princeton University
Princeton, New Jersey

Introductory Remarks.

Over the last decade, a new branch of quantum field theory has been created, constructive quantum field theory. This theory has a feature in common with so-called axiomatic field theory: in it a result is a result only if given a precise mathematical statement and proof. Barring human fallibility, its results are therefore truth. (Their significance for physics is something that each person has to judge for himself.)

Constructive quantum field theory differs from axiomatic field theory in that it attempts to construct the solutions of specific concrete model theories, typically simplified analogues of self-coupled meson theories or Yukawa theories of meson-baryon couplings. On the other hand, axiomatic field theory customarily attempts to make statements about all theories satisfying certain quite general assumptions.

Although in the early 1960's axiomatic theory was a very active subject, by 1970 most work on the mathematical foundations of quantum field theory was constructive quantum field theory. It is easy to understand this shift in emphasis. By the mid 1960's most of the easy general results of the general theory of quantized fields seemed to have been found, and no general structure theory of fields had emerged. It seemed necessary to study special cases more deeply to get the insight that would make possible a general theory. This

*Supported in part by AFOSR

evolution is a typical example of the interplay of general theory and the study of specific examples in the development of a deep and complicated subject.

Of course, the idea of using explicitly soluble models as a guide for general theory has been a recurring theme in quantum field theory. The Lee and Thirring models immediately come to mind. However, the models under discussion are ones for which it seems very unlikely that an explicit solution will ever be found. One has to attack them more indirectly.

At this point in the development of the subject, most of the work has been done on super-renormalizable theories, the most detailed results having been obtained in two dimensional space-time. In this review we will confine ourselves almost entirely to that case.

Two archetypical models which will be discussed in detail in the following by L. Rosen and R. Schrader are $\mathcal{P}(\phi)_2$ and Y_2. These symbols are shorthand for models with the following formal Lagrangean densities and Hamiltonians

$$\mathcal{L} = 1/2(\partial^\mu\phi\ \partial_\mu\phi - m_0^2\ \phi^2) - \mathcal{P}(\phi) \tag{1}$$

$\boxed{\mathcal{P}(\phi)_2}$

$$H_{un} = \int[1/2(\pi^2 + (\nabla\phi)^2 + m_0^2\ \phi^2) + \mathcal{P}(\phi)]\ dx \tag{2}$$

Here $\mathcal{P}$ is a real polynomial bounded below i.e. it is of even degree with a positive coefficient for the largest term. ϕ is a hermitean scalar field.

$$\mathcal{L} = \bar{\psi}(i\gamma^\mu\partial_\mu - M)\psi - \lambda\bar{\psi}\psi\phi + 1/2(\partial^\mu\phi\ \partial_\mu\phi - m_0^2\ \phi^2) \tag{3}$$

$\boxed{Y_2}$

$$H_{un} = \int dx[1/2(\pi^2 + (\nabla\phi)^2 + m_0^2\ \phi^2) + \bar{\psi}(-i\gamma^1\partial_1 + M)\psi + \lambda\bar{\psi}\psi\phi] \tag{4}$$

ϕ is again a hermitean scalar field. ψ is a two component (spinor) field.

$$\bar{\psi} = \psi^{+}\gamma^{0} \qquad \gamma^{0} = \begin{pmatrix} 0 & 1 \\ 1 & 0 \end{pmatrix} \qquad \gamma^{1} = \begin{pmatrix} 0 & 1 \\ -1 & 0 \end{pmatrix} \quad . \tag{5}$$

The subscript 2 on $\mathcal{P}(\phi)_2$ and Y_2 refers to two dimensions of space-time. The Y is for Yukawa a current notation which will have to be elaborated when Yukawa theories with other than scalar mesons and scalar coupling are considered as they have not been up to now. The subscript un stands for unrenormalized. At the present stage of the exposition these expressions for Hamiltonians should be regarded as purely formal manipulations on the classical Lagrangeans.

A. OBJECTIVES

The first main point of this introduction is an answer to the question: what does it mean to "solve" a theory. Of course, one can answer by saying one wants to construct a field satisfying the usual axioms of quantum field theory that somehow solves the equations of the model, but in fact, one wants much more than that. The objectives will be listed in a somewhat arbitrary order. The first is to prove the existence of the

1. Renormalized Hamiltonian: H_{ren} in the space of physical states. How one recognizes that it is the renormalized Hamiltonian for the somewhat illdefined model one starts out from is a question about which much will be said later.

2. Positivity of the Energy: $H_{ren} \geq 0$ or more generally the relativistic spectral condition

$$H_{ren}^2 - \vec{P}^2 \geq 0 \quad . \tag{6}$$

3. Vacuum: $H_{ren}\Omega = 0, \ \vec{P}\Omega = 0$ (7)

4. Fields: The basic fields of the theory ought to be constructed so that they transform under space and time translations. For example, for a hermitean scalar field ϕ

$$\exp(iH_{ren}t)\phi(f)\exp(-iH_{ren}t) = \phi(\{t,0,1\}f) \tag{8}$$

$$\exp(-i\vec{P}.\vec{a})\phi(f)\exp(i\vec{P}.\vec{a}) = \phi(\{0,\vec{a},1\}f) \quad . \tag{9}$$

Here

$$(\{t,\vec{a},1\}f)(x^0,\vec{x}) = f(x^0-t,\vec{x}-\vec{a}) \quad . \tag{10}$$

The fields ought to be local

$$[\phi(f),\phi(g)] = 0 \tag{11}$$

if the supports of f and g are space-like separated. This reflects the relativistic propagation of influence.

5. Lorentz Invariance: The invariance of the theory under the Poincaré group requires the existence of a continuous unitary representation of the Poincaré group in the space of physical states satisfying

$$U(a_1,\Lambda_1)U(a_2,\Lambda_2) = U(a_1+\Lambda_1 a_2,\Lambda_1\Lambda_2) \tag{12}$$

where $a_{1,2}$ are space-time translations and $\Lambda_{1,2}$ are Lorentz transformations. This implies in addition to the existence of H_{ren} as the infinitesimal time translation operator and $\vec{P}$ as the infinitesimal space translation operator, the existence of a boost operator $\vec{N}$, the infinitesimal operator of pure Lorentz transformations (≡ boosts).

The vacuum should be invariant under $U(a,\Lambda)$

$$U(a,\Lambda)\Omega = \Omega \tag{13}$$

and the field should transform, for a scalar field

$$U(a,\Lambda)\phi(f)U(a,\Lambda)^{-1} = \phi(\{a,\Lambda\}f) \tag{14}$$

where

$$(\{a,\Lambda\}f)(x) = f(\Lambda^{-1}(x-a)) \tag{15}$$

6. Vacuum Expectation Values and Green's Functions; Their Temperedness and Lorentz Invariance:

To make sense of the vacuum expectation value

$$(\Omega,\phi(f_1)\ldots\phi(f_n)\Omega) \tag{16}$$

it suffices to prove that $\phi(f)$ can be defined on a domain such that

$$\phi(f)D\subset D \tag{17}$$

i.e. such that applying $\phi(f)$ to a vector in the domain, D, yields another vector in D, and to prove that

$$\Omega\varepsilon D\ . \tag{18}$$

To be sure that (16) is a multilinear functional in $f_1\ldots f_n$ i.e. that it is linear in each f_j with the others held fixed, it suffices to show that $\phi(f)\Psi$ is linear in f for each fixed $\Psi\ \varepsilon\ D$. If it can be further shown that for each fixed pair of vectors $\Phi,\Psi\ \varepsilon\ D$

$$(\Phi,\phi(f)\Psi) \tag{19}$$

is continuous in f in the sense of the theory of distributions, then a general theorem (the kernel or nuclear theorem[20]), assures us that there is a unique distribution F_n in all variables $x_1\ldots x_n$ together, defined for test functions $f(x_1\ldots x_n)$ which gives meaning to

$$F_n(f) = \int\ldots\int dx_1\ldots dx_n\ f(x_1\ldots x_n)(\Omega,\phi(x_1)\ldots\phi(x_n)\Omega) \tag{20}$$

and reduces to (16) when f is a product $f(x_1\ldots x_n) = \prod_{j=1}^{n} f_j(x_j)$. If (19) is a tempered distribution so is (20).

Finally, the Lorentz invariance of F_n

$$F_n(f) = F_n(\{a,\Lambda\}f) \tag{21}$$

with

$$(\{a,\Lambda\}f)(x_1 \ldots x_n) = f(\Lambda^{-1}(x_1 - a), \ldots \Lambda^{-1}(x_n - a)) \tag{22}$$

follows from the transformation laws (13) and (14), if it can be established that D is invariant under the representation of the Poincaré group

$$U(a,\Lambda)D \subset D \quad . \tag{23}$$

All these arguments for vacuum expectation values are standard in the general theory of quantized fields (see, for example,[21]). For Green's functions, i.e. time ordered vacuum expectation values, further more detailed discussion is necessary. The most elementary definition of the n-point Green's function for a scalar field uses the formula

$$G_n(x_1, x_2, \ldots x_n) = \sum_{\text{Perm}} \Theta(x^0_{i_1} - x^0_{i_2})\Theta(x^0_{i_2} - x^0_{i_3}) \ldots \Theta(x^0_{i_{n-1}} - x^0_{i_n}) \tag{24}$$

$$(\Omega, \phi(x_{i_1})\phi(x_{i_2}) \ldots \phi(x_{i_n})\Omega)$$

smeared with test functions, $g_1 \ldots g_n$, in space alone. For this definition to be unambiguous it suffices if the space-smeared vacuum expectation values

$$(\Omega, \phi(x^0_1, g_1) \ldots \phi(x^0_n, g_n)\Omega) \tag{25}$$

are functions of $t_1 \ldots t_n$ and not generalized functions near $t_j = t_k$, $j \neq k$ i.e. (25) should contain no singular functions like $\delta(t_1 - t_2)$ $\delta'(t_3 - t_4)$ etc. Of course, if such singularities are present it does not mean that a reasonable definition of Green's function cannot be found, but only that (24) would not have an unambiguous meaning as it stands. Until recently there were no convenient conditions on the operators $\phi(t,g)$ which guaranteed the existence of the Green's functions. An important contribution of Nelson seems to offer a reasonable definition under quite general assumptions[22].

As far as the Lorentz invariance of the Green's functions

is concerned, it does not follow, in general, from the Lorentz invariance of the vacuum expectation values. It requires a further analysis of singularities. If the vacuum expectation values contain vector rather than scalar fields it is notorious that the time-ordered vacuum expectation values are not tensors but differ from them by non-covariant terms involving equal time commutators. Thus to get to covariant Green's functions for such fields one has to subtract from the analogue of the definition certain non-covariant quantities formed from vacuum expectation values. It seems to be still an open problem to organize these phenomena in a precise mathematical way. Thus, when it comes to the construction of covariant Green's functions in the models it is likely that special arguments based on the detailed behavior of the model will be necessary.

There is one peculiar feature of two-dimensional space-time which should be noted. Because the homogeneous Lorentz group has a continuous family of one-dimensional representations it can happen that the dynamics causes the transformation law of a field to vary with the coupling constant. This was first pointed out by Dubin[23], I believe.

7. Currents; Heisenberg Picture Equations of Motion:

The formal equations of motion that would follow from the Lagrangeans (1) and (3) are respectively

$$(\Box + m^2)\phi = -\mathcal{P}'(\phi) \tag{26}$$

and

$$(\Box + m^2) \quad = -\lambda\bar{\psi}\psi \tag{27}$$

$$(i\gamma^\mu\partial_\mu - M)\psi = \lambda\phi\psi \quad . \tag{28}$$

The right-hand sides of these equations will be for brevity called currents; they are ill-defined as they stand. It is an important part of the theory of a model to give well-defined expressions for the currents in terms of ϕ, ψ and $\bar{\psi}$, in such a way that the resulting equations are valid as operator identities, the Heisenberg picture field equations, when the fields have transformation laws under time and space translation given by (8) and (9) for ϕ (and analogous equations for ψ and $\bar{\psi}$.)

As we will see corrections to the masses of particles arising from the interaction are implicit in the prescription for the currents.

8. Uniqueness of the Vacuum; Dynamical Instability; Broken Symmetry:

The phrase "uniqueness of the vacuum" can mean quite different things in different contexts. In one meaning standard in axiomatic field theory, it requires that only one vector exist satisfying the invariance property

$$U(a,\Lambda)\Omega = \Omega$$

under the representation of the Poincaré group. On the other hand, "uniqueness of the vacuum" can be taken to mean that starting from a heuristically formulated Hamiltonian such as (2) or (4), there is an essentially unique theory for each choice of coupling constants (including bare masses as coupling constants). The phrase "essentially unique" is used to mean unique up to a certain kind of unitary equivalence. Namely two theories labeled

Hilbert Space	$\mathcal{H}\,1$	$\mathcal{H}\,2$
Representation of Poincaré	U_1	U_2
Vacuum	Ω_1	Ω_2
Field	ϕ_1	ϕ_2

are equivalent if there exists a unitary mapping V of the $\mathcal{H}_1$ onto $\mathcal{H}_2$ such that

$$U_1 = V^{-1}U_2V \qquad \Omega_2 = V\Omega_1$$
$$\phi_1 = V^{-1}\phi_2V \quad . \tag{29}$$

When there are two or more inequivalent theories for

given coupling constants and bare masses, the construction procedures to be described shortly may yield a vacuum state which is not pure; it is a density matrix with non-vanishing probabilities of different pure vacuum states belonging to inequivalent theories. When this happens it is customary to say that the vacuum of these theories is degenerate or non-unique. I would prefer to use the terminology introduced by Nambu and Jona-Lasinio and say the theories are dynamically unstable. The theory defined by the density matrix is then a mixed theory and the theories into which it decomposes are pure theories. In each of the pure theories the vacuum is unique in the first sense mentioned above.

The preceding discussion has to be modified slightly when, as in Y_2, infinite renormalizations are necessary. Then one expects to parametrize the theory using renormalized constants in place of bare constants whenever renormalization is necessary.

Dynamical instability may or may not be accompanied by broken symmetry. When it is, the various pure theories lack an invariance group possessed by the Lagrangean. The action of the invariance group then carries one pure theory into another rather than a pure theory into itself. Later in this section some circumstances under which dynamical instability and/or broken symmetry may be expected in the models under discussion will be described.

In Figure 1, I have repeated a schematic description of these phenomena which makes them easy to remember.

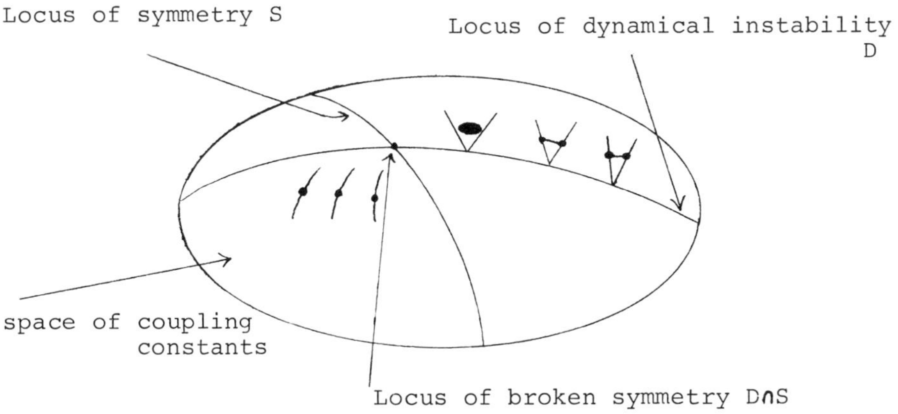

Fig. 1 The space of theories is a bundle of vacuum states over the space of coupling constants. A theory is determined by a pair consisting of a point in coupling-constant space and a vacuum state. For a general value of the coupling constants there is a unique theory whose vacuum state is indicated on the diagram by a bristle (one dimensional) attached at the point. For those special values of the coupling constants lying on the locus of dynamical instability D the theory is not uniquely determined by the values of the coupling constants. Then the cone of vacuum states has two or more dimensions. (In the diagram the normalized vacuum states have been indicated by spots on the bristles, line segments on the two-dimensional cones and by a disc on the three-dimensional cones.)

9. Existence of Single Particle States; Mass Gap and Goldstone Bosons:

For the purposes of the present discussion any state of definite mass which is not a vacuum state is a single particle state. (This is the usage of the general theory of collisions where any stable subsystem which can appear as a constituent of a colliding beam is regarded as a single particle.) Once the question of dynamical stability or instability is settled for a theory, the next natural question is the structure of the low lying mass states, and in particular the existence of single particle states. The simplest situation from the point of view of existing general theory is one in which there is a mass gap: there are no zero mass particles, so that the spectrum would look as in Figure 2 for the case of just one single particle state. As we will see this situation is possible even when the theory is dynamically unstable.

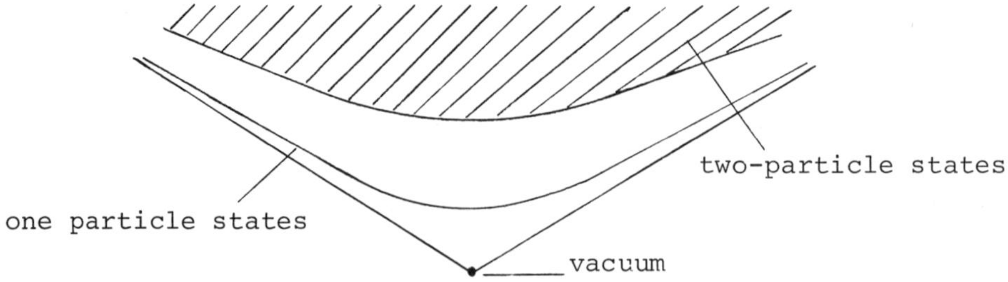

Fig. 2 Energy Momentum Spectrum for a Theory with one Stable Particle.

However, there are also situations in which dynamical instability implies the existence of zero mass particles and, therefore, the absence of a mass gap. That is the case first pointed out by Goldstone when there is breaking of a continuous internal symmetry group[24]; the zero mass particles are Goldstone bosons.

Although the notation we are using does not distinugish composite from elementary particles, it is easy to see how distinctions might arise in the theory which would lead one to regard some particles as composite. The essential point is to consider the coupling constant dependence of the single particle masses. Suppose that when an appropriate subset of the coupling constants becomes small some of the masses disappear into continua. Then it would be natural to regard those single particle states as composites of the remaining single particle states. Roughly speaking, one would say that the composite particles are produced by the interaction of the remaining particles, when the interaction is sufficiently strong. This situation would be somewhat analogous to n-body Schrödinger theory where strengthening an interaction potential can produce new bound states. Of course, in Schrödinger theory one has uniquely identifiable elementary particles making up the bound states in contrast to the field theory case where no such absolute identification seems possible. In fact, it is quite possible that by making a different set of coupling constants small one would be led to regard a different subset of particles as composite.

10. Asymptotic Completeness:

If it can be established that a theory has a mass gap, and a non-empty set of single particle states, then the general theory of collisions comes into operation[25] (Haag-Ruelle, Jost book). It shows that there are collision states corresponding to any number of single particles in the incoming beams. The remaining big question is then asymptotic completeness: is every state a superposition of collision states? The unitarity of the S matrix is then an immediate consequence of asymptotic completeness.

It should be said that the question of asymptotic completeness is not a trivial problem at all. Evidence for this statement is the fact that in the Schrödinger n-body problem asymptotic completeness has not been established for $n \geq 4$ even for attractive Yukawa forces.

Despite a voluminous literature the theory of collisions in the presence of massless particles has not reached a stage where general statements can be made analogous to those just mentioned. A substitute for the S-matrix has not yet been found. Thus at the moment no analogue of the problem of proving the unitarity of the S-matrix can be stated.

11. Validity of Perturbation Theory:

The renormalized perturbation series for the S-matrix elements and Green's functions of Lagrangean field theories are the core of the existing general wisdom about field theory. Hence it is not unreasonable to regard a field theory as the solution of a given heuristically formulated Lagrangean theory if it gives the same renormalized perturbation series.

To determine the perturbation series of a Green's function of a theory one has to show that the Green's function is differentiable to all orders in the coupling constant in some interval which has the origin either in its interior or as an end point. The terms of the perturbation series are then just the derivatives ($\times \frac{(\text{coupling constant})^n}{n!}$) evaluated either at the origin or evaluated as limits as the coupling constant approaches zero. To show that all such derivatives exist is showing that the Green's function has an asymptotic series in the coupling constant.

Once this has been established it is natural to ask whether the exact solution can be recovered from the perturbation series. If the Green's function is analytic in the coupling constant at zero coupling, the series converges to the Green's function in some interval, but there is good evidence that analyticity does not hold. Thus the most that one can hope for is that some summability procedure will enable one to recover the Green's function from the series. Barry Simon will describe to you how this can be done in some detail.

These problems having to do with the perturbation series are special cases of a more general question. Where are the Green's functions analytic as a function of the coupling constant and what is the nature of their singularities? In particular, can it happen as it does in the theory of phase transitions that there is a singularity for a real coupling constant such that the theory above and below are qualitatively different. As you will hear from Barry Simon this phenomenon cannot occur in $\mathcal{P}(\phi)_2$ until one removes the space cutoff, but in Y_2 there is the possibility that the vacuum becomes degenerate at some particular value of the coupling constant even with a space cutoff. Another related question: when two couplings are present how is the perturbation series for one affected by the presence of the other? An answer to this question might resolve some of the apparent paradoxes which arise when the electromagnetic mass splittings of particles multiplets are calculated in the presence of strong coupling using the electromagnetic interaction as a perturbation.

In addition to the questions involving analyticity in the coupling constant there are questions about the analyticity of Green's functions in momentum variables. To what extent are the analyticity properties in the momenta of the terms of the perturbation series a reliable guide to the analyticity properties of the Green's functions themselves? There is some evidence that natural boundaries occur in the latter which do not appear in the former[26]. If that is the case, it is a fact to be reckoned with in dispersion theory and analytic S-matrix theory.

That is an imposing list of objectives but what makes the present prospects of field theory so interesting is that, as a result of the work of Glimm, Jaffe, et al. it is not unreasonably optimistic to think that many of them can be reached in the next few years at least for super-renormalizable theories.

B. SOME MATHEMATICAL BACKGROUND

The purpose of this section is to review a few technical points which could, if they were not understood, make difficult an understanding of what follows. The reader who has had a course in methods of modern mathematical physics[27,28] should pass directly to C.

Forms and Operators

The various operators appearing in the models are usually presented not as operators in the precise mathematical sense defined below but as forms. A sesquilinear form (sometimes called bilinear form) with domain $D_1 \times D_2$ is a complex valued function F of two variables: $F(\Phi,\Psi)$ defined for $\Phi \in D_1$ $\Psi \in D_2$, linear in Ψ and anti-linear in Φ:

$$\begin{aligned} F(\Phi,\alpha\Psi_1 + \beta\Psi_2) &= \alpha F(\Phi,\Psi_1) + \beta F(\Phi,\Psi) \\ F(\alpha\Phi_1 + \beta\Phi_2,\Psi) &= \bar{\alpha}F(\Phi_1,\Psi) + \bar{\beta}F(\Phi_2,\Psi) \ . \end{aligned} \tag{30}$$

Here D_1 and D_2 are supposed to be dense linear subsets of the Hilbert space of states, $\mathcal{H}$.

Particular examples of sesqui-linear forms are provided by operators. If A is an operator defined on D_2

$$F(\Phi,\Psi) = (\Phi,A\Psi) \tag{31}$$

is a sesquilinear form, and so is

$$F(\Phi,\Psi) = (B\Phi,\Psi) \tag{32}$$

if B is an operator defined on D_1. By Schwarz' Inequality, forms of this special kind satisfy

$$|F(\Phi,\Psi)| \leq ||\Phi|| \ ||A\Phi|| \tag{33}$$

and

$$|F(\Phi,\Psi)| \leq ||B\Phi|| \ ||\Psi|| \tag{34}$$

respectively. Thus in the first case F is continuous in its first argument and can be extended by continuity from $D_1 \times D_2$ to $\mathcal{H} \times D_2$. Similarly, in the second argument. A converse holds

Lemma

A sesqui-linear form F with domain $D_1 \times D_2$ arises from an operator, A, defined on D_2, according to (31) if for each $\Psi \in D_2$ there exists a constant $C(\Psi)$ such that

$$|F(\Phi,\Psi)| \leq C(\Psi)||\Phi|| \tag{35}$$

for all $\Phi \in D_1$. Similarly if for each $\Phi \in D_1$ there exists a constant $D(\Phi)$ such that

$$|F(\Phi,\Psi)| \leq D(\Phi)||\Psi|| \tag{36}$$

for all $\Psi \in D_2$, then F is of the form (32). The proof is left to the reader.

Examples (constructed with annihilation operators a(k))

$$F(\Phi,\Psi) = (\Phi,a(k)\Psi) \tag{37}$$

$$F(\Phi,\Psi) = (a(k)\Phi,\Psi) \tag{38}$$

$$F(\Phi,\Psi) = (a(k)\Phi,a(k)\Psi) \quad . \tag{39}$$

All three are definable on dense sets of vectors Φ,Ψ. (37) satisfies (35) but not (36), (38) satisfies (36) but not (35), (39 satisfies neither (35) nor (36).

As a consequence of the Lemma an annihilation operator, a(k), is a densely defined operator, even without being smeared with a test function of k. The same holds for products $a(k_1)\ldots a(k_n)$. On the other hand, the same does not hold for creation operators. The rule defining Wick ordering: bring all annihilation operators to the right and creation operators to the left, exploits these facts. The creation operators on the left can be converted to annihilation operators when one computes a form.

The Wick ordered powers in a field $:\phi^n:(x)$ satisfy neither of the continuity restrictions (35) and (36). However, once they are smeared with a test function in space decreasing rapidly at infinity they satisfy both on dense domains and therefore define operators. (True in two but not larger dimensional space-time!) This provides a starting point for the discussion of the operator properties of expressions like

$$\int dx\ g(x):\phi^n:(x) \quad . \tag{40}$$

It is natural to define hermiticity for a form by mimicking the definition

$$(\Phi,A\Psi) = \overline{(\Psi,A\Phi)}$$

for an operator. A form F is <u>hermitean</u> on a domain D x D if

$$F(\Phi,\Psi) = \overline{F(\Psi,\Phi)}$$

for all $\Phi,\Psi \in D$. (40) turns out to be hermitean in this sense on a suitable dense domain.

Now I turn to the discussion of operators. An operator defined on all of $\mathcal{H}$ is <u>bounded</u> if its norm is finite

$$||A|| = \sup_{\Phi\neq 0} \frac{||A\Phi||}{||\Phi||} < \infty \quad . \tag{41}$$

It is elementary to show that $||A||$ can also be defined by

$$||A|| = \sup_{\Phi,\Psi\neq 0} \frac{|(\Phi,A\Psi)|}{||\Phi||\ ||\Psi||} \quad . \tag{42}$$

Thus, a bounded operator has bounded expectation values in all normalized states. Since many quantities in physics are unbounded, for example the energy, we will have much to do with unbounded as well as bounded operators.

If an operator is defined on a dense set of vectors, D(A) the <u>domain</u> of A, and satisfies

$$||A\Phi|| \leq C\ ||\Phi|| \tag{43}$$

for all $\Phi \in D(A)$ with a fixed constant C then A can be extended by continuity to all of $\mathcal{H}$ and so extended is a bounded operator with bound $\leq$ C. Thus unbounded operators must fail to have the continuity property (43), and in all the cases we consider will not be defined for every vector in $\mathcal{H}$.

In the detailed arguments of constructive quantum field theory, one has constantly to deal with the operations of adjoining elements to the domain of an unbounded operator or deleting elements from that domain. For that reason the notion of extension and restriction is introduced. An operator B whose domain is D(B) is an <u>extension</u> of an operator A whose domain is D(A) if

$$D(B) \supset D(A) \tag{44}$$

and

$$B\Phi = A\Phi \tag{45}$$

for all $\Phi \in D(A)$ i.e. if B is defined on every vector that A is, and they agree wherever both are defined. A is called a <u>restriction</u> of B and one writes $A \subset B$ or $B \supset A$ and $A = B\restriction D(A)$ (read A is B restricted to the domain of A).

The notion of extension is expressed in transparent geometrical form if one introduces the idea of the graph of a transformation. Let A be a transformation from a Hilbert space $\mathcal{H}_1$ to a Hilbert space $\mathcal{H}_2$. (We admit the possibility $\mathcal{H}_1 \neq \mathcal{H}_2$ temporarily for clarity in what follows.) The pairs $\{\Phi, A\Phi\}$, $\Phi \in D_1(A)$ are vectors in the direct sum, $\mathcal{H}_1 \oplus \mathcal{H}_2$, of the two Hilbert spaces. The set of all such pairs with Φ running over D(A) is called the <u>graph</u>, $\Gamma(A)$, of A. It is a linear subset of $\mathcal{H}_1 \oplus \mathcal{H}_2$ since A is linear. The relation A B for two operators is expressible as

$$\Gamma(A) \subset \Gamma(B) \tag{46}$$

where $\subset$ in (46) means inclusion as sets.

Is every linear subset, Γ, of $\mathcal{H}_1 \oplus \mathcal{H}_2$ the graph of some linear transformation? No, only if $\{0, \Psi\} \in \Gamma$ implies $\Psi = 0$. This is a matter of importance if one considers the closure $\overline{\Gamma(A)}$ in $\mathcal{H}_1 \oplus \mathcal{H}_2$ of the graph, $\Gamma(A)$, of a transformation A. Is it the graph of a transformation? Not necessarily, but if so, the transformation is called $\bar{A}$ the closure of A. Clearly $A \subset \bar{A}$ if $\bar{A}$ exists. (An example of a densely defined operator without a closure is the unsmeared annihilation operator a(k), as will be evident from the discussion below.)

The notion of graph also permits one to define a notion of adjoint of a transformation succinctly and geometrically. Since A has been assumed to be a transformation for $\mathcal{H}_1$ to $\mathcal{H}_2$, A* turns out to be a transformation from $\mathcal{H}_2$ to $\mathcal{H}_1$. Of course, if $\mathcal{H}_1 = \mathcal{H}_2 = \mathcal{H}$ both transformations act within $\mathcal{H}$. Since $\Gamma(A)$ is a linear subset of $\mathcal{H}_1 \oplus \mathcal{H}_2$ so is its orthogonal complement, $\Gamma(A)^{\perp}$, the set of all vectors orthogonal to every vector of $\Gamma(A)$. If $\Gamma(A)^{\perp}$ is the graph of a transformation from $\mathcal{H}_2$ to $\mathcal{H}_1$, that transformation is defined to be -A*, where A* is the adjoint of A. In other words, since the pair $\{x,\Psi\} \in \Gamma(A)^{\perp}$ if for all $\Phi \in D(A)$

$$(x,\Phi) + (\Psi,A\Phi) = 0 \tag{47}$$

we have that $\Psi \in D(A^*)$ if there exists an x satisfying (47) for all $\Phi \in D(A)$, and then $x = -A^*\Psi$ so

$$-(A^*\Psi,\Phi) + (\Psi,A\Phi) = 0 \quad . \tag{48}$$

It is not difficult to see that if D(A) is dense in $\mathcal{H}_1$, $\Gamma(A)^{\perp}$ is the graph of a linear transformation so A* exists. Applying this condition to A* itself one gets that, if D(A*) is dense, the closure $\bar{A}$ of A exists and is A**.

An unbounded operator A mapping a Hilbert space $\mathcal{H}$ into itself is hermitean if

$$(\Phi,A\Psi) = (A\Phi,\Psi) \tag{49}$$

for every pair of vectors Φ,Ψ in its domain, which we assume to be dense and linear. Expressed in terms of the above notion of adjoint this says

$$A \subset A^* \quad \text{or} \quad \Gamma(A) \subset \Gamma(A^*) \quad . \tag{50}$$

Passing to orthogonal complements we have

$$\Gamma(A^*)^{\perp} \subset \Gamma(A)^{\perp} \quad \text{or} \quad A^{**} \subset A^* \tag{51}$$

so a hermitean operator is the restriction of its closure which in turn is the restriction of its adjoint

$$A \subset A^{**} \subset A^* = A^{***} \quad . \tag{52}$$

If $A = A^*$, A is said to be <u>self-adjoint</u>. If $A^{**} = A^*$, or what is the same thing A has a closure which is self-adjoint, then A is said to be <u>essentially self-adjoint.</u>

Notice that the same argument which leads from (50) to (51) yields that

$$A \subset B \tag{53}$$

implies

$$B^* \subset A^* \quad . \tag{54}$$

Thus, if a hermitean operator B is an extension of hermitean operator A

$$A \subset B \subset B^* \subset A^* \quad . \tag{55}$$

Thus, a self-adjoint operator possesses no hermitean extensions except itself.

In a typical practical case one will be given an operator, A, which is hermitean on some convenient domain. One can always extend it without losing hermiticity by passing to the closure, $A = A^{**}$. (This is a sort of analogue of extending a bounded operator by continuity.) Will it not then be true in all practical cases that the closure is self-adjoint so that $A^{**} = A^*$ i.e. will it not be so that in all practical cases the hermitean operators one meets are essentially self-adjoint? The answer is a resounding no. There are numerous non-pathological cases in physics in which a hermitean operator is not essentially self-adjoint either because it has <u>no</u> self-adjoint extensions or because it has <u>several</u> and additional boundary conditions are necessary to distinguish them. Usually, the hard part of the theory of a hermitean operator is precisely to prove that it has a unique self-adjoint extension, i.e. to prove that its closure is self-adjoint.

<u>Stone's Theorem and the Trotter Product Formula</u>

And what is so great about an operator being self-adjoint? There are several answers. First of all, it is the self-adjoint operators, A, that have purely real spectrum and resolvents, $[A-z]^{-1}$, which are bounded operators analytic in z except when z belongs to the spectrum. (For a hermitean but not self-adjoint operator A, vectors of the form $(A-z)\Phi$, z fixed, Φ running over the domain of A, can fail to be dense in $\mathcal{H}$ even for z off the real axis; such z are said to belong to the residual spectrum.) Second, when A is self-adjoint

$$U(t) = \exp(iAt)$$

is a unitary one parameter group continuous in t. Third, the converse of this statement holds (Stone's Theorem).

<u>Theorem</u>

Let $U(t)$ $-\infty < t < \infty$ be a one parameter continuous unitary group of operators i.e. a family which satisfies

$$U(0) = 1 \tag{56}$$

$$U(t)U(t') = U(t+t')$$

with $(\Phi, U(t)\Psi)$ continuous in t for each fixed pair Φ, Ψ of vectors of $\mathcal{H}$. Define an operator A by: $\Phi \in D(A)$ if

$$d/dt(U(t)\Phi)\Big|_{t=0}$$

exists, and then by definition

$$d/dt(U(t)\Phi)\Big|_{t=0} = iA\Phi \quad . \tag{57}$$

A so defined is self-adjoint and

$$U(t) = \exp(iAt) \quad . \qquad (58)$$

Stone's theorem shows that a continuous one parameter group is uniquely determined by its infinitessimal operator, A. The significance of Stone's theorem for quantum mechanics ought to be evident. It shows that a dynamics is uniquely specified by a self-adjoint Hamiltonian.

Let me give a concrete example of this kind of problem which appears in $\mathcal{P}(\phi)_2$. There the Hamiltonian in a box consists of two parts: the free part H_0 and the interaction parts $H_I(g)$. It is not difficult to treat these separately and prove them self-adjoint in respective domains $D(H_0)$ and $D(H_I(g))$. The sum $H(g) = H_0 + H_I(g)$ is evidently hermitean on the domain $D(H_0) \cap D(H_I(g))$. However, it is a decidedly non-trivial problem to prove H(g) self-adjoint or even essentially self-adjoint on this domain. That accounts for the hullabaloo raised by the paper[11] of Glimm and Jaffe in which one of several results was a proof of self-adjointness for the case $(\phi^4)_2$.

An important mathematical tool in coping with sums of operators such as $H_0 + H_I(g)$, a tool indispensable for the proof of the theorem of Segal discussed in section C, below, is the Trotter Product Formula. It permits one to express the unitary operator of a one parameter group $\exp it(\overline{A+B})$ whose infinitesimal generator is the closure of the sum A + B as the limit of a product whose factors are of the form exp it'A, exp it"B. The formula has been widely used in quantum mechanics for a long while. In the specific case of Schrödinger theory where A is the kinetic energy operator and B the potential, it is the operator form of the so-called Feynman-Kac formula used by Feynman in his treatment of history integrals[29]. For an elegant proof of the theorem (see para. 8 of Ref. 30).

Theorem (Trotter)

Let A and B be self-adjoint operators such that A + B is essentially self-adjoint on the domain $D(A) \cap D(B)$. Then

$$\exp[it(\overline{A+B})] = \operatorname*{s-lim}_{n\to\infty} [\exp(i\ t/n\ A)\exp(i\ t/n\ B)]^n \quad . \qquad (59)$$

Notice that the closure $(\overline{A+B})$ is self-adjoint so the left-hand side makes sense. (59) means

$$\lim ||\exp[it(\overline{A+B})]\Phi - [\exp(i\ t/n\ A)\exp(i\ t/n\ B)]^n\Phi|| = 0$$

for all $\Phi \in \mathcal{H}$ i.e. the limit in (59) is a strong limit.

The importance of the Trotter Product Formula is that it permits one to disentangle A and B and to exploit the commutativity properties of exp itA and exp itB with other operators separately. (For an example of its power see[32].)

Convergence of Operators especially Unbounded Self-Adjoint Operators

One of the most important technical devices in what follows is the approximation of a self-adjoint Hamiltonian by self-adjoint cut-off Hamiltonians or alternatively the definition of a self-adjoint Hamiltonian as a limit of self-adjoint cut-off Hamiltonians. To discuss this subject one needs to know what notion of limit is appropriate.

The question has a considerable variety of answers depending on the circumstances. To begin with let me recall the standard notions of convergence for bounded operators. A sequence of bounded operators B_n, n=1,2... converges weakly if for each pair of vectors $\Phi,\Psi \in \mathcal{H}$, the matrix elements $(\Phi,B_n\Psi)$ form a convergent sequence of complex numbers. The sequence B_n converges strongly if for each $\Phi \in \mathcal{H}$ the sequence of vectors, $B_n\Phi$, n=1,2,... converges strongly i.e. if for each $\varepsilon > 0$ there is an integer N such that

$$||B_n\Phi - B_m\Phi|| < \varepsilon \tag{60}$$

for all $n,m > N$. The sequence of operators B_n converges uniformly ($\equiv$ converges in norm) if for each $\varepsilon > 0$ there exists an integer N such that

$$||B_n - B_m|| < \varepsilon \tag{61}$$

for all $n,m > N$.

There is a fine point connected with strong convergence that has to be mentioned here, I regret to say, since an analogous point causes lots of trouble later on in the construction of the physical vacuum. A sequence of bounded operators B_n converging in the sense of (60) has a unique limit B, in the sense

$$\lim_{n\to\infty} || B_n \Phi - B \Phi || = 0 \quad .$$

B is everywhere defined but need not be a bounded operator. On the other hand, if the B_n are uniformly bounded: $||B_n|| \leqslant C$ (the constant C is independent of n), then B is bounded, and that will often suffice for our purposes. The standard way of defining strong convergence using neighborhoods, found in the books, has the property that a strong limit point of bounded operators is a bounded operator. I will not write it down here. For the notion of uniform convergence, no such difficulty arises. If the sequence B_n n = 1,2,... converges uniformly, it converges to a bounded operator B, in the sense that $||B_n - B|| \to 0$.

If one is given a sequence of <u>unbounded</u> operators A_n n=1,2,..., one has to face the fact that in general the domains of the different A_n do not coincide. Thus if one attempts to extend the above definitions of weak and strong convergence it is unclear for what vectors one should require the definitions to hold. To extend the notion of uniform convergence seems hopeless because the norms occurring in the definition will not exist in general.

In dealing with self-adjoint operators there is one simple and natural way to get around this difficulty. It is to work with resolvents instead of the unbounded operators themselves i.e. to define A_n n = 1,2,... as convergent if the resolvents $[A_n - z]^{-1}$ converge in one of the above three senses for some z not on the real axis.

It will turn out that the resolvents of the self-adjoint cut-off Hamiltonians $H_{K,V}(g)$, defined below, converge in norm to the resolvent of the self-adjoint Hamiltonian H(g) so let us confine our attention to sequences of operators whose resolvents converge in norm. One simple fact about such sequences $[A_n - z]^{-1}$ is that if they converge for one non-real z they converge for all non-real z. (To see this note that $||(A_n-z)^{-1}|| \leqslant |\frac{1}{\mathrm{Im} z}|$. The usual elementary argument which is used to give that $[A_n-z]^{-1}$ is analytic in z, then yields

$$[A_n-w]^{-1}\,[1-(w-z)\,[A_n-z]^{-1}] = [A_n-z]^{-1} \quad .$$

Since for w sufficiently close to z

$$|w-z|\; ||[A_n-z]^{-1}|| < 1$$

the operator $[1 - (w-z)\,[A_n-z]^{-1}]^{-1}$ can be defined by a norm convergent power series in (w-z) and so defined is continuous in $[A_n-z]^{-1}$. Hence

$$[A_n-w]^{-1} = [1 - (w-z)\,(A_n-z)^{-1}]^{-1}\,[A_n-z]^{-1}$$

and the norm convergence of $[A_n-z]^{-1}$ implies the norm convergence of the function of $[A_n-z]^{-1}$ on the right hand side and hence of the left hand side for all w sufficiently close to z. One proceeds in this way throughout the half plane in which z lies and passes to the opposite half plane by $[[A_n - z]^{-1}]^* = [A_n - \bar{z}]^{-1}$.)

To use norm convergence of resolvents to define a limiting self-adjoint operator one must be sure that the norm limit of $[A_n - z]^{-1}$ is the resolvent $[A - z]^{-1}$ of a self-adjoint operator A. That is <u>not</u> always the case. Take, for example, $A_n = n.\mathbb{1}$, then $||[A_n - z]^{-1}|| = 1/|n-z| \to 0$ as $n \to \infty$ and 0 is not the resolvent of any self-adjoint operator. This example provides the key to the problem. The solution is contained in the following

<u>Lemma</u>

Let A_n n=1,2,... be a sequence of self-adjoint operators whose resolvents $[A_n - z]^{-1}$ converge for some non-real z (and therefore all non-real z) to an operator R(z). Then R(z) is the resolvent of a self-adjoint operator A:

$$R(z) = [A - z]^{-1}$$

iff $R(z)\Phi = 0$ implies $\Phi = 0$. (For a proof see Ref.[33] pp427-8).

The relation between the norm convergence of the resolvents of a sequence of self-adjoint operators $A_n, n=1,2,\ldots$ and convergence of the unitary groups they generate is given by the following theorem.

Theorem

Let $A_n, n=1,2,\ldots$ be a sequence of self-adjoint operators, A_n, whose resolvents converge uniformly to the resolvent $[A - z]^{-1}$ of a self-adjoint operator A. Then the corresponding one parameter unitary groups

$$\exp itA_n \quad , \; -\infty < t < \infty$$

converge strongly to

$$\exp itA$$

uniformly on each bounded interval of t (see Ref.33 pp502-3).

This theorem assures us that if we succeed in defining a self-adjoint Hamiltonian as the limit of self-adjoint cut-off Hamiltonians by proving uniform convergence of resolvents then the corresponding dynamical laws of evolution will converge too.

ΦOK Space and ΦOK Representations of the CCR and CAR

The constructions which serve as a starting point in $\mathcal{P}(\phi)_2$ and Y_2 are carried out in ΦOK space, and the basic operators are constructed from the ΦOK representations of the CCR and CAR. It is the last task of this section to define these notions.

Let $\mathcal{H}$ be the Hilbert space of single particle wave functions. For both $\mathcal{P}(\phi)_2$ and Y_2 it can be taken (and usually is so in elementary field theory) as the space $L^2(R)$ of all square integrable wave functions on momentum space with the scalar product

$$(\Phi,\Psi) = \int_R dk \; \overline{\Phi(k)}\Psi(k) \quad . \tag{62}$$

Then the n-particle wave functions lie in the n-fold product space $\mathcal{H}^{\otimes n}$ which is nothing but the space $L^2(R^n)$ of square integrable functions defined on n dimensional momentum space with the scalar product

$$(\Phi^{(n)},\Psi^{(n)}) = \int_{R^n} dk_1 \ldots dk_n \; \overline{\Phi(k_1 \ldots k_n)}\Psi(k_1 \ldots k_n) \quad . \quad (63)$$

For bosons one restricts to the subspace of functions symmetrical under permutations; for fermions to those antisymmetric under permutations. The <u>Φοκ space over $\mathcal{H}$</u> is obtained by taking the direct sum of all these spaces, the zero particle space being taken as the one-dimensional Hilbert space, the complex numbers equipped with the scalar product $\overline{\Phi^{(0)}}\Psi^{(0)}$. This statement is expressed symbolically

$$\mathcal{F}_\varepsilon(\mathcal{H}) = \bigoplus_{n=0}^{\infty} \mathcal{H}^{(n)} = \bigoplus_{n=0}^{\infty} (\mathcal{H}^{\otimes n})_\varepsilon \quad (64)$$

where the subscript ε is s for the symmetric case and a for the antisymmetric case. The elements of $\mathcal{F}_\varepsilon(\mathcal{H})$ are therefore sequences

$$\Phi = \{\Phi^{(0)}, \Phi^{(1)}, \Phi^{(2)}, \ldots\} \quad \text{where } \Phi^{(n)} \varepsilon \mathcal{H}^{(n)} \quad (65)$$

with the scalar product

$$(\Phi,\Psi) = \overline{\Phi^{(0)}}\Psi^{(0)} + \sum_{n=1}^{\infty} (\Phi^{(n)},\Psi^{(n)})_{\mathcal{H}^{(n)}} \quad . \quad (66)$$

While the realization (62) is the most practical one for most of the computations of constructive field theory, it does not make the relativistic transformation law simple for free particles. For that purpose it is natural to introduce single particle wave functions which have the scalar product

$$(\Phi,\Psi) = \int \frac{dk^1}{\sqrt{(k^1)^2+m_0^2}} \overline{\Phi(k^1)} \; \Psi(k^1) \quad (67)$$

and to regard $\Phi(k^1)$ as a function of the energy momentum vector $\{\sqrt{(k^1)^2 + m^2}, k^1\}$ defined on the positive energy mass shell. Then the transformation law becomes

$$(U(a,\Lambda)\Phi)(k) = \exp(ik\cdot a)\Phi(\Lambda^{-1}k) \quad (68)$$

where $k = (k^0, k^1)$.

Clearly passage from the amplitudes which yield (62) to those which yield (67) corresponds to multiplying the wave function by $\sqrt[4]{(k^1)^2 + m^2}$. In both cases, one can pass from the wave function to its Fourier transform and get another realization of the space of single particle states. The four possible realizations may be indicated as follows

$$\begin{array}{ccccc} & L^2(\mathbb{R}, dk) & \xrightarrow{\text{FOURIER TRANSFORM}} & L^2(\mathbb{R}, dx) & \\ \mathcal{H}^{(1)} = & \Big\downarrow \begin{array}{l}\text{MULTIPLY} \\ \sqrt[4]{(k^1)^2 + m_0^2}\end{array} & & \Big\downarrow \begin{array}{l}\text{MULTIPLY} \\ \sqrt[4]{-\frac{d^2}{dx^2} + m_0^2}\end{array} & \\ & L^2\left(\mathbb{R}, \frac{dk^1}{\sqrt[2]{(k^1)^2 + m^2}}\right) & \xrightarrow[\text{FOURIER TRANSFORM}]{} & H_{-\frac{1}{2}}(\mathbb{R}) & \end{array} \quad (69)$$

The upper right realization is appropriate for the discussion of localization in the sense of Newton and Wigner while the lower right is to be used when one has to deal with localization in the sense of the Dirac or Klein-Gordon equation. We will stick to the upper left realization in most of the following.

The annihilation operators of the Φοκ representation are defined on $\mathcal{H}^{(n)}$ by

$$(a(k)\Phi)^{(n-1)}(k_1 \ldots k_{n-1}) = \sqrt{n}\, \Phi^{(n)}(k, k_1 \ldots k_{n-1}) \,. \quad (70)$$

$a(k)$ maps $\mathcal{H}^{(n)}$ into $\mathcal{H}^{(n-1)}$. It is a densely defined operator because if $\Phi^{(n)}$ is smooth the right-hand side $\Phi^{(n)}(k, k_1 \ldots k_{n-1})$ is square integrable in $k_1 \ldots k_{n-1}$ for each fixed k, provided $\Phi^{(n)}$ is square integrable in all its n variables. On the other hand, when rough $\Phi^{(n)}$ are admitted the formula will not make sense for some k's. The creation operators $a^*(k)$ are in fact not operators in the sense defined above but only sesquilinear forms. They are defined on $\mathcal{H}^{(n)}$ by

$$(a^*(k)\Phi)^{(n+1)}(k_1 \dots k_{n+1}) = \frac{1}{\sqrt{n+1}} \sum_{j=1}^{n+1} \begin{Bmatrix} (-1)^{j+1} \\ 1 \end{Bmatrix} \delta(k-k_j) \qquad (71)$$

$$\Phi^{(n)}(k_1 \dots \hat{k}_j \dots k_{n+1}) \quad .$$

The upper sign is for the anti-symmetrical case, the lower for the symmetric case. The occurrence of the δ function in the formula indicates that the expression (71) does not define an operator. However, if we smear (71) with any square-integrable function to obtain $a^*(f) = \int dk\, a^*(k) f(k)$ we get the formula

$$(a^*(f)\Phi)^{(n+1)}(k_1 \dots k_{n+1}) = \frac{1}{\sqrt{n+1}} \sum_{j=1}^{n+1} \begin{Bmatrix} (-1)^{j+1} \\ 1 \end{Bmatrix} f(k_j) \qquad (72)$$

$$\Phi^{(n)}(k_1 \dots k_j \dots k_{n+1})$$

which defines an operator on all of $\mathcal{H}^{(n)}$, mapping it into $\mathcal{H}^{(n+1)}$. Similarly, one can define a smeared annihilation operator everywhere on $\mathcal{H}^{(n)}$:

$$(a(f)\Phi)^{(n-1)}(k_1 \dots k_{n-1}) = \sqrt{n} \int f(k)\, dk\, \Phi^{(n)}(k, k_1 \dots k_{n-1}) \,. \qquad (73)$$

These definitions then extend by linearity to yield for each $f \in L^2(\mathbb{R})$ an $a(f)$ and an $a^*(f)$ defined on the dense subset of $\mathcal{F}_\varepsilon(\mathcal{H})$ consisting of all vectors, Φ, for which only a finite number of $\Phi^{(n)}$ are different from zero. On this domain they satisfy

$$[a(f), a(g)]_- = 0, \quad [a(f), a^*(g)]_- = \int dk\, f(k) g(k) \qquad (74)$$

for the symmetric $\mathcal{F}_s(\mathcal{H})$, the C(anonical) C(ommutation) R(elations) and

$$[a(f), a(g)]_+ = 0, \quad [a(f), a^*(g)]_+ = \int dk\, f(k) g(k) \qquad (75)$$

for the anti-symmetric case in $\mathcal{F}_a(\mathcal{H})$, the C(anonical)

A(nti-commutation) R(elations). We have also by a straightforward computation

$$a^*(f) = a(\bar{f})^* \tag{76}$$

and

$$a(f)\Omega_0 = 0 \tag{77}$$

for the <u>ΦοΚ vacuum</u>

$$\Omega_0 = \{1,0,0,\ldots\} \quad . \tag{78}$$

The remarkable feature of the CAR is that they imply

$$||a(f)^*\Phi||^2 + ||a(f)\Phi||^2 = ||\Phi||^2 \int |f(k)|^2 dk \tag{79}$$

so

$$||a(f)^*\Phi|| \leq ||f||_{L^2}||\Phi|| \qquad ||a(f)\Phi|| \leq ||f||_{L^2}||\Phi|| \tag{80}$$

and therefore $a(f)^*$ and $a(f)$ are continuous on their dense domain in $\mathcal{F}_a(\mathcal{H})$. They therefore can be extended by continuity to all of $\mathcal{H}$ and so extended are bounded operators with bounds

$$||a(f)^*|| = ||a(f)|| = ||f||_{L^2} \tag{81}$$

(80) yields $||a(f)^*|| \leq ||f||_{L^2}$; inserting $\Phi = \Omega_0$ in (79) yields $||a(f)^*|| = ||a(f)|| \geq ||f||_{L^2}$.)

For the CCR no such boundedness holds and the best that one can say is that restricted to $\mathcal{H}^{(n)}$ $a(f)$ is a bounded operator of bound

$$||a(f)\restriction\mathcal{H}^{(n)}|| = \left(||f||_{L^2}\right)^{\sqrt{n}} \quad . \tag{82}$$

The unboundedness of the $a(f)$ and $a^*(f)$ in the symmetric case arises from these $\sqrt{n}$ and in fact one can show that the closure of $a(f)$ as we have defined it has precisely the domain of the operator $\sqrt{N+1}$ where N is the operator of the total number of particles defined by

$$(N\Phi)^{(n)}(k_1\ldots k_n) = n\Phi^{(n)}(k_1\ldots k_n) \quad . \tag{83}$$

Using the annihilation and creation operators of the CCR, one can define the free scalar field of mass m

$$\Phi(x) = [4\pi]^{-\frac{1}{2}} \int_{\mathbb{R}} dk^1 [\mu(k^1)]^{-\frac{1}{2}} [a(k^1)\exp(-ik\cdot x) + a(k^1)^*\exp(ik\cdot x)] \tag{84}$$

(Here $k = \{k^0,k^1\}$, $x = \{x^0,x^1\}$, $k^0 = \mu(k^1) = \sqrt{(k^1)^2 + m^2}$, and $k\cdot x = k^0x^0 - k^1x^1$.) $\phi(x)$ is defined initially as a sesquilinear form on the domain $D \times D$ where D are the <u>nice vectors</u> in $\mathcal{F}_s(\mathcal{H})$, namely the vectors, Φ, for which $\Phi^{(n)}$ is infinitely differentiable, rapidly decreasing and different from zero only for a finite number of distinct n. When one smears $\phi(x)$ with a square integrable function, g, in space, one obtains $\phi(x^0,g)$ an operator on D

$$\phi(x^0,g) = a([2\mu]^{-\frac{1}{2}}\hat{g}e^{-i\mu x^0}) + a^*([2\mu]^{-\frac{1}{2}}\tilde{g}e^{i\mu x^0}) \tag{85}$$

where

$$\hat{g}(k) = [2\pi]^{-\frac{1}{2}} \int dx e^{ikx} g(x) \text{ and } \tilde{g}(k) = \hat{g}(-k) \quad . \tag{86}$$

The corresponding canonically conjugate field, π, is obtained by differentiating (84) with respect to x^0:

$$\pi(x) = i^{-1}[4\pi]^{-\frac{1}{2}} \int dk^1 [\mu(k^1)]^{\frac{1}{2}} [a(k^1)\exp(-ik\cdot x) - a(k^1)^*\exp(ik\cdot x)] \quad . \tag{87}$$

When smeared in space π is also defined on D the domain of nice vectors.

Both $\phi(x^0,g)$ and $\pi(x^0,g)$ are essentially self-adjoint if g is real so that one can pass from them to the bounded operators

$$U(f) = \exp i\phi(0,f) \tag{88}$$

$$V(g) = \exp i\pi(0,g) \quad .$$

U and V satisfy the relations

$$U(f_1)U(f_2) = U(f_1 + f_2)$$

$$V(g_1)V(g_2) = V(g_1 + g_2) \tag{89}$$

$$U(f)V(g) = \exp(-i\int dx\ f(x)g(x))V(g)U(f)$$

valid for all real f,g in $L^2(R)$.

The relations (89) are usually referred to as <u>Weyl's form of the CCR</u> because in studying the canonical variables $q_1 \ldots q_n$, $p_1 \ldots p_n$ for a system with a finite number of degrees of freedom, Weyl proposed to replace the Heisenberg commutation relations

$$[q_j, q_k]_- = 0 = [p_j, p_k]_-$$

$$[q_j, p_k]_- = i\delta_{jk} \tag{90}$$

by

$$U(\alpha)U(\alpha^1) = U(\alpha+\alpha^1) \quad V(\beta)V(\beta^1) = V(\beta+\beta^1)$$

$$U(\alpha)V(\beta) = \exp -i(\sum_{j=1}^{n} \alpha_j\beta_j)V(\beta)U(\alpha) \tag{91}$$

where

$$U(\alpha) = \exp(i\sum_{j=1}^{n} \alpha_j q_j)$$

and
$$V(\beta) = \exp(i\sum_{j=1}^{n} \beta_j p_j) \quad .$$

It is easy to extend this form of Weyl's relations to the

case where n is infinite and the sequences α and β satisfy $\sum_{j=1}^{\infty} \alpha_j^2 < \infty$ and $\sum_{j=1}^{\infty} \beta_j^2 < \infty$. The q_j and p_j can be regarded as arising from annihilation and creation operators a_j and $a_j{}^*$ which satisfy

$$[a_j, a_k]_- = 0, \ [a_j, a_k{}^*] = \delta_{jk} \tag{92}$$

$$a_j = \frac{1}{\sqrt{2}}(q_j + ip_j) \quad a_j^* = \frac{1}{\sqrt{2}}(q_j - ip_j) \quad .$$

The a_j's in turn can be regarded as arising from the above defined a(p) by choice of a particular orthonormal basis ϕ_j in the single particle space $L^2(\mathbb{R})$

$$a_j = \int dk \ \overline{\phi_j(k)} \ a(k) \quad . \tag{93}$$

In section C, the discrete form of the CCR, (92) will be used to construct free fields on a torus.

It is worth noting that the formulation (89) of the CCR makes evident a fact which is sometimes overlooked. The CCR depend on what scalar product is adopted in the real Hilbert space of test functions. However, even after the scalar product indicated in (89) is fixed, there is a one parameter family of representations determined by the mass m and all called ΦOK representations. To see this compute $(\Omega_0, \exp i\phi(f)\Omega_0)$ where Ω_0 is the ΦOK vacuum (78). The result is $\exp - \int dx \ f[4\mu]^{-1} f$ which clearly depends on m. Thus there are really many ΦOK representations.

The free field ϕ given in (84) has been constructed to satisfy the free wave equation

$$(\Box + m^2)\phi(x) = 0 \quad . \tag{94}$$

Its temporal evolution, which is given explicitly in (85), can be unitarily implemented

$$\phi(t,g) = \exp(iH_0 t)\phi(0,g)\exp(-iH_0 t) \quad . \tag{95}$$

The self-adjoint operator, H_0, is defined symbolically by

$$H_0 = \int \mu(k) a^*(k) a(k)\, dk \tag{96}$$

an expression which expressed in Φoκ space is

$$(H_0\Phi)^{(n)}(k_1 \ldots k_n) = [\sum_{j=1}^{n} \mu(k_j)]\Phi^{(n)}(k_1 \ldots k_n) \tag{97}$$

on nice vectors Φ. More precisely H_0 is the closure of this operator defined on nice vectors.

In Y_2 the fermions have anti-particles. Thus, the state space for the theory is a product of a Φoκ space of bosons, a Φoκ space of fermions, and a Φoκ space of anti-fermions.

$$\mathcal{H} = \mathcal{F}_s(\mathcal{H}^{1\ \text{boson}}) \otimes \mathcal{F}_a(\mathcal{H}^{1\ \text{fermion}}) \otimes \mathcal{F}_a(\mathcal{H}^{1\ \text{anti-fermion}}) \tag{98}$$

The boson field then acts only in the first factor i.e. it is of the form

$$\phi(x) \otimes 1 \otimes 1 \tag{99}$$

where $\phi(x)$ is the field discussed above. The baggage $\otimes\ 1 \otimes 1$ will be dropped in most of the following, for simplicity. Similarly, the annihilation creation operators for particles act in the second factor of (98) while those of the anti-particles act in the third. The only trick is that to make the annihilation operators of the particles anti-commute with those of the anti-particles one includes an extra factor $(-1)^{N\ \text{part}}$ in the definition of the anti-particle annihilation operators.

The total free field hamiltonian for Y_2 is defined by

$$H_0 = \int \mu(k) a^*(k) a(k)\, dk$$

$$+ \int \omega(p) \sum_{\varepsilon=\pm 1} b^*(p,\varepsilon) b(p,\varepsilon)\, dp \tag{100}$$

where $b(p,+1)$ is an annihilation operator for fermions and $b(p,-1)$ for anti-fermions and $\omega(p) = [p^2 + M^2]^{\frac{1}{2}}$.

The free fermion field itself is written

$$\psi(x) = [4\pi]^{-\frac{1}{2}} \int dp[\omega(p)]^{-\frac{1}{2}}[b(p,+1)u(p)\exp(-ip\cdot x) + b^*(p,-1)u^c(p)\exp(ip\cdot x)] \quad . \tag{101}$$

Here u(p) and $u^c(p)$ are standard two-component solutions of the Dirac equation in momentum space, chosen so

$$[i(\gamma^0 \frac{\partial}{\partial x^0} + \gamma^1 \frac{\partial}{\partial x^1}) - M]\psi(x) = 0 \quad . \tag{102}$$

It is easy to verify that H_0 for Y_2 defined as the closure of (100) defined on nice vectors is self-adjoint.

C. <u>APPROACH THROUGH CUT OFF THEORIES</u>

What stands in the way of a direct attack on $\mathcal{P}(\phi)_2$ and Y_2 made by the insertion of a standard (ΦOK) representation of the CCR or CAR in the Hamiltonian? For example, what is wrong with

$$\int dx\ [\tfrac{1}{2}\colon (\pi^2 + (\nabla\phi)^2 + m_0^2)\colon + \colon\mathcal{P}(\phi)\colon] \quad , \tag{103}$$

the standard representation (84) of the canonical variables, ϕ, π, having been inserted? The answer is two-fold

a) <u>Haag's Theorem</u> for $\mathcal{P}(\phi)_2$

b) <u>Haag's Theorem</u> and <u>Ultra-violet Divergences</u> for Y_2.

Let me begin then with $\mathcal{P}(\phi)_2$ and Haag's Theorem. The theorem guarantees that H will be meaningless if any ΦOK representation is inserted. If any representation of the CCR makes (103) meaningful it must be non-ΦOK. However, it has so far been impossible to characterize directly which non-ΦOK representation should be used. The rather indirect procedure adopted in constructive field theory and discussed in the following leads to an appropriate representation; it can be regarded as a round about way of choosing a representation of the CCR to make (103) meaningful.

The procedure is, in part, the traditional one in quantum field theory. The first step involves modifying the Hamiltonian by ad hoc box cut-offs. This modified Hamiltonian makes sense when a ΦOK representation of the CCR is inserted.

This is possible because the modified Hamiltonian is no longer Euclidean invariant. Consequently, Haag's theorem does not apply. The second step in the theory involves using the results of the first step to find a new representation of the CCR which makes the original non-cutoff Hamiltonian meaningful.

Two Kinds of Boxes

The first kind of box used in the construction is the traditional one defined by periodic boundary conditions. Described geometrically, it is a torus or in space of one dimension, a circle. Thus the space-time used in this cutoff theory is a circular cylinder as shown in Figure (3).

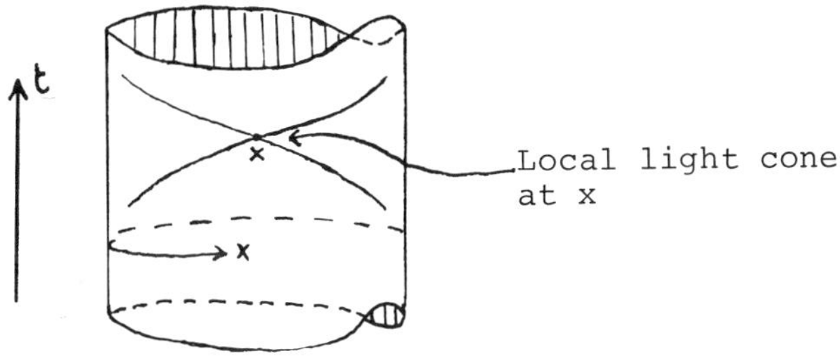

Fig. 3 Space-time for the torus world (≡ box with periodic boundary conditions).

Locally, the propagation of waves on the cylinder is indistinguishable from that in Minkowski space. However, in the large it is quite different; reverberations caused by waves running around the world again and again are a phenomenon absent in Minkowski space but present on the torus. The existence of reverberations is related to the fact that the future light cone intersects itself an infinite number of times.

It is constructive to look at the free scalar field, ϕ_V, on the torus. It satisfies the equation

$$(\Box + m^2)\ \phi_V(x) = 0,\ [\phi_V(x), \phi_V(y)]_- = \frac{1}{i}\ \Delta^V(m,\ x-y). \quad (104)$$

Here sub and superscripts v have been used to indicate that

the quantities are for the torus of perimeter $|V|$. The field ϕ_V and the distribution Δ^V are expanded in Fourier series on the circle. For example, at t=0,

$$\phi^V(x) = [2|V|^{-\frac{1}{2}}] \sum_{k\varepsilon\Gamma_V} [\mu(k)]^{-\frac{1}{2}} [a_V(k) + a_V^*(-k)] \exp(ikx) \quad (105)$$

where Γ_V is the lattice.

$$k = \frac{2\pi}{|V|}n \qquad n = 0, \pm 1, \pm 2, \ldots \quad (106)$$

on the real line $\mathbb{R}$. Here as promised in the preceding section, we have used the discretely labeled annihilation and creation operators

$$[a_V(k), a_V(\ell)]_- = 0, [a_V(k), a_V^*(\ell)]_- = \delta_{k\ell}\ k, \ \ell\varepsilon\Gamma_V. \quad (107)$$

Thus the Fourier transform of $\phi(x)$ in space-time is non-vanishing only for a discrete set of points of the mass shell. See Figure (4).

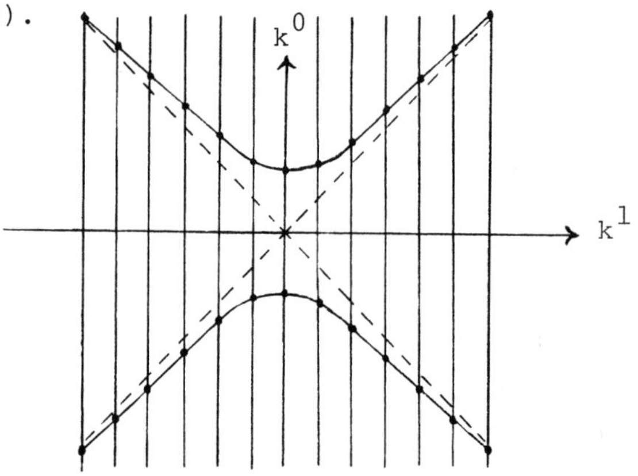

Fig. 4 The Fourier transform of the field is non-vanishing only at the indicated points.

The field, $\phi^V(x)$, transforms in the standard way under translations of the space time cylinder. There is a unitary representation of the translation group, U(a) such that

$$U(a)\ \phi^V(x)\ U(a)^{-1} = \phi^V(x + a) \quad . \quad (108)$$

(Of course vectors like x and x + a have to be defined

modulo $|V|$ in their space components.) The energy momentum spectrum of $U(a)$ is a discrete set obtained by adding multiples of energy momentum vectors lying on the positive energy mass shell in Figure (4).

The explicit form of the energy and momentum operators is

$$H_{o,V} = \sum_{k\varepsilon\Gamma_V} \mu(k) \quad a_V^*(k) \; a_V(k) \tag{109}$$

$$P_V = \sum_{k\varepsilon\Gamma_V} k \; a_V{}^*(k) \; a_V(k) \quad . \tag{110}$$

There are theorems quite analogous to those quoted above for H_o which guarantee that when defined on a suitable domain $H_{o,V}$ and P_V are essentially self-adjoint.

As for Δ^V, it is the sum of $\Delta^{V(+)}$ and $\Delta^{V(-)}$ where

$$\Delta^{V(+)}(x) = - \Delta^{V(-)}(-x) \tag{111}$$

and $\Delta^{V(+)}$ is defined by the Fourier series

$$\Delta^{V(+)}(x) = \frac{i}{2|V|} \sum_{k\varepsilon\Gamma_V} \mu(k)^{-\frac{1}{2}} \exp(-ik\cdot x) \quad . \tag{112}$$

Notice that the only vestige of Lorentz invariance which remains in this theory on the torus is the fact that the ground state has zero energy momentum and that the single particle states (here a discrete set) lie on the mass shell. There is no boost operation which carries states of energy momentum k into those of energy momentum Λk, Λ being a Lorentz transformation. The absence of such a boost is not surprising since there is no analogue of the homogeneous Lorentz group acting on the cylinder of space-time and mapping it into itself with preservation of light cones.

Starting with this formalism for a free field, one can undertake to define and study $\mathcal{P}(\phi)_2$ on the torus. This was first done by Glimm and Jaffe for $(\phi^4)_2$[11] and later by Rosen for general $\mathcal{P}(\phi)_2$[34]. The interaction Hamiltonian is taken as

$$H_{r,V} = \int_V dx :\mathcal{P}(\phi_V): \tag{113}$$

and the full Hamiltonian as

$$H_V = H_{o,V} + H_{I,V} \quad . \tag{114}$$

Although this cutoff Hamiltonian is of considerable interest, it has not so far played as important a role in the development of the theory as the cutoff Hamiltonian for the second kind of box, an object which will be discussed at length in the following. However, before leaving the torus, I want to make one point which seems of general interest. It is the answer to the question: is it so that the property of the single particle states noted above for the free field theory in the torus persists in $\mathcal{P}(\phi)_2$ on the torus. That is, after a shift has been made to adjust the vacuum to zero energy in $\mathcal{P}(\phi)_2$ do all the single particle states lie on a mass hyperboloid $p^2=m^2$ for some suitable m?

One might be led to hope that the answer is yes because for the field ϕ_V that solves $\mathcal{P}(\phi)_2$ on the torus the differential equation

$$(\Box + m_o^2)\ \phi_V(x) = - :\mathcal{P}'(\phi_V(x)): \tag{115}$$

can be proved. Thus, $\phi_V(x)$ is the solution of a <u>locally</u> Lorentz invariant equation. If Ω_V is the ground state of the theory, one can study

$$(\Box + m_o^2)(\Omega_V, \phi_V(x)\Psi) = -(\Omega_V, :\mathcal{P}'(\phi_V(x)):\Psi) \tag{116}$$

and ask whether one can write

$$- :\mathcal{P}'(\phi_V(x)): = J(x) + \delta m^2 \phi_V(x) \tag{117}$$

in such a way that

$$(\Omega_V, J(x)\Psi) = 0 \tag{118}$$

for all the single particle states. If so then all the single particle states have mass given by $m^2 = m_o^2 - \delta m^2$. The simplest way to convince oneself that this does not happen is to calculate the energy shifts of the ground and single particle states to lowest order in perturbation theory in a representative theory. In fact, the energy shifts depart from a hyperboloid by amounts which go to zero as $|V| \to \infty$. So there is no exact remnant of Lorentz invariance for the energy momentum spectrum of $\mathcal{P}(\phi)_2$ for the torus.

The second kind of box is defined for fields on Minkowski space. It is introduced by replacing the formal Hamiltonian (2) of $\mathcal{P}(\phi)_2$ by

$$H(g) = H_o + \int dx\ g(x) : \mathcal{P}(\phi(x)) : \qquad (119)$$

and analogously for Y_2

$$H(g) = H_o + \int dx\ g(x) \lambda : \bar{\psi}\psi : (x) \phi(x) \quad . \qquad (120)$$

Here g(x) is a positive function typically chosen to be 1 in bounded region and to fall to zero outside. (For various parts of the theory various assumptions suffice. For example, in part, $g \in L^1 \cap L^2$, i.e. integrable and square integrable suffices. In other parts of the theory some smoothness of g has to be imposed.) Evidently, this way of introducing a box is not invariant under translations in space. It expresses the physical idea that the theory be modified so that interaction take place only in the bounded region where $g \neq 0$, and that where $g = 1$ the interaction is not modified at all. Outside, where $g = 0$, the particles of the butchered theory move freely.

An important fact about this second way of introducing the box is that it is a local operation in the coordinates. As will be seen shortly, it is this local property which can be exploited to construct a theory independent of cutoff in the Φοκ representation.

What has been achieved by the introduction of the two kinds of box? The answer is that the guaranteed meaningless expression (103) has been replaced by (114) and (119) which make sense when the appropriate $\Phi o\kappa$ representation of ϕ_V, π_V or ϕ, π respectively is inserted. The proof of this statement is one of first problems solved in constructive field theory and it's precisely one of the points of Rosen's talk to give an idea how this is done. Here I want merely to prepare the notation and to relate some of the ideas involved to the general set of definitions found in B.

It first is shown that $H_{I,V}$ and $H_I(g)$ are essentially self-adjoint on the domain of nice vectors. Since the same is true of H_o and $H_{o,V}$ the sums H_V and $H(g)$ are certainly densely defined Hermitean operators. But are they essentially self adjoint? No general theorem says they have to be. Nevertheless, we have

<u>Theorem</u>

$H_V = H_{o,V} + H_{I,V}$ is essentially self-adjoint in $D(H_{o,V}) \cap D(H_{I,V})$.

$H(g) = H_o + H_I(g)$ is essentially self-adjoint on $D(H_o) \cap D(H_I(g))$.

Two useful pieces of machinery that go into the Glimm-Jaffe-Rosen proof are the <u>Hamiltonian with ultra violet cut off on the torus</u> $H_{K,V}$ and the <u>Hamiltonian with two boxes and ultra violet cutoff</u> $H_{K,V}(g)$. The first is defined as

$$H_{K,V} = H_{o,K,V} + H_{I,K,V} \tag{121}$$

where

$$H_{o,K,V} = \sum_{k\epsilon\Gamma_{K,V}} \mu(k)\; a_V^*(k)\; a_V(k) \tag{122}$$

and

$$H_{I,K,V} = \int_V dx\; :\mathcal{P}(\phi_{K,V}(x)): \tag{123}$$

with

$$\phi_{K,V}(x) = [2|V|]^{-\frac{1}{2}} \sum_{k\epsilon\Gamma_{K,V}} [\mu(k)^{-\frac{1}{2}} [a_V(k)+a_V^*(k)]e^{ikx} \quad . \tag{124}$$

Here $\Gamma_{K,V}$ is the finite set of k's consisting of those elements Γ_V that satisfy $|k| \leq K$. Clearly these operators are defined by dropping all annihilation and creation operators labeled by wave numbers larger than K in magnitude.

The second Hamiltonian $H_{K,V}(g)$ is obtained by imposing both an ultra violet cutoff $|k| \leq K$ and a torus cutoff on the Hamiltonian H(g). It relates the two kinds of boxes we have introduced and, in fact, leads to an embedding of the Φοκ space of the theory on the torus in the Φοκ space of the theory on Minkowski space. To define it, define a set of $a_V(k)$ for $k \varepsilon \Gamma_V$ by averaging the a(k)

$$a_V(k) = \left(\frac{|V|}{2\pi}\right)^{\frac{1}{2}} \int_{-\pi/|V|}^{\pi/|V|} a(k+\ell)\,d\ell \quad . \tag{125}$$

Then the $a_V(k)$ and $a_V{}^*(k) = a_V(k)^*$ satisfy the discrete form of the CCR for annihilation and creation operators (107), and the field $\phi_{K,V}(x)$ is given a meaning acting on the Φοκ space $\mathcal{F}_s(\mathbb{L}^2(\mathbb{R}))$. The same is true of the cut off Hamiltonian

$$H_{o,K,V} = \int_{|k|\leq K} dk\ \mu([k]_V)a^*(k)a(k) \tag{126}$$

$[k]_V$ is the point of the lattice Γ_V nearest to k, and

$$H_{I,K,V}(g) = \int dx\ g(x) : \mathcal{P}(\phi_{K,V}(x)) : \tag{127}$$

and

$$H_{K,V}(g) = H_{o,K,V} + H_{I,K,V}(g) \quad . \tag{128}$$

The simple and useful device of embedding all these theories on the torus in the Φοκ space for the field theory on Minkowski space was introduced by Glimm and Jaffe.[11]

In the proof of the preceding theorem discussed by Rosen it is shown that the $H_{K,V}(g)$ are self adjoint and their resolvents converge to the resolvent of H(g) as K and $|V| \to \infty$.

Ultra-Violet Divergences and Y_2

The introduction of boxes does not suffice to make the Hamiltonian of Y_2 into a well defined operator. That is to be expected if one takes the indications provided by perturbation theory seriously. There is an infinite boson mass renormalization and an infinite vacuum energy renormalization required because the fermion bubble diagrams diverge:

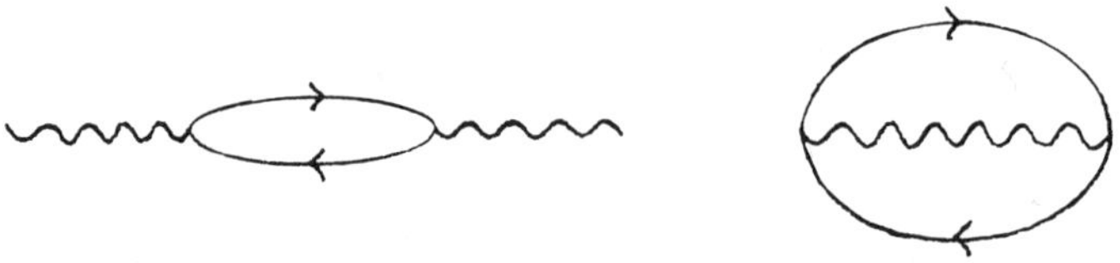

However, this, according to perturbation theory is the only ultra violet divergence so one might hope that by modifying the Hamiltonian by the mass renormalization counter term and a vacuum energy counter term determined from second order perturbation theory one might get a well defined result.

Although technically the work done by Glimm and Jaffe was a tour de force, it does not contain any new ideas simple enough to explain in this introduction. You will hear from Schrader how the argument goes later.

Guenin's Idea and Segal's Theorem: The Right (Algebraic) Theory in the Wrong (Φοκ) Representation

The idea that influence propagates more slowly than light when the dynamics is governed by a butchered Hamiltonian such as H(g) makes it plausible that the definition

$$\phi(x^0,x^1) = \exp(iH(g)x^o)\phi(0,x^1)\exp(-iH(g)x^0) \qquad (129)$$

should be independent of g provided the diamond in space time associated with the region where g = 1 includes $\{0,x^1\}$ and $\{x^0,x^1\}$. (See Figure (5)). The reason for the independence of g is that H(g) acts correctly on those operators that can't feel the effects of the wrong interactions in the regions where $g \neq 1$. The amazing thing about this proposal is that, if it works, it gives an exact solution of the equation of motion (115), even though the basic Hamiltonians used are the butchered H(g). This solution is, in an

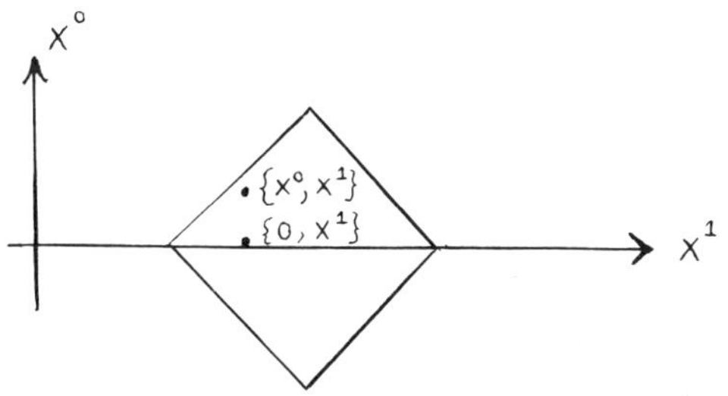

Fig. 5 The diamond is the dependence domain of the region on the line $x^0 = 0$, where $g(0,x^1) = 1$. The point $\{x^0,x^1\}$ is obtained from the point $\{0,x^1\}$ by a time translation.

algebraic sense, the solution of $\mathcal{P}(\phi)_2$ but it is very different in character from the solutions declared acceptable in axiomatic field theory. For example, although the temporal evolution of the field given by (129) is visibly locally unitarily implemented by $\exp(i\, H(g)t)$ there is no single Hamiltonian H_{ren} which gives the temporal development according to (129) for all regions of space-time. Thus, there is no energy operator and the spectrum condition makes no sense. Nevertheless, since the algebraic structure of the theory has been gotten exactly right one can look for another representation of it in which there is a Hamiltonian and in which the spectral condition is satisfied.

The idea of this two step construction of the theory to evade the troubles arising from Haag's Theorem was Guenin's[6] if I read the history correctly, but he did not apply it to any of the relativistic model field theories under discussion here. It was Irving Segal[10] who showed that the proposal (129) indeed yields a $\phi(x^0, x^1)$ independent of g provided the essential self-adjointness of $H(g)$ on $D(H_0) \cap D(H_I(g))$ can be established. The whole scheme fitted in beautifully with general algebraic ideas that Segal had been urging on reluctant physicists for a long time. Finally, Glimm and Jaffe[11] proved the decisive essential self-adjointness for $(\phi^4)_2$. (In fact, they proved self-adjointness,a sharper result). The gold rush was then on.

To state the result of Segal simply and to describe later developments, it is necessary to introduce some of the ideas of the C* algebra approach to quantum field theory. For each bounded open set $\mathcal{O}$ on the line t = 0, we define an algebra of operators, $\mathfrak{A}(\mathcal{O})$, obtained by taking the algebra generated from the bounded operators exp i $\phi(f)$ and exp i $\pi(g)$ where the supports of f and g lie in $\mathcal{O}$ by applying the operations of addition, multiplication, multiplication by complex numbers, hermitean adjoint and passage to the weak limit. (Such a weakly closed algebra is called a <u>von Neumann algebra</u>). $A(\mathcal{O})$ is called the <u>local algebra of $\mathcal{O}$</u>. The algebra $\mathfrak{A}$ obtained by taking the union over all bounded open $\mathcal{O}$, $\bigcup_{\mathcal{O}} \mathfrak{A}(\mathcal{O})$ and then taking the uniform closure of the algebra so obtained is called the <u>quasi-local algebra.</u> The operators belonging to $\mathfrak{A}$ are "essentially" localized in some bounded region; they either belong in fact to some $\mathfrak{A}(\mathcal{O})$ for a bounded region or are norm limits of such. Normed closed algebras of operators such as $\mathcal{O}$ are called C* algebras. (They can also be characterized abstractly but that is irrelevant for our purposes.)

Given a C* algebra, $\mathfrak{A}$, an <u>automorphism</u> of $\mathfrak{A}$ is a one to one mapping of $\mathfrak{A}$ into itself which preserves all algebraic operations:

$$\alpha(AB) = \alpha(A)\,\alpha(B) \qquad \alpha(A+B) = \alpha(A) + \alpha(B)$$

$$\alpha(aA) = a\,\alpha(A) \qquad \alpha(A^*) = \alpha(A^*) = \alpha(A) \tag{130}$$

A <u>one parameter group of automorphisms</u> is a family α_t, $-\infty < t < \infty$ such that α_0 is the identity automorphism and

$$\alpha_{t_1}(\alpha_{t_2}(A)) = \alpha_{t_1+t_2}(A) \quad . \tag{131}$$

Notice that a simple example of a group of automorphisms is

$$\alpha_t(A) = \exp(i\,H_o t)A\exp(-i\,H_o t) \quad . \tag{132}$$

This example was chosen because it plays a role in the

following. The action of the automorphism can be determined from the solution of the initial value problem for the free wave equation. We know that $\phi(t,x)$ is expressible in terms of $\phi(0, y)$ and $\pi(0,y)$ for $\{0, y\}$ in the influence domain of $\{t, x\}$ i.e. for $\{0, y\}$ in the intersection of the past light cone from $\{t, x\}$ with the line $t = 0$. Thus $\phi(t,g)$ is expressible as a linear combination of $\phi(0,g_1)$ and $\pi(0, g_2)$ with test functions g_1, g_2 whose supports lie in the support of g enlarged to the left by subtracting t and to the right by adding t. From this it is easy to see that α_t maps $\mathfrak{A}(\mathcal{O})$ into $\mathfrak{A}(\mathcal{O}_t)$ where $\mathcal{O}_t$ is obtained from $\mathcal{O}$ by an operation analogous to that just described.

<u>Theorem</u> (Segal)

If $H(g) = H_0 + H_I(g)$ is essentially self-adjoint on $D(H_0 \cap D(H_I(g))$ then there is a uniquely determined one parameter group of automorphisms α_t such that

$$\alpha_t(A) = \exp(i\, H(g)\, t)A \exp(-i\, H(g)\, t) \tag{132}$$

for all $A \in \mathfrak{A}(\mathcal{O})$, $\mathcal{O}$ being an open interval, provided $g = 1$ on $\mathcal{O}_t$, the open interval which is $\mathcal{O}$ enlarged by $-t$ to the left and t to the right. Furthermore $\alpha_t(A) \in \mathfrak{A}(\mathcal{O}_t)$.

The idea of the proof is to use the Trotter Product Formula which, for the case at hand, reads

$$\exp(i\, H(g)t) = \underset{n\to\infty}{\text{s-lim}}\, [\exp(i\tfrac{t}{n}H_0)\exp(i\tfrac{t}{n}H_I(g))]^n \,. \tag{133}$$

Thus the action of $H(g)$ on A as given on the right hand side of (132) is the limit as $n \to \infty$ of the n-fold repetition of the operation

$$A \to \exp(i\tfrac{t}{n}H_0)\, \exp(i\tfrac{t}{n}H_I(g))A \exp(-i\tfrac{t}{n}H_I(g))\, \exp(-i\tfrac{t}{n}H_0). \tag{134}$$

Now

$$\exp(i\tfrac{t}{n}H_I(g))A \exp(-i\tfrac{t}{n}H_I(g)) \tag{135}$$

is independent of what g does outside $\mathcal{O}$ because $H_I(g)$ can be split into $H_I(g_1) + H_I(g_2)$ where the support of g_2 lies outside $\mathcal{O}$ and the support of g_1 is in $\bar{\mathcal{O}}$. $\exp(i\frac{t}{n}H_I(g_2))$ commutes with A so only $H_I(g_1)$ has any effect on A. (In their original paper, Glimm and Jaffe only proved self-adjointness of $H_0 + H_I(g)$ where g is smooth so their g_1 had support arbitrarily close to $\mathcal{O}$. This complicated the later details. Since one now knows that g_1 need only be L^1 and L^2 $e^{iH_I(g_1)} \in \mathfrak{A}(\mathcal{O})$ when supp $g_1 \subset \bar{\mathcal{O}}$ one can avoid these complications). Thus (135) lies again in $\mathfrak{A}(\mathcal{O})$. On the other hand, the second part of the operation in (134), the part involving exp(i t/n H_0) maps an operator in $\mathfrak{A}(\mathcal{O})$ into one in $\mathfrak{A}(\mathcal{O}_{t/n})$. Repetition of (134), n times thus yields an operator in $\mathfrak{A}(\mathcal{O}_t)$ which is independent of what g does outside the region where it is 1.

The theorem of Segal as we have stated it describes the action of the automorphism on bounded operators. It is then a technical enterprise to show that the unbounded operators obtained when the field is smeared in space or in space and time with real test functions are in fact essentially self-adjoint on a suitable domain and satisfy the local commutativity assumption. I will not say more about that but instead make a few remarks preparatory to Schrader's discussion of Y_2.

The reason that the extension of the preceding results to Y_2 is difficult is that an analogue of the statement H(g) essentially self-adjoint on $D(H_0) \cap D(H_I(g))$ does not exist. The singularity of the interaction and the consequent necessity of mass renormalization make it impossible to separate H_0 and $H_I(g)$ + counter terms in Y_2. To apply the Trotter Formula one has to introduce ultra-violet cut-offs which destroys the local character of the interaction unless further special measures are taken. In the end when all cut-offs are removed, there is an automorphism of the quasi-algebra and there are fields satisfying a (renormalized) field equation. It is this kind of thing that makes the papers of Glimm and Jaffe meaty.

The Physical Vacuum as an Invariant State on the Quasi-local Algebra; Construction of the Physical Representation by GNS

As a consequence of the results described in the

preceding paragraphs, one has the quasi-local algebra $\mathcal{A}$ with its family of subalgebras $\mathcal{A}(\mathcal{O})$ and on $\mathcal{A}$ the one parameter group of automorphises which describes the dynamics of $\mathcal{P}(\phi)_2$ or Y_2 as the case may be.

The construction of the physical representation starting from this information uses one of the most important general methods in the representation theory of C* algebras: the G(elfand) N(aimark) S(egal) construction. The first task of this section is to provide an introduction to this subject.

For the purposes of the construction one needs a more general notion of state than the one used in elementary quantum mechanics. There a state is given either by a vector in Hilbert space or a density matrix. In the theory of C* algebras this notion is generalized: a <u>state</u> on a C* algebra $\mathcal{A}$ is a complex valued function, ω, on $\mathcal{A}$ satisfying three conditions. It is

a) linear

$$\omega(aA+bB) = a\ (A) + b\ (B) \tag{135}$$

b) positive

$$\omega(A^*A) \geqslant 0 \tag{136}$$

c) normalized

$$\omega(\mathbb{1}) = 1 \quad . \tag{137}$$

Notice that a normalized vector, Ψ, defines a state in this sense

$$\omega(A) = (\Psi, A\Psi) \quad . \tag{138}$$

A state on $\mathcal{A}$ of this form is called a <u>vector state</u>. A density matrix, ρ, also defines a state

$$\omega(A) = \mathrm{tr}\ (\rho A) \tag{139}$$

customarily referred to as a <u>density matrix state</u> or <u>normal state</u>.

Thus the usage in C* algebra theory shortens the phrase

"expection value in a state" to "state". (138) is evidently a special case of (139) in which ρ is a projection operator onto a one dimensional subspace. C* algebras like the quasi-local algebra have lots of states that are not normal.

We will be particularly interested in states invariant under an automorphism group α_t i.e. states ω satisfying

$$\omega(\alpha_t(A)) = \omega(A) \quad . \tag{140}$$

Given a state ω in a C* algebra, $\mathcal{A}$, the GNS construction yields a Hilbert space, $\mathcal{H}_\omega$, a representation, π_ω, of $\mathcal{A}$ by operators in $\mathcal{H}_\omega$, and a cyclic vector, Ω_ω, for π_ω such that

$$\omega(A) = (\Omega_\omega, \pi_\omega(A)\Omega_\omega) \quad . \tag{141}$$

(That Ω_ω is <u>cyclic</u> means that the vectors $\pi_\omega(A)\Omega_\omega$, with A running over $\mathcal{A}$, are dense in $\mathcal{H}_\omega$.)

If the state ω is invariant under a one parameter group of automorphisms α_t, then the construction also yields a representation $\alpha_t \to U(t)$ of the group α_t by unitary operators in $\mathcal{H}_\omega$ such that

$$U(t)\Omega_\omega = \Omega_\omega \tag{142}$$

and

$$\pi_\omega(\alpha_t(A)) = U(t)\pi_\omega(A)U(t)^{-1} \quad . \tag{143}$$

If α_t is continuous in t in the sense that, for each $A \varepsilon \mathcal{A}$ $||\alpha_t(A) - \alpha_{t'}(A)||$ is arbitrarily small for sufficiently small $|t - t'|$, then U(t) is continuous in t and by Stone's theorem is of the form exp iHt for a unique self adjoint H.

Thus the GNS construction offers a method of constructing a representation in which there <u>is</u> a Hamiltonian, H, as well as a candidate for the vacuum state vector, Ω_ω.

Some of the ideas of the construction will now be sketched. The main device is to introduce a Hilbert space structure on $\mathcal{A}$ itself using the given state ω to define the scalar product. We tentatively define the scalar product between two elements A, B and $\mathfrak{A}$ as $\omega(A^*B)$. This ansatz clearly is sesqui-linear i.e. linear in B and anti-linear in A. The positivity property of ω makes $\omega(A^*A)$ non-negative. On the other hand, it might happen that

$$\omega(A^*A) = 0 \tag{144}$$

even though $A \neq 0$. If no such A occurs in $\mathfrak{A}$, then the tentatively assigned scalar product can be used as an honest scalar product because the corresponding norm satisfies

$$||A||_\omega = \sqrt{(A,A)_\omega} = \sqrt{\omega(A^*A)} = 0 \tag{145}$$

implies $A = 0$. If there are A's $\neq 0$ satisfying (144), they form a closed left ideal N_ω in $\mathfrak{A}$. That means first of all that N_ω is a (linear) subspace of $\mathfrak{A}$. (To see this compute $\omega((aA+bB)^*(aA+bB))$. It vanishes by virtue of $\omega(A^*A) = 0 = \omega(B^*B)$ and the Schwarz inequality

$$|\omega(A^*B)|^2 \leqslant \omega(A^*A)\omega(B^*B) \quad . \tag{146}$$

The Schwarz inequality holds for any non-negative sesqui-linear form, the proof being the standard one.) Second, N_ω is invariant under left multiplication i.e. if $A \in N_\omega$ and $B \in \mathfrak{A}$, then $BA \in N_\omega$. (This is again a consequence of the Schwarz inequality $\omega((BA)^*BA) \leqslant \omega(((BA)^*B)^*(BA)^*B), \omega(A^*A)=0.$) Finally, N_ω is closed in norm i.e. if A_n $n = 1,2,\ldots$ is a sequence of elements of N_ω which converges to an element $A \in \mathfrak{A}$, then $A \in N_\omega$. (This follows from the fact that any state ω on a C* algebra A is necessarily norm continuous, satisfying, in fact, the inequality $|\omega(A)| \leqslant ||A||$. To obtain a proof of this last inequality notice that if $||x||<1$, the element $1 - x$ has a square root defined by a norm convergent power series in x, and the square root is hermitean if x is hermitean. Thus, if $||y||<1$

$$\omega(1-y^*y) = 1 - \omega(y^*y) \geqslant 0 \quad . \tag{147}$$

Apply this inequality to $y = A/||A||(1+\varepsilon)$ for $\varepsilon > 0$. The result is

$$\omega(A^*A) \leqslant (1+\varepsilon)||A||^2 \tag{148}$$

for every $\varepsilon > 0$ and therefore for $\varepsilon = 0$. Hence

$$|\omega(A)| = |\omega(1\ A)| \leqslant \sqrt{\omega(1)}\ \sqrt{\omega(A^*A)} \leqslant ||A|| \ . \tag{149}$$

Form now the quotient space $\mathcal{A}/N_\omega$ whose elements consist of the equivalence classes $A + N_\omega$ i.e. two elements A,B lie in the same equivalence class if their difference A-B lies in N_ω.

The space $\mathcal{A}/N_\omega$ plays the role that $\mathcal{A}$ does when N_ω consists of nothing but the operator 0. It has the structure of a prehilbert space with the scalar product defined by

$$(A,B)_\omega = \omega(A^*B) \quad . \tag{150}$$

Here we have indulged in a standard abuse of notation, the use of the symbols A and B to denote both equivalence classes in $\mathcal{A}/N_\omega$ as on the left hand side, and representative operators from the equivalence classes on the right hand side. The Schwarz inequality and the properties of N_ω discussed above assure us that the right hand side is independent of which representative operators from the equivalence classes of A and B are used in its evaluation.

That $\mathcal{A}/N_\omega$ is a prehilbert space means that $(\ ,\ )_\omega$ is an honest scalar product such that the corresponding norm $||\ ||_\omega$ satisfies $||A||_\omega = 0$ implies $A = 0$. (Here $A = 0$ means the equivalence class of A is N_ω.) Thus, $\mathcal{A}/N_\omega$ has all the properties of a Hilbert space except possibly completeness. If it is not complete, complete it. (This is a standard operation for metric spaces which requires the introduction of equivalence classes of Cauchy sequences. It will not be described here. See ref.36). The resulting set of vectors is $\mathcal{H}_\omega$. Thus, the vectors of the representation space $\mathcal{H}_\omega$ are either equivalence classes modulo N_ω or limits in the $||\ ||_\omega$ norm of such.

Next we want to define the representation π_ω of $\mathcal{A}$ on $\mathcal{H}_\omega$.

We define it first on $\mathfrak{A}/N_\omega$ by: $\pi_\omega(A)B$ is the equivalence class of AB. Since

$$||\pi_\omega(A)B||_\omega = ||AB||_\omega \leq ||A||_\omega \; ||B||_\omega \leq ||A|| \, ||B||_\omega \; . \tag{151}$$

The operator $\pi_\omega(A)$ is bounded on the dense set $\mathfrak{A}/N_\omega$ in $\mathcal{H}_\omega$ and therefore can be extended by continuity to all of $\mathcal{H}_\omega$. So continued it satisfies

$$||\pi_\omega(A)\Psi|| \leq ||A|| \quad ||\Psi||_\omega \tag{152}$$

and is therefore continuous in A in the norm $||A||$ also. $\pi_\omega(A)$ is a representation of $\mathfrak{A}$

$$\pi_\omega(AB) = \pi_\omega(A)\pi_\omega(B)$$

$$\pi_\omega(aA + bB) = a\pi_\omega(A) + b\pi_\omega(B)$$

$$\pi_\omega(A^*) = \pi_\omega(A)^* \tag{153}$$

(only the last needs comment. It is a result of the computation $(C,\pi_\omega(A^*)B)_\omega = (C,A^*B)_\omega = \omega((AC)^*B) = (AC,B)_\omega = (\pi_\omega(A)C,B)_\omega$, valid for C, $B\varepsilon \; \mathfrak{A}/N_\omega$, which implies by continuity $(\Phi, \pi_\omega(A^*)\Psi)_\omega = (\pi_\omega(A)\Phi,\Psi)_\omega$ for arbitrary $\Phi, \Psi \; \varepsilon \, \mathcal{H}_\omega$.)

To establish the correspondence (141), we have now only to give the equivalence class of the operator 1 the name Ω_ω. Then

$$(\Omega_\omega, \pi_\omega \; (A)\Omega_\omega)_\omega = \omega(1 \; \pi_\omega(A)1) = \omega(A) \tag{154}$$

as required. Ω_ω is clearly cyclic for π_ω by construction.

It remains to define the U(t) which represents the automorphism α_t. For A ε $\mathfrak{A}/N_\omega$ we define

$$U(t)A = \alpha_t(A) \qquad . \tag{155}$$

The right hand side is linear in A because α_t is an

automorphism. Furthermore, it is independent of the representative A in the equivalence class because α_t maps N_ω into itself. (This is a consequence of the invariance of ω: $\omega(\alpha_t(A^*A)) = \omega(A^*A))$. The same invariance implies $||\alpha_t(A)||_\omega = ||A||_\omega$. U(t) is a bounded operator since α_t preserves the $||\ ||_\omega$ norm so it can be extended by continuity to all of $\mathcal{H}_\omega$. Furthermore, it is unitary because it has the group property

$$U(t+t') = U(t)\ U(t') \tag{156}$$

and

$$U(t)^* = U(-t) \quad . \tag{157}$$

(Commute $(A,U(t)B)_\omega = \omega(A^*\alpha_t(B)) = \omega(\alpha_{-t}(A^*\alpha_t(B)))$

$= \omega(\alpha_-(A^*)B) = (U(-t)A,B)_\omega$. Extend by continuity.)

The invariance of the vacuum (142) is trivial

$$U(t)\Omega_\omega = \alpha_t(1) = 1 = \Omega_\omega \tag{158}$$

and the transformation law (143) of π_ω under U(t) follows by straightforward computation on $\mathfrak{A}/N_\omega$

$$U(t)\pi_\omega(A)B = \alpha_t(AB) = \alpha_t(A)\alpha_t(B) = \pi_\omega(\alpha_t(A))U(t)B \ . \tag{159}$$

It remains to discuss the continuity of U(t) in t. First, note that if we had continuity of α_t in t in the sense that for each $B \in \mathfrak{A}$, $||\alpha_t(B) - \alpha_{t'}(B)||$ is small when $|t-t'|$ is small we could conclude that $(A,U(t)B)_\omega$ is continuous in t:

$$|(A,\ (U(t) - U(t')B)_\omega| = |\omega(A^*(\alpha_t(B) - \alpha_{t'}(B)))|$$

$$\leq ||A||_\omega\ ||\alpha_t(B) - \alpha_{t'}(B)||_\omega \leq ||A||\ ||\alpha_t(B) - \alpha_{t'}(B)|| . \tag{160}$$

However, the assumption of this kind of continuity for α_t is too strong; it would fail to be true in elementary examples.

To prove the continuity of U(t) and hence by Stone's theorem the existence of a renormalized Hamiltonian satisfying $U(t) = \exp it H_{ren}$ it is necessary to use more information about ω, information which arises from particular features of its construction.

This completes a lengthy introduction to the GNS construction. For further information about C* algebras and their representations consult for example[35]. The GNS construction reduces the problem of constructing a Hilbert space, and a representation of the quasi-local algebra to the construction of an appropriate state ω. Let me now turn to the construction by which Glimm and Jaffe obtained such a state[36,37].

They succeeded in showing that $(\phi^4)_2$ has a non-degenerate ground state Ωg. This result was later extended to $\mathcal{P}(\phi)_2$[34]. The physical vacuum according to Glimm and Jaffe is to be obtained as a limit

$$\omega = \lim_{g\to 1} \omega g \qquad (161)$$

where ω_g is the state in the quasi-local algebra defined by

$$\omega_g(A) = (\Omega g,\ A\ \Omega g) \quad . \qquad (162)$$

For Y_2, the program is not so clear cut, because the non-degeneracy of the vacuum for H(g) has only been established for sufficiently small values of the coupling constant. However a general theorem of Kato assures us that if the vacuum ever did become degenerate it would be as a consequence of level crossing in which each of the levels involved is analytic in the coupling constant[40]. There can be no branch points of the eigenvalues and eigenfunctions of H(g) on the real λ axis for Y_2. Beyond the value of λ where the hypothetical crossing has occurred the ground state is again non-degenerate unless it is permanently degenerate. (Such permanent degeneracy is not unheard of for sytems of fermions in space time of four dimensions; the spin of the ground state of K^{40} is 4, so it is nine-fold degenerate.) In any case, the only additional complication in applying the Glimm-Jaffe method to the case in which the ground state is degenerate is that one has to decide which Ωg to choose.

Notice that the Glimm-Jaffe proposal when coupled with the GNS construction is a precise concrete mathematical realization of what has been folklore in theoretical physics for two decades: the quantities which are required to have limits as $g \to 1$ are the expectation values in the ground state of local quantities in space. It is not required that every matrix element of $\exp iH(g)t$ have a limit but only matrix elements between "quasi-local states":

$$(A\Omega_g, \exp(iH(g)t)B\Omega_g) \to \omega(\pi_\omega(A^*)U(t)\pi_\omega(B)) \quad . \tag{163}$$

To prove the actual convergence of ω_g seems to be a difficult problem. What Glimm and Jaffe established was the convergence of a subsequence ω_{gn} after a spatial averaging process which insures that the limit state is invariant under space translations. They did not exclude the possibility that a different choice of subsequence would lead to a different limit. An analogous result was obtained for Y_2 by R. Schrader[41]. Is the possibility of several distinct limiting states real or only a reflection of the limitations of current techniques of proof? To obtain some perspective on this question it is useful to look at its analogue in statistical mechanics. There it can happen that the thermodynamic limit $N, |V| \to \infty, \frac{N}{|V|} = \rho$ fixed can be made to yield distinct states depending on the boundary conditions used in the finite system. Consider for example an Ising model in zero magnetic field at temperature, T, below the Curie temperature. If for the finite systems all spins on the boundary are forced to be up, the thermodynamic limit will yield a state with magnetization up, and analogously for down spins. The same effect, the existence of two distinct limiting states, can be obtained by taking the limit with a nonvanishing magnetic field and afterwards passing to the limit $H \to 0_+$

and $H \to 0_-$ respectively. At temperature T there are two distinct pure equilibrium states. Do analogous phenomena occur in $\mathcal{P}(\phi)_2$ or Y_2? It seems unlikely. The only analogue for field theory of the boundary conditions would appear to be the arbitrariness in the choice of box g and it does not seem to favor one theory over another, even when as discussed in the next section one expects two (or more) inequivalent theories. What is much more likely when there are two inequivalent limiting theories is that one gets a unique mixture of the two. Varying the magnetic field in the Ising model is analogous to varying the coefficients of the polynomial in $\mathcal{P}(\psi)_2$ and it indeed ought to have an effect. (Bogolubov's Principle ought to apply in field theory as well as in statistical mechanics. See the next section for its statement and significance.) However, it does not seem to lead to several distinct limits for (161) for a given polynomial. Thus, it is likely that improved technique ought to lead to a uniqueness proof of consequences of the limit.

How in fact did Glimm and Jaffe get their result? To begin with they chose a sequence g_n, n = 1,2,...

$$g_n(x) = g(\frac{x}{n}) \tag{164}$$

and then averaged ω_{gn} over translations using an auxiliary function h

$$\omega_n(A) = \frac{1}{n} \int h\ (\frac{x}{n})\ \omega_{gn}\ (\beta_x(A))\ dx \tag{165}$$

where h is positive and infinitely differentiable, has support in the interval (-1, 1) and

$$\int h(x)\,dx = 1 \ . \tag{166}$$

β_x is the automorphism induced on the quasi local algebra by space translation by x.

The first step in the argument was a general fact about the dual $\mathfrak{A}^*$ (= the set of continuous linear functionals) of a C* algebra : the unit ball in the dual $\mathfrak{A}^*$ of a C* algebra is compact in the w* topology. Let me explain what this statement means. The continuous linear functionals, F, on a C* algebra can be given a norm

$$||F|| = \sup_{\substack{A \in \mathcal{A} \\ ||A|| = 1}} |F(A)| \tag{167}$$

and the unit ball in $\mathcal{A}^*$ consists of all F such that $||F|| \leq 1$. By virtue of (149) every state of a C* algebra, $\mathcal{A}$, lies in the unit ball of $\mathcal{A}$. To give a topology for $\mathcal{A}^*$, it suffices to give a neighborhood base for its family of open sets. A neighborhood of a functional F in the w* topology is labeled by a finite number of elements of $\mathcal{A}$ and a real number $\varepsilon > 0$:

$$N(F;A_1 \ldots A_n;\varepsilon) = \{\mathcal{A};\ |\mathcal{A}(A_j) - F(A_j)| < \varepsilon,\ j = 1,\ldots n\}. \tag{168}$$

The open sets of the w* topology are then those sets $\mathcal{O}$ such that if a functional $F \in \mathcal{O}$ there exists a neighborhood of the form (168) containing F and contained in $\mathcal{O}$. So the w* topology is one in which two functionals are near each other if they agree closely on a finite number of elements of $\mathcal{A}$. Finally, that the unit ball is compact means that every covering of it by sets open in the w* topology contains a finite open subcovering. For a proof of the theorem see for example[27].

The importance of this general theorem for the study of the ω_n is that as a consequence of it the set of states $\{\omega_n;\ n = 1,2,\ldots\}$ has at least one limit point in the w* topology. Since it is easy to see that such a limit point, F, again satisfies $F(1) = 1$ and $F(A^*A) \geq 0$ it is also a state, the desired ω.

As is usual with such gloriously general arguments, it turns out that relatively little can be said about the limiting state on this basis alone. For example, because of the peculiarities of the w* topology one cannot even be sure there is a subsequence of ω_n that converges to ω. (There is a so-called generalized subsequence or sub-net, however). In any case, what Glimm and Jaffe did was to dig deeper into the properties of the ω_n and to show that even though the general theorem does not guarantee the existence of convergent subsequences the special properties of the ω_n do.

The key technical point in the discussion is the proof that $\omega_n \restriction \mathcal{A}(\mathcal{O})$ for each bounded region is not only w* compact but is also compact in the norm topology. (The norm topology on $\mathcal{A}(\mathcal{O})^*$ is defined by giving the neighborhoods $N(F;\varepsilon) = \{\mathcal{A}; ||\mathcal{A}-F||<\varepsilon\}$ of the functional F. $\mathcal{S}$ is then an open set if for each $F \in \mathcal{S}$ there is an $N(F,\varepsilon)$ contained in $\mathcal{S}$). If ω is a state on $\mathcal{A}$ that is a w* limit point of ω_n, then it turns out that its restriction $\omega \restriction \mathcal{A}(\mathcal{O})$, where $\mathcal{O}$ is any bounded open set, is normal i.e. $\omega \restriction \mathcal{A}(\mathcal{O})$ is given by a density matrix. From the fact that $\omega \restriction \mathcal{A}(\mathcal{O})$ is normal, Glimm and Jaffe deduce that the corresponding representation $\pi_\omega \restriction \mathcal{A}(\mathcal{O})$ is unitarily equivalent to the ΦOK representation

$$\pi_\omega(A) = V(\mathcal{O})\ A\ V(\mathcal{O})^{-1} \tag{169}$$

where $V(\mathcal{O})$ is a unitary mapping of $\mathcal{F}_s(L^2(R))$ onto $\mathcal{H}_\omega$. (Warning!: The ΦOK representations of distinct masses are unitarily <u>in</u>equivalent, but their restrictions to a bounded region are unitarily equivalent. Thus it makes sense to talk about <u>the</u> ΦOK representation <u>locally</u>.)

The validity of this equation for every bounded open set $\mathcal{O}$ is what is meant by the representation π_ω being <u>locally ΦOK</u>. The usefulness of this result is indicated by the fact that from it one can easily deduce that $V(t)$ is continuous in t and hence by Stone's theorem

$$V(t) = \exp\, itH_{ren} \tag{170}$$

where, by definition, H_{ren} is the self adjoint infinitesimal <u>generator</u> of the continuous unitary one parameter group. Furthermore it is easy to prove that ω is the limit point of a subsequence of the ω_n and from that the spectral property of the $H(g)$: $H(g) \geqslant 0$, persists for H_{ren}:

$$H_{ren} \geqslant 0 \quad . \tag{171}$$

Even with these remarkable results in hand, the representation π_ω is rather inaccessible and an intense effort is currently being made to prove that the limiting state, ω, is unique. Results in this direction are encouraging[42]. As to when this limit describes a mixed theory and when a pure theory, that question will be discussed in the following section.

D. PROPERTIES OF THE PHYSICAL REPRESENTATION

The Physical Representation

Most of the discussion in B and C and in the talks of Rosen and Schrader to which they are an introduction is concerned with the first few of the objectives listed in A. One wants to prove the existence of the renormalized Hamiltonian and fields satisfying the general requirements of field theory. In the present section I will talk about what little is known concerning the other objectives on the list and connect them with some of the insights that have been obtained over the past two decades in statistical mechanics and particle physics. These insights give partial answers to such questions as: when does a theory as given by a Lagrangian have a unique solution for given values of the coupling constants and when is it dynamically unstable? When is dynamical instability accompanied by broken symmetry and when not? When does one expect the interaction to produce composite particles and when not? Many of the results obtained have the following character. They say: if you are given a theory with such and such properties it will have such and such other properties. Sometimes the conclusion is supported by a firm chain of reasoning and sometimes only by poetic argument but whichever it is the skeptic can say: how do you know there are any? Any theories, that is, satisfying the hypotheses. The point I want to make is that constructive field theory has a potentially important role to play here even at the stage where it does not solve theories describing the real world. To obtain qualitative control over the solutions of $\mathcal{P}(\phi)_2$ and Y_2 including the answers to the above questions would, in my opinion, provide important support for the conceptual scheme currently used in particle physics. To say this is not to minimize the hard problems that remain so far untouched e.g. the extension of constructive field theory to models that are renormalizable but not super-renormalizable. Nor would I want to argue that there will not be completely new phenomena in four dimensional space-time.

The Conventional Wisdom About Dynamical Instability, Broken Symmetry and the Mass Gap

More than a decade ago, as a result of work by Goldstone, Heisenberg, Jona-Lasinio, Nambu and others, there appeared a lore on what is to be expected as the conditions of dynamical instability in a number of model theories. In their simplest form, these conditions are based on a classical description. Let me describe what they say about $\mathcal{P}(\phi)_2$.

According to this conventional wisdom the main qualitative features of the solution of $\mathcal{P}(\phi)_2$ can be read off from the properties of the polynomial

$$P_1(\xi) = \frac{1}{2} m_0^2 \xi^2 + P(\xi) \tag{172}$$

of the real variable ξ. By assumption this polynomial is bounded below so it has a minimum value.

a) If the minimum value of P_1 is taken on for a single $\xi = \xi_1$, then the theory is dynamically stable and has an essentially unique vacuum state. The resulting theory has a single particle state whose mass squared, m^2, is given by the curvature of P_1 at ξ_1:

$$m^2 = P_1^{(2)}(\xi_1) = m_0^2 + P^{(2)}(\xi_1) \quad . \tag{173}$$

If $m^2 > 0$, there will be a mass gap, if $m^2 = 0$, not. The expectation value of the field is ξ_1

$$(\Omega, \phi(x)\Omega) = \xi_1 \quad . \tag{174}$$

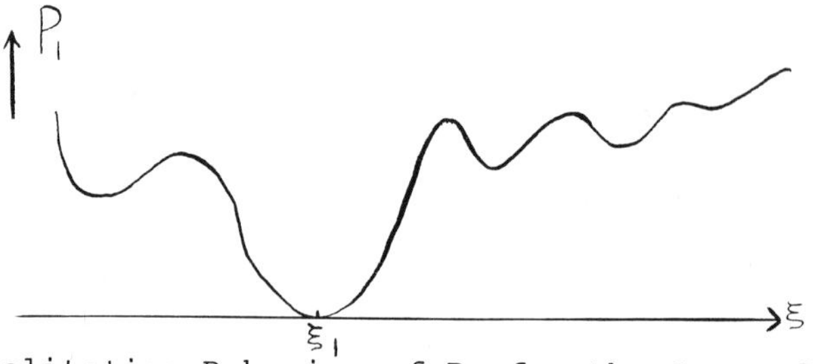

Figure 6. Qualitative Behavior of P_1 for the Case of Dynamical Stability

b) If the polynomial P_1 takes its minimum value at several distinct points $\xi_1 \ldots \xi_n$ the theory is dynamically unstable. There will be a pure theory for each such ξ_j having a single particle state of mass

$$m_j^2 = m_o^2 + P^{(2)}(\xi_j) \tag{175}$$

and an expectation value of the field

$$(\Omega_j, \phi(x)\Omega_j) = \xi_j \quad .$$

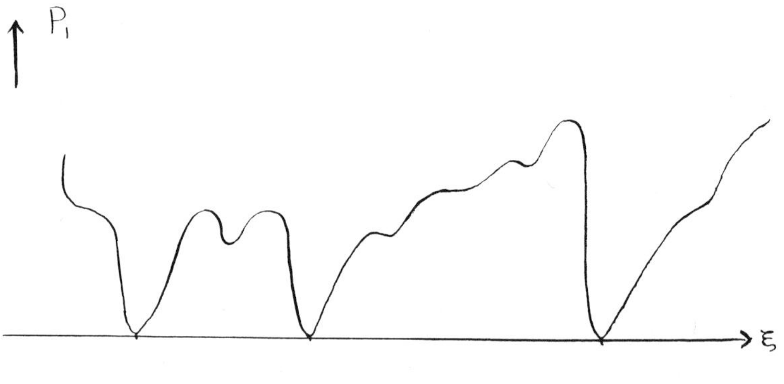

Figure 7. Qualitative Behavior of P_1 for the Case of Dynamical Instability

Goldstone argued that the quantized field, ϕ, should be written as $\phi = \xi + \delta\phi$ where $\delta\phi$ is a field quantized in the Φoκ representation with mass equal to $m_o^2 + P^{(2)}(\xi)$. In physical terms, $\delta\phi$ describes the fluctuations about ξ. For brevity a) and b) will be referred to as the Goldstone Criterion for Dynamical Instability.

One thing Goldstone left unsettled in his account was the status of the relative minima other than ξ. It would be in the spirit of a) and b) to state tentatively:

c) If $\mathcal{P}(\phi)_2$ is dynamically stable the values of ξ for which P_1 has a relative minimum rather than an absolute minimum, correspond to excited states in which $(\Omega, \phi(x)\Omega) = \xi$. Building on such a state, there will be a series of metastable states, quasi-particle excitations with masses approximately given by

$$m^2 = m_o^2 + P^{(2)}(\xi) \quad . \tag{176}$$

If $P(\phi)_2$ is dynamically unstable, the same holds also for the values of ξ at absolute minima different from the distinguished value described in b).

Notice that according to the Goldstone Criterion dynamical instability in $\mathcal{P}(\phi)_2$ is in general not accompanied by broken symmetry. The only available internal symmetry is reflection about some average field $(\phi-\alpha) \to -(\phi-\alpha)$. The Hamiltonian has no such symmetry in the case shown in Figure 7. However, for polynomials of the fourth degree the most general polynomial leading to dynamical instability has P_1 of the form

$$\begin{aligned} P_1(\xi) &= c(\xi-a)^2 (\xi-b)^2 \\ &= c[(\xi-(\tfrac{a+b}{2}))^2 - (\tfrac{a-b}{2})^2]^2 \quad . \end{aligned} \tag{177}$$

The Lagrangian density then possesses the internal symmetry.

$$(\phi - (\tfrac{a+b}{2})) \to -(\phi - (\tfrac{a+b}{2})) \quad . \tag{178}$$

Thus for polynomials $P(\phi)$ of the fourth degree $\mathcal{P}(\phi)_2$ always has dynamical instability accompanied by broken symmetry according to the Goldstone Criterion.

Since there is no continuous internal symmetry group in $\mathcal{P}(\phi)_2$, there are no Goldstone bosons accompanying the breaking of continuous internal symmetry. Of course, by slightly complicating the model, introducing n scalar fields $\phi_1 \ldots \phi_n$, $n \geq 2$ one gets in the standard way models with O(n) as a continuous symmetry group. In such a case dynamical instability may be accompanied by broken symmetry which may be accompanied by Goldstone bosons. See Figure 8.

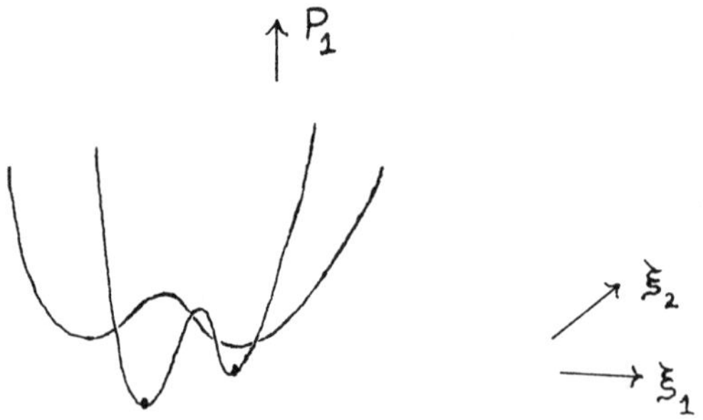

a) P_1 has two passes and takes its minimum at two points. There is dynamical instability but no Goldstone bosons. There may be broken symmetry or not.

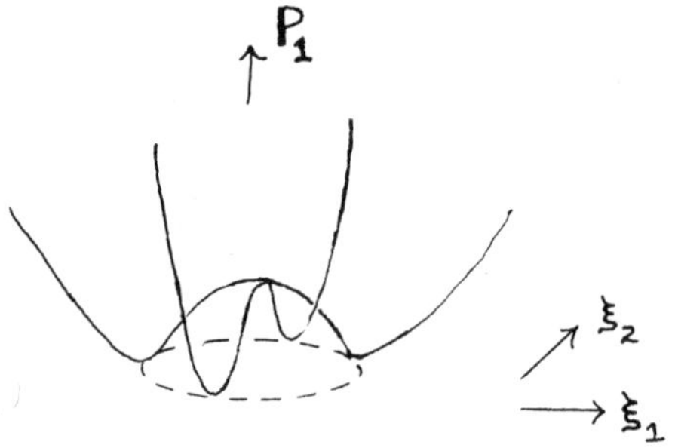

b) P_1 takes its minimum on a circle. Excitations along the valley floor are massless Goldstone bosons. There is dynamical instability and broken symmetry.

Figure 8: Dynamical instability for a theory with two scalar fields. Here the indicator polynomial is

$$P_1(\xi_1,\xi_2) = \frac{m_{o1}^2}{2}\,\xi_1^2 + \frac{m_{o2}^2}{2}\,\xi_2^2 + P(\xi_1,\xi_2) \quad .$$

In its original form, this wisdom gave no indication whether there should exist quantum corrections to the Goldstone criterion. As a result of further work by many authors including Jona-Lasinio, B.W. Lee, and Symanzik, a systematic theory developed, in which there are formulae for the determination of the ground state energy as a function of coupling constants. The formulae can be expanded systematically in the so-called closed loop expansion and the first term yields precisely the Goldstone criterion provided that one applies a fundamental principle suggested by Bogolubov as a result of his studies of the BCS model of super conductivity.

Bogolubov's Principle says that the various distinct pure theories corresponding to the same set of coupling constants for a dynamically unstable theory can be obtained as limits of dynamically stable theories obtained by small variations of coupling constants. For example, if $P(\xi) = \lambda\xi^4 - \lambda_1\xi^2$, Goldstone's Criterion predicts dynamical instability for $\lambda_1 > m_o^2/2$. The addition of a term $\lambda_2\,\xi$ will yield dynamical stability for $\lambda_2 \neq 0$ since the right hand minimum of P_1 is lower than the left if $\lambda_2 > 0$ and vice versa if $\lambda_2 < 0$. If one takes the limits $\lim\limits_{\lambda_2\to 0+}$ and $\lim\limits_{\lambda_2\to 0-}$ one expects to get two distinct theories corresponding to

$$(\Omega_1,\phi(x)\Omega_1) = \xi_1 = \pm\sqrt{\frac{2\lambda_1-m_o^2}{4\lambda}},(\Omega_2,\phi(x)\Omega_2) = \xi_2 = -\sqrt{\frac{2\lambda_1-m_o^2}{4\lambda}}. \tag{179}$$

It is essential that the limits in the coupling constants used in Bogolubov's Principle be taken <u>after</u> all box and ultraviolet cutoffs have been removed because one expects finite systems of bosons to show no discontinuities in the coupling constant.

The net effect of all this work is to indicate that there must, in general, be quantum corrections to the Goldstone Criterion, but, nevertheless, it may be a good guide and a source of interesting conjectures. In the remainder of this section, I will describe several results which I think are helpful in understanding the status of the Goldstone Criterion in constructive quantum field theory.

As a first step, consider the problem of finding the choice of $f \in L^2(\mathbb{R})$, that minimizes the expression

$$E(f) = (\exp(-i\pi_V(f))\Phi_{O,V}, H \exp(-i\pi_V(f))\Phi_{O,V}) \qquad (180)$$

where $\Phi_{O,V}$ is the ΦOK vacuum for the theory on the torus, V. Here and throughout the following H stands for $H_{O,V} + H_{I,V}$ while H_V is $H-E_V$ and therefore has spectrum beginning at 0. Using the standard result

$$(\exp(-i\pi_V(f)\Phi_{O,V}, :\phi^n:(x) \exp(-i\pi_V(f))\Phi_{O,V}) = [f(x)]^n \qquad (181)$$

we get

$$E(f) = \int_V dx \; [\tfrac{1}{2}((\nabla f)^2 + m_O^2 f^2) + P(f)] \; . \qquad (182)$$

The minimum of this expression is obtained by taking f constant, say = ξ, and finding the value or values of ξ that minimize P_1 as given in (172). (To see this, note that if f is a given differentiable function on the torus, V, the form $\frac{1}{2} m_O^2 f^2 + P(f)$ takes on a minimum value on V, say at ξ. The constant function g with value = $f(\xi)$ will yield a value for $E(g) \leq E(f)$.) Thus, the Goldstone Criterion gives exactly the condition on f to insure that the translated ΦOK vacuum gives the minimum energy. The corresponding state

$$\exp(-i\sqrt{|V|}\,\xi p_O)\Phi_{O,V} \qquad (183)$$

with

$$p_O = \frac{1}{\sqrt{|V|}} \int_V dx \; \pi_V(x) \qquad (184)$$

is a sort of caricature of the true ground state Ω_V.

When the polynomial P_1 has several relative minima occuring at $\xi_1 \ldots \xi_n$, it is natural to consider wave functions of the form

$$\Psi = \Sigma_{j=1}^{n} c_j \exp(-i\sqrt{|V|}\,\xi_j p_O)\Phi_{O,V} \qquad (185)$$

and to attempt to minimize $(\Psi, H\Psi)/||\Psi||^2$, by a suitable choice of the real numbers c_j. The normalization integral in the denominator is

$$||\Psi||^2 = \sum_{j=1}^{n} c_j^2 + 2 \sum_{j<k} c_j\, c_k\, (\Phi_{o,v}, \exp(i\sqrt{|V|}\,(\xi_g-\xi_k) p_o \Phi_{o,v})$$

$$= \sum_{j=1}^{n} c_j^2 + 2 \sum_{j<k} c_j\, c_k\, \exp(-\frac{|V|}{4}(\xi_j-\xi_k)^2) \qquad (186)$$

while the numerator is

$$(\Psi, H\Psi) = \sum_{j=1}^{n} \sum_{k=1}^{n} c_j\, c_k\, (\Phi_{o,v}, \exp(i\sqrt{|V|}\ \xi_j p_o) H \exp(-i\sqrt{|V|}\ \xi_k p_o) \Phi_{o,v})$$

$$= \sum_{j=1}^{n} c_j^2\ |V|\ (\Phi_{o,v}, : P_1((q_o/\sqrt{|V|}) + \xi_j) : \Phi_{o,v}) \qquad (187)$$

$$+ 2 \sum_{j<k} c_j\, c_k\ |V|\,(\Phi_{o,v}, \exp(i\sqrt{|V|}\,\xi_j p_o) : P_1(\frac{q_o}{\sqrt{|V|}}) : \exp(-i\sqrt{|V|}\,\xi_k p_o) \Phi_{o,v}).$$

The off-diagonal terms in both the denominator (186) and the numerator (187) contain a factor exp $[-|V|(\xi_j-\xi_k)^2]$ and go to zero as $|V| \to \infty$. If the minimum of P_1 is taken on for a single value of ξ, say ξ_j, one clearly gets the lowest energy for large $|V|$ by taking $c_j = 1$ and all other $c_k = 0$. If the minimum value of P_1 is taken on at $\xi_1 \ldots \xi_n$, one evidently gets n linearly independent wave functions of the corresponding energy in the limit $|V| \to \infty$. A rough idea of the behavior of these states may be obtained as follows. For finite $|V|$, the lowest state will be nodeless and therefore may be expected to have all its c_j of the same sign. The excited states orthogonal to this approximate ground state may be expected to be nearly degenerate differing from the ground state in energy by an amount of the order of magnitude of the diagonal elements and therefore $O(\exp[-|V|\ \text{const}])$.

The preceding calculation is quite instructive in indicating the qualitative behavior of the low lying states in dynamically unstable theories, but it proves nothing about the actual behavior of the solutions of the problem. There is a traditional way of deriving necessary conditions on the vacuum expectation values of the theory by replacing $\Phi_{0,V}$ in the wave functions (185) by the true vacuum Ω_V. (See, for example, G. Baym [44]). Although these conditions do not throw much light on the nature of the energy momentum spectrum let us write them down for completeness.

Consider the evident inequality

$$(\exp(-i\pi(f)\Omega_V, H_V \exp(-i\pi(f)\Omega_V) \geqslant 0 \tag{188}$$

i.e.

$$\int dx \; (\Omega_V, \{\tfrac{1}{2}(:(\nabla(\phi+f))^2: + m_0^2 : (\phi+f)^2: +: P(\phi+f):\}\Omega_V) - E_V \geqslant 0$$

i.e. using invariance under space translations

$$\int_V dx \; \{\tfrac{1}{2}(\nabla f)^2 + \frac{m_0^2}{2}(f^2 + 2f\,(\Omega_V, \phi(x)\Omega_V))$$

$$+ \sum_{j=0}^{2n} (\Omega_V, : P^{(j)}(\phi): f^j (j!)^{-1}\Omega_V)\} \geqslant 0 \quad . \tag{190}$$

The coefficient of the linear term in f has to vanish

$$m_0^2 \; (\Omega_V, \phi(x)\Omega_V) + (\Omega_V, :P^{(1)}(\phi(x)):\Omega_V) = 0 \quad . \tag{191}$$

(The same condition follows immediately without all this fol de rol if one takes the expectation value of the equation of motion in the state Ω_V.) Again there is nothing to be gained by considering f other than constant in space. What then remains is a form of degree 2n-1

$$\frac{m_0^2}{2}\xi^2 + \sum_{j=2}^{2n} (\Omega_V, : P^{(j)}(\phi):\Omega_V)\,\frac{\xi^j}{j!} \geqslant 0 \quad . \tag{192}$$

For example, for n = 4 and $P(\xi) = \sum_{n=1}^{4} c_n \xi^n$ (192) becomes

$$[\frac{m_0^2}{2} + c_2 + 3c_3\,(\Omega_V, \phi\Omega_V) + 6\,c_4\,(\Omega_V, :\phi^2:\Omega_V)]\,\xi^2$$

$$+ [c_3 + 4\,c_4\,(\Omega_V, \phi\Omega_V)]\,\xi^3 + c_4\,\xi^4 \geqslant 0 \tag{193}$$

i.e. since a factor ξ^2 can be removed leaving a quadratic form

$$c_4 \geqslant 0 \ , \ [\frac{m_o^2}{2} + c_2 + 3c_3\ (\Omega_V,\phi\Omega_V) + 6\ c_4\ (\Omega_V,:\phi^2:\Omega_V)] \geqslant 0$$

$$[c_3 + 4\ c_4\ (\Omega_V,\phi\Omega_V)]^2 \leqslant 4c_4\ [\frac{m_o^2}{2} + c_2 + 3\ c_3\ (\Omega_V,\phi\Omega_V) + 6\ c_4\ (\Omega_V,:\phi^2:\Omega_V)] \ . \qquad (194)$$

Possibly useful but not earthshaking.

Now let me turn to the basic question of the occurence or non-occurence of a gap in the mass spectrum. In molecular physics it is standard to estimate such gaps by Rayleigh-Ritz and Weinstein-Bazeley variational techniques. At first sight, it might seem hopeless to use such techniques to bound the gap, $E_{1V} - E_V$, between the ground state, E_V, and the first excited state E_{1V} of H. Since the eigenvalues of H are expected to be of the form

$$E_{nV} = |V|\,(e_o + 0(\frac{1}{|V|}\ \alpha)) \qquad (195)$$

where $\alpha > 0$, it would seem that to avoid having an error proportional to $|V|$, in the bound on the gap, the trial functions will have to yield e_o exactly, and that seems very difficult. In fact, that is precisely what can be done if one uses the true ground state as a constituent of the trial wave function. Since the Founding Fathers of constructive field theory have shown us how to estimate the expectation values of various operators in the true vacuum, the problem is not quite impossible as it would seem to first sight[45].

In fact, consider the state

$$\Psi = (q_o - \langle q\rangle)\Omega_V \qquad . \qquad (196)$$

(Throughout the following calculation $\langle A\rangle = (\Omega_V,\ A\Omega_V)$.) It is orthogonal to the vacuum, Ω_V. Thus

$$\frac{(\Psi,H\Psi)}{||\Psi||^2} \geqslant E_{1V} \qquad . \qquad (197)$$

Now the momentum is

$$\langle (q_o-\langle q_o\rangle) H (q_o-\langle q_o\rangle)\rangle = \langle [(q_o-\langle q_o\rangle), H](q_o-\langle q_o\rangle)\rangle$$
$$+ \langle H(q_o-\langle q_o\rangle)^2\rangle \tag{198}$$
$$= E_v \, ||\Psi||^2 + \frac{1}{2} \; .$$

Thus,

Proposition (R. Baumel[45])

$$E_{1v} - E_v < \frac{1}{2\langle (q_o-\langle q_o\rangle)^2\rangle} \; . \tag{199}$$

Here is an upper bound on the gap which depends on the fluctuation in the quantity, q_o, in the ground state. The bound goes to zero if the fluctuation becomes infinite.

How do we expect this quantity to behave? For purposes of orientation consider the case $P(\xi) = \lambda\xi^4 - \lambda_1\xi^2 \; \lambda, \lambda_1 > 0$. Then for $\lambda_1 > \frac{m_o^2}{2}$, $P(\xi)$ has two minima at

$$\xi_{1,2} = \pm \sqrt{\frac{\lambda_1 - m_o^2/2}{2\lambda}} \; . \tag{200}$$

Instead of the fluctuation in the true ground state, take it in the caricature of the ground state wave function provided by

$$\chi = \frac{1}{\sqrt{2}} \left(e^{-i\sqrt{|V|}\,\xi_1 p_o} + e^{-i\sqrt{|V|}\,\xi_2 p_o}\right)\Phi_{o,V} \; . \tag{201}$$

The norm of χ is given by

$$||\chi||^2 = 1 + \exp\left(-\frac{|V|}{4}(\xi_1-\xi_2)^2\right) \; . \tag{202}$$

The expectation value of q_o is zero, while the expectation of q_o^2 is

$$(\chi, q_o^2\chi) = |V|(\xi_1^2+\xi_2^2) + \text{lower order} \; . \tag{203}$$

Thus, if this approximate calculation is any guide dynamical instability will be accompanied by infinite fluctuations and a vanishing mass gap in the limit $|V| \to \infty$. It should be emphasized that this phenomenon leads to a degenerate vacuum but need not be accompanied by any zero mass particles; there may be a gap in the physical spectrum above the degenerate vacuum.

The fluctuation in q_o approaching infinity as $|V| \to \infty$ is the analogue for our models of the critical point fluctuations occuring in liquid-gas or ferromagnetic phase transitions. In quantum field theory the phenomenon is really a product of the approximation methods used. $\mathcal{P}(\phi)_2$ for a finite volume must have a unique ground state. Dynamical instability shows itself in the limit $|V| \to \infty$ by the fact that the limiting theory is a mixture of pure theories with different values of $(\Omega, \phi(x)\Omega)$. The infinite fluctuations of q_o are caused by $(\Omega, \phi(x)\Omega)$ having to take on the different values appropriate to the different pure theories.

According to the Goldstone Criterion the occurence of zero mass particles ought also to give a vanishing mass gap, the physical reason being that the curvature of P_1 vanishes at the minimum. Baumel also has an inequality for that .

Proposition (R. Baumel[45])

$$E_{1V} - E_V < [m_o^2 + \frac{1}{|V|} \int dx <: P^{(2)}(\phi(x)):>]^{\frac{1}{2}} \quad . \quad (204)$$

This inequality is derived by means almost equally simple to those used for the first inequality. The right hand side is precisely the quantity that one would expect to have when all quantum corrections to the curvature occuring in the Goldstone Criterion are included.

If one can get an upper bound on the fluctuation

$$<(q_o - <q_o>)^2>_o \leqslant B$$

one can get lower bounds on the mass gap using the spectral representation of the two point function

$$(\Omega_V, \phi^V(x)\phi^V(y)\Omega_V) = \sum_{k\varepsilon\Gamma_V} \sum_{\omega_j(k)} \mu(\omega_j(k),k)$$

$$\exp\,[-i(\omega_j(k)(x^o-y^o) - k(x^1-y^1))] \ .$$

Here the μ are non-negative numbers and the $\omega_j(k)$ are the differences between the ground state energy and that of the jth state of momentum k

$$\omega_j(k) = E_{jV}(k) - E_V \quad .$$

The canonical commutation relations then require for each $k \ \varepsilon \ \Gamma_V$

$$|V|[\sum_{\omega_j(k)} \omega_j(k)\mu(\omega_j(k),k) + \sum_{\omega_j(-k)} \omega_j(-k)\mu(\omega_j(-k), -k)] = 1 \ .$$

In particular, for k = 0

$$2|V| \sum_{\omega_j(0)} \omega_j(0)\mu(\omega_j(0),0) = 1 \ .$$

Now the field strength renormalization constant is defined as the contribution to this sum from the first excited state Ω_{1V}

$$2|V| \ \omega_1(0)\mu(\omega,(0),0) = Z \quad .$$

On the other hand

$$<(q_o-<q_o>)^2> \ > \mu(\omega,(0),0)|V| = \frac{Z}{2\omega_1(0)}$$

so[46]

$$E_{1V} - E_V > \frac{Z}{2B} \quad .$$

Whether this is the most useful lower bound is not yet clear. Baumel has others, but neither he nor I is prepared at this moment to make a statement on the mass gap of the theory without box cutoff. Nevertheless I hope I have at least convinced you that for $\mathcal{P}(\phi)_2$ the intuition provided by the Goldstone Criterion can be made quantitative in the form of inequalities which look promising in the study of the spectral condition. Regrettably, for Y_2 I know of no analogous wisdom. Does anyone know any reason why Y_2 should show the phenomenon of dynamical instability? Does it happen in Y_2 that for some large values of the coupling constant a bound state of zero mass appears? If so what happens for larger values of the coupling constant?

GLOSSARY

C* Algebra

A C* algebra, $\mathcal{A}$, is an associative algebra over the complex numbers with an involution* and a norm $\|\ \|$. It is complete in the norm and satisfies the identity

$$\|A^*A\| = \|A\|^2 \quad .$$

An associative algebra is a set with three operations: addition, multiplication, and multiplication by complex scalars. They satisfy the usual associative laws

$$A(BC) = (AB)C \qquad (A+B) + C = A + (B+C) \quad .$$

The distributive laws

$$A(B+C) = AB + AC \qquad (A+B)C = AC + BC \quad .$$

The commutative law of addition

$$A + B = B + A$$

and the laws of scalar multiplication

$$a(bC) = (ab)C \ , \ 1A = A$$

$$a(B+C) = aB + aC \quad .$$

An involution* is a mapping of $\mathcal{A}$ onto itself

satisfying

$$(A + B)^* = A^* + B^* \qquad (AB)^* = B^*A^*$$

$$(aA)^* = \bar{a}\, A^* , \qquad (A^*)^* = A \quad .$$

An associative algebra with involution is called a **algebra*.

A *norm* on a algebra, $\mathcal{A}$, is a real valued function $|| \quad ||$ on satisfying

$$||A+B|| \leq ||A||+||B|| \qquad ||AB|| \leq ||A|| \; ||B||$$

$$||aA|| = |a| \; ||A|| \quad , \quad ||x|| \geq 0 \; , \quad ||x||=0 \text{ implies } x=0 \; .$$

A *normed algebra* is an associative algebra with norm. If it is complete in the norm i.e. if each Canchy sequence has a limit $\mathcal{A}$ is called a *Banach algebra*. A C* algebra is a Banach *algebra whose norm satisfies $||A^*A|| = ||A||^2$.

Graph $\Gamma(A)$ (of an operator A on a Hilbert space $\mathcal{H}$). The set of pairs $\{\Phi, A\Phi\}$ where Φ runs over the domain, D(A), of A. The graph is a linear mainfold in $\mathcal{H} \otimes \mathcal{H}$.

Extension (of an operator A in a Hilbert space $\mathcal{H}$). An operator A in a Hilbert space $\mathcal{H}$). An operator B is an extension of an operator A if

$$\Gamma(A) \subset \Gamma(B)$$

this usually written symbolically

$$A \subset B \quad .$$

Closure (of an operator A on a Hilbert space $\mathcal{H}$). The closure $\Gamma(A)$ of the graph $\Gamma(A)$ of an operator A may or may not be the graph of an operator. If it is, that operator is called $\bar{A}$, the closure of A.

Adjoint A^* (of an operator A on a Hilbert space $\mathcal{H}$). As a linear manifold in $\mathcal{H} \oplus \mathcal{H}$ the graph $\Gamma(A)$ of A has an orthogonal complement, $\Gamma(A)^{\perp}$, the set of all pairs $\{\chi, \Psi\}$

satisfying $(\{\chi,\Psi\}, \{\Phi,A\Phi\}) = (\chi,\Phi) + (\Psi,A\Phi) = 0$ for all $\Phi \in D(A)$. If the set of $\{\chi,\Psi\}$ is of the form $\{-A^*\Psi\}$ where A^* is a linear operator, then A^* is called the adjoint of A. Since $(\Gamma(A)^{\perp})^{\perp} = \overline{\Gamma(A)}$, the closure of the graph of A, it follows that $(A^*)^* = \bar{A}$ whenever the closure exists.

Hermitean (property of an operator A in a Hilbert space $\mathcal{H}$).
Expressed in terms of the graph of A, the hermiticity of A is

$$\Gamma(A) \subset \Gamma(A^*) \quad .$$

This is usually written symbolically $A \subset A^*$. Expressed more directly, A is hermitean if for every Φ and Ψ in the domain of A, D(A)

$$(\Phi,A\Psi) = (\Psi,A\Phi) \quad .$$

In some books and papers the word symmetric is used as a substitute for hermitean.

Self-Adjoint (property of an operator A in a Hilbert space $\mathcal{H}$).
An operator, A, is self-adjoint if

$$A = A^* \quad .$$

Essentially Self-Adjoint
An operator is essentially self-adjoint if its closure exists and is self-adjoint, or expressed in symbols

$$A^{**} = A^* \quad .$$

Note that if A is hermitean it always has a closure.

CCR (canonical commutation relations).
For a system of a finite number of degrees of freedom described by canonical variables $q_1 \ldots q_n \; p_1 \ldots p_n$ the canonical commutation relations are

$$[q_j,q_k]_- = 0 = [p_j,p_n]$$
$$[q_j,p_k] = i\,\delta_{jk}$$

or alternatively in terms of the annihilation and creation operators

$$a_j = \frac{1}{\sqrt{2}}(q_j + ip_j) \qquad a_j^* = \frac{1}{\sqrt{2}}(q_j - ip_j)$$

$$[a_j, a_k]_- = 0 \qquad [a_j, a_k^*]_- = \delta_{jk} \quad .$$

For a system of an infinite number of degrees of freedom described by a field $\phi(f)$ and conjugate field $\pi(g)$ with f and g real test functions, the canonical commutation relations are

$$[\phi(f), \phi(f_2)]_- = 0 = [\pi(g_1), \pi(g_2)]_-$$

$$[\phi(f), \pi(g)]_- = i(f,g)$$

where (,) in general is some real scalar product but in $\mathcal{P}(\phi)_2$ is

$$(f,g) = \int dx\ f(x)\ g(x) \quad .$$

See also Weyl system.

CAR (canonical anti-commutation relations).

For a system of a finite number of degrees of freedom described by annihilation and creation operators a_j $j = 1,\ldots n$ and a_j^*, $j = 1,\ldots n$

$$[a_j, a_k]_+ = 0 \qquad [a_j, a_k]_+ = \delta_{jk} \quad .$$

For a two component field $\psi(x)$ with conjugate field $\psi^+ = \psi^*\gamma^o$, the CAR are

$$[\psi(f), \psi^+(g)] = \int dx\ f(x)\gamma^o g(k) \quad .$$

ΦοΚ Space ((= Fock space) over a Hilbert space $\mathcal{H}$).

There are two such spaces

$$\mathcal{F}_\varepsilon(\mathcal{H}) = \bigoplus_{n=0}^{\infty} \mathcal{H}_\varepsilon^{(n)} \qquad \varepsilon = \begin{cases} s(\text{ymmetrical}) \\ a(\text{ntisymmetrical}) \end{cases}$$

$\mathcal{H}_\varepsilon^{(o)}$ = one dimensional Hilbert space

$\mathcal{H}_\varepsilon^{(n)} = (\mathcal{H}^{\otimes n})_\varepsilon$ the n-fold symmetrical or anti-symmetrical product space.

<u>ΦOK Representation</u> ((= Fock Representation) of the CCR or CAR)

The phrase ΦOK representation is used to describe several somewhat distinct objects. If $\mathcal{H}$ is the Hilbert space of square integrable functions on the real line, $L^2(\mathbb{R})$, the ΦOK representation of the annihilation and creation operators are defined on $\mathcal{F}_\varepsilon(\mathcal{H})$ by

$$(a(k)\Phi)^{(n-1)}(k_1 \ldots k_{n-1}) = \sqrt{n}\, \Phi^{(n)}(k\ k_1 \ldots k_{n-1})$$

$$(a^*(k)\Phi)^{(n+1)}(k_1 \ldots k_{n+1}) =$$

$$\frac{1}{\sqrt{n+1}} \sum_{j=1}^{n+1} \{ {1 \atop (-1)}^{j+1} \}\ \delta(k-k_j)\ \Phi^{(n)}(k_1 \ldots k_j \ldots k_{n+1}) \quad .$$

Smeared with a square integrable function f, these becomes operators a(f) and a*(f)

$$(a(f)\Phi)^{(n-1)}(k_1 \ldots k_{n-1}) = \sqrt{n} \int f(k)dk\Phi^{(n)}(k\ k_1 \ldots k_{n-1})$$

$$(a^*(f)\Phi)^{(n+1)}(k_1 \ldots k_{n+1}) =$$

$$\frac{1}{\sqrt{n+1}} \sum_{j=1}^{n+1} \{ {1 \atop (-1)}^{j+1} \}\ f(k_j)\ \Phi^{(n)}(k_1 \ldots k_j \ldots k_{n+1}) \quad .$$

On the other hand, using this ΦOK representation of the annihilation and creation operators one can construct a one parameter family of fields ϕ and conjugate fields π

$$\phi(x) = [4\pi]^{-\frac{1}{2}} \int dk\ [\mu(k)]^{-\frac{1}{2}} [a(k)\exp(-i(\mu(k)x^o - kx^1))$$
$$+ a^*(k)\exp(i(\mu(k)x^o - kx^1))]$$

$$\pi(x) = [4\pi]^{-\frac{1}{2}}\, i^{-1} \int dk\ [\mu(k)]^{\frac{1}{2}} [a(k)\exp(-i(\mu(k)x^o - kx^1))$$
$$- a^*(k)\exp(i(\mu(k)x^o - kx^1))]$$

where $\mu(k) = [k^2+m^2]^{\frac{1}{2}}$. They satisfy the CCR

$$[\phi(x), \phi(y)]\Big|_{x^0=y^0} = 0 = [\pi(x), \pi(y)]\Big|_{x^0=y^0}$$

$$[\phi(x), \pi(y)]\Big|_{x^0=y^0} = i\,\delta(x^1-y^1)$$

A one parameter family of representations of the CAR is defined by

$$\psi(x) = [4\pi]^{-\frac{1}{2}} \int dp\, [b(p,+1)u(p)\exp(-ip.x) + b^*(p,-1)u^c(p)\exp(ip.x)]\ \omega(p)^{-\frac{1}{2}} .$$

Here b(p,+1) and b(p,-1) are annihilation operators for the fermions and anti-fermions respectively and u(p) and $u^c(p)$ are standardized solutions of the Dirac equation in momentum space

$$(-\not p + M)\ u(p) = 0 \qquad (+\not p + M)\ u^c(p) = 0$$

normalized so that

$$u(p)^+\ u(p) = M$$

and

$$\omega(p) = [p^2 + M^2]^{\frac{1}{2}} .$$

$\psi(x)$ and its adjoint $\psi^+(x) = \psi(x)^*\gamma^0$ satisfy the CAR

$$[\psi(x), \psi(y)]_+\Big|_{x^0=y^0} = 0$$

$$[\psi(x), \psi^+(y)]_+\Big|_{x^0=y^0} = \gamma^0\delta(x^1-y^1) .$$

Locally ΦOK ((= locally Fock) of a representation of the CCR or CAR)

The adjective applies to representations π in the form in which the test functions are defined on a Euclidean space. When the test functions are restricted to have support in a fixed bounded open set $\mathcal{O}$, one gets a representation of the local algebra for that region $\pi \restriction \mathcal{A}(\mathcal{O})$. (The symbol $\restriction$ means restricted to. $\mathcal{A}(\mathcal{O})$ is the weakly closed algebra generated by the bounded functions of fields smeared with test functions whose supports lie in the region.) If for every bounded open $\mathcal{O}$, $\pi \restriction \mathcal{A}(\mathcal{O})$ is unitarily equivalent to the ΦOK representation restricted to the region, then π is said to be locally ΦOK.

State (on a C* algebra, $\mathcal{A}$)

A state is a complex valued function F defined on $\mathcal{A}$, which is linear

$$F(aA + bB) = aF(A) + bF(B)$$

positive on positive elements

$$F(A^*A) > 0$$

and normalized

$$F(1) = 1 \quad .$$

Normal State (on a C* algebra, $\mathcal{A}$)

A normal state is a state that has an additional continuity property (ultra weak continuity). It can equivalently be defined as a state that is given by a density matrix ρ:

$$\omega(A) = \mathrm{tr}(\rho A) \quad .$$

N_τ (τ a real number)

This is an operator on ΦOK space defined by

$$(N_\tau \Phi)^{(n)}(k_1 \ldots k_n) = \sum_{j=1}^{n} [k_j^2 + m^2]^{\tau/2} \, \Phi^{(n)}(k_1 \ldots k_n) \ .$$

Thus $N_0 = N$, the number operator, and

$N_1 = H_0$, the Hamiltonian for free particles of mass m.

N_τ Estimates

Estimates of Wick-ordered monomials

$$W = \int \ldots \int dk_1 \ldots dk_n \; w(k_1, k_2, \ldots k_n) \; a^*(k_1) \ldots a(k_n)$$

of the form

$$||N_\tau^{-\alpha} W N_\tau^{-\beta}|| \leq \text{const } ||w||$$

where $||w||$ is some norm of w as a function of n variables and α and β are some real numbers.

Linear Estimate

Generally refers to estimates of the form

$$A \leqslant B$$

such as

$$H_0 \leqslant a(H(g) + b)$$

or

$$\pm \phi(f) \leqslant a(H(g) + b)$$

in contrast to

Higher Order Estimate

Generally refers to estimates of the form

$$A^k \leqslant B^\ell$$

for example

$$H_0^2 < a(H(g)^2 + b) \quad .$$

Pull Through Formula

A formula which relates the product of a creation or annihilation operator and the resolvent to the product in reverse order with the resolvent translated plus a correction term. For example

$$a(k)\ R(z) = R(z-\mu(k))a(k) - R(z-\mu(k))[a(k),\ H_I]\ R(z)\ .$$

<u>Local Number Operators</u>

These are operators obtained from the operators N or N_τ by localizing in space. For example

$$N_{\tau,\mathcal{O}} = \iint E_{\mathcal{O}}(x)\ c(x,y) E_{\mathcal{O}}(y) A(x) A(y)\, dx\ dy$$

where

$$A(x) = [2\pi]^{-\frac{1}{2}} \int a(p)\ \exp(-\ ipx)\, dp\ .$$

$E_{\mathcal{O}}$ is the characteristic function of the set $\mathcal{O}$ (= 1 on the set and $\mathcal{O}$ outside), and c(x,y) is the kernel

$$c(x,y) = [2\pi]^{-\frac{1}{2}} \int \mu(p)^{\tau}\ \exp -ip(x-y)\ dp\ .$$

<u>Resolvent Convergence</u> (A notion of convergence for self-adjoint operators, bounded or unbounded on a Hilbert space $\mathcal{H}$). A sequence of self-adjoint operators A_n n = 1,2,... is convergent in this sense if the resolvents $[A_n - z]^{-1}$ converge for some non-real z. This convergence of the resolvents may be in the weak strong or uniform sense. It turns out that weak resolvent convergence is equivalent to strong. In $\mathcal{P}(\phi)_2$ and Y_2 uniform resolvent convergence of the cutoff Hamiltonians $H_{K,V}(g)$ to H(g) holds.

<u>Weyl System</u>

Two families of unitary operators U(f) and V(g) satisfying

$$U(f_1)\ U(f_2) = U(f_1 + f_2)$$

$$V(g_1)\ V(g_2) = V(g_1 + g_2)$$

$$U(f)\ V(g) = \exp(-i\ (f,g))V(g)\ U(f)$$

where the f's and g's are real test functions

$$(f,g) = \int dx\ f(x)\ g(x)$$

under the assumption that $U(tf)$ and $V(tg)$ are continuous in t, one recovers the canonical fields $\phi(f)$ and $\pi(g)$ as the self-adjoint infinitesimal generators

$$U(tf) = \exp i t \phi(f)$$

$$V(tg) = \exp i t \pi(g) \quad .$$

<u>Q-Space</u>

A realization of a representation of the CCR in which the Abelian algebra generated by the operators $\exp i \phi(f)$ is diagonalized. For a system of a finite number of degrees of freedom this corresponds to diagonalizing the oscillator coordinates.

REFERENCES

Historical and General References

Among the historical antecedents of current constructive field theory, there should be mentioned at least

1. E. Nelson, Interaction of Non-relativistic Particles with a Quantized Scalar Field, J. Math. Phys. 5, 1164 (1964) in which a renormalized Hamiltonian for a model requiring a mass renormalization was constructed for the first time, and a number of papers on cutoff field theories.
2. A.M. Jaffe, Dynamics of a Cutoff $\lambda\phi^4$ Field Theory, Princeton Thesis 1965 where the cutoff ϕ^4 theory is studied in any number of space-time dimensions and the existence of Green's functions is shown.
3. Y. Kato, Some Converging Examples of Perturbation Series in Quantum Field Theory, Prog. Theoret. Phys. 26, 99 (1961) in which the self-adjointness of the Hamiltonian for certain cutoff Yukawa interactions is proved.
4. O. Lanford, Construction of Quantum Fields Interacting by a Cutoff Yukawa Coupling, Princeton Thesis 1966 which extended Kato's results and also proved the existence of the vacuum expectation values of products of fields.

A review article giving background for the early stages of these developments is

5. A.S. Wightman, Introduction to Some Aspects of the Relativistic Dynamics of Quantized Fields in High Energy Electromagnetic Interactions and Field Theory, Cargese Lectures in Theoretical Physics 1964, Ed. M. Levy.

A paper whose ideas were very important for later developments although the results obtained with them were soon superceded was

6. M. Guenin, On the Interaction Picture, Comm. in Math. Phys. 3, 120 (1966).

The new era began with the papers by Nelson, Glimm, Segal, and Glimm and Jaffe.

7. E. Nelson, A Quartic Interaction in Two Dimensions in Mathematical Theory of Elementary Particles, R. Goodman

and I. Segal, Ed. 1966 in which the Hamiltonian H(g) of $(\phi^4)_2$, defined below, was shown to be bounded below.

8. J. Glimm, Yukawa Coupling of Quantum Fields in Two Dimensions I, Comm. in Math. Phys. 5, 343 (1967); II, ibid. 6, 61 (1967).
9. Boson Fields with Non-linear Self Interaction in Two Dimension, ibid. 8, 12 (1968).
10. I. Segal, Notes Toward the Construction of Non-linear Relativistic Quantum Fields I The Hamiltonian in Two Space-time Dimensions as the Generator of a C*-automorphism Group, Proc. Nat. Acad. Sci. U.S.A. 57, 1178 (1967).
11. J. Glimm and A. Jaffe, A $\lambda\phi^4$ Quantum Field Theory Without Cutoffs I, Phys. Rev. 176, 1945 (1968).

References to the further papers will be quoted later in connection with the detailed results. To complete the present list I quote some of the reviews which can provide a general orientation on the subject.

12. J. Glimm, Models for Quantum Field Theory in Local Quantum Theory, Ed. R. Jost, Varenna Lectures, 1968.
13. A. Jaffe, Constructing the $\lambda(\phi^4)_2$ Theory in Local Quantum Theory, Ed. R. Jost, Varenna Lectures 1968.
14. A. Jaffe, Whither Quantum Field Theory?, Rev. of Mod. Phys. 41, 576 (1969).
15. J. Glimm, The Foundations of Quantum Field Theory, Advances in Math. 1, 101 (1969).
16. K. Hepp, Theorie de la Renormalisation, Springer, Lecture Notes in Physics 1970.
17. J. Glimm and A. Jaffe, Quantum Field Theory Models, Les Houches Lectures 1970, Ed. C. DeWitt and R. Stora, Gordon and Breach, 1971.
18. B. Simon, Studying Spatially Cutoff $(\phi^{2n})_2$ Hamiltonians Haifa Lectures 1971.
19. J. Glimm and A. Jaffe, Boson Quantum Field Models London Lectures 1971.

References to Section A

For the nuclear or kernel theorem see

20. B. Simon, Distributions and their Hermite Expansions, J. Math. Phys. 12, 140 (1971).

Its application in quantum field theory is described in

21. R.F. Streater and A.S. Wightman PCT Spin and Statistics and All That, W.A. Benjamin, 1964.
22. E. Nelson, Time Ordered Products of Sharp-time Quadratic Forms, J. Functional Analysis, to appear.
23. D.A. Dubin, The Group-Theoretical Structure of Free Quantum Fields in Two Dimensions, Nuovo Cimento (10), 67B, 39 (1970).
24. J. Goldstone, Field Theories with Super Conductor Solutions Nuovo Cim. (10), 9, 154 (1961).

For the Haag-Ruelle theory of collisions see

25. R. Jost, The General Theory of Quantized Fields, Amer. Math. Soc. 1964.
26. R. Gudmansson, The Analytic and Boundedness Properties of a Three Point Function in a Quantum Electrodynamic Like Model, Nuclear Phys. 40, 202 (1963).

References to Section B

A general account of functional analysis oriented to the needs of constructive quantum field theory is the forthcoming book

27. M. Reed and B. Simon, Methods of Modern Mathematical Physics, Academic Press I, 1972, II and III in preparation.

A more specially aligned account is

28. B. Simon, Topics in Functional Analysis, Lectures at the LMS/NATO Instructional Conference "Mathematics of Contemporary Physics", Aug/Sept, 1971.
29. R. Feynman, A. Hibbs, Quantum Mechanics and Path Integrals, McGraw Hill 1965. The Trotter Product Formula for finite dimensional matrices goes back to S. Lie in the 19th century. It was Trotter who recognized its validity under the hypothesis that A + B is essentially self-adjoint and proved the theorem stated.
30. E. Nelson Topics in Dynamics I Flows Princeton Lecture Notes.
31. H. Trotter, On the Product of Semi-Groups of Operators, Proc. Amer. Math. Soc. There is an elegant account in 10, 595 (1959).

32. J. Ginibre, Existence of Phase Transitions for Quantum Lattice Systems, Commun. in Math. Phys. 14, 205 (1969).

33. T. Kato, Perturbation Theory for Linear Operators, Springer 1966.

References to Section C

34. L. Rosen, A $\lambda\phi^{2n}$ Field Theory Without Cutoffs, Comm. in Math. Phys. 16, 157 (1970).

35. R. Kadison, Cargese Lectures 1965.

36. J. Glimm and A. Jaffe, The $(\lambda\phi^4)_2$ Quantum Field Theory Without Cutoffs III The Physical Vacuum, Acta Math. 125, 203 (1970).

37. J. Glimm and A. Jaffe, The $(\lambda\phi^4)_2$ Quantum Field Theory Without Cutoffs II The Field Operators and the Approximate Vacuum, Annals of Math. 91, 362 (1970).

38. J. Glimm and A. Jaffe, Positivity and Self-Adjointness of the $\mathcal{P}(\phi)_2$ Hamiltonian (to appear).

39. B. Simon, On the Glimm Jaffe Linear Lower Bound in $\mathcal{P}(\phi)_2$ Field Theories (to appear).

40. T. Kato, Perturbation Theory of Linear Operators, Springer 1965.

41. R. Schrader, Yukawa, Quantum Field Theory in Two Space-Time Dimensions Without Cutoffs, Ann. of Phys. (to appear).

42. K. Osterwalder and R. Schrader, On the Uniqueness of the Energy Density in the Infinite Volume Limit for Quantum Field Models (to appear).

References to Section D

The standard references to the work on dynamical instability and broken symmetry can be traced backwards from

43. K. Symanzik Renormalizable Models with Simple Symmetry Breaking I Symmetry Breaking by a Source Term, Comm. in Math. Phys. 16, 48 (1970).

There is a standard formal argument which uses the condition (118) to show that the Hamiltonian of the $(\phi^3)_{S+1}$ theory is unbounded below. See, for example,

44. G. Baym, Inconsistency of Cubic Boson-Boson Interactions Phys. Rev. 117, 886 (1960).

See also[5], where Baym's argument is repeated with some attention paid to the problem of Wick ordering.

The discussion of bounds on the gap in the spectrum of $\mathcal{H}_V$ is based in part on

45. R. Baumel Princeton thesis in preparation.

46. This bound was pointed out to us by B. Simon whom we thank for this and other useful remarks.

THE $P(\phi)_2$ MODEL

Lon Rosen*
Princeton University
Princeton, New Jersey

1. Introduction

The model for which the most complete results have been obtained is the $P(\phi)_2$ model describing self-interacting massive scalar bosons in two space-time dimensions. It is known that this model satisfies all of the Haag-Kastler axioms and many of the Wightman axioms. In terms of the program outlined by Arthur Wightman in the previous talk, progress on $P(\phi)_2$ has been carried forward to about item 7 and thus provides a good laboratory for investigating the questions of a physical nature contained in items 8-10.

The central work on $P(\phi)_2$ has been done by Glimm and Jaffe in a series of papers[1-8] to which the serious student is referred. In the last two years many other people have contributed towards extending and clarifying what we know with the result that $P(\phi)_2$ has now attained a certain level of simplicity and elegance[9-22]. Since several excellent reviews of this model are available [23-25], I shall in this talk make no attempt at comprehensiveness, but shall focus on some aspects of the construction which illustrate the techniques used.

Let me begin by describing some of the principles that have emerged so far in the program.

a) Since the perturbation series diverge[26], we must use nonperturbative methods. Nevertheless it would be rash to reject perturbation theory altogether, for as in quantum electrodynamics there is a great deal of information contained

*Supported in part by AFOSR under contract No.AF49(638)1545.

in lower order perturbation theory. In particular perturbation theory seems to be an infallible guide to the nature and magnitude of renormalization counter terms[27]. With $P(\phi)_2$ there are no ultraviolet divergences in perturbation theory and indeed the Hamiltonian requires only an infinite vacuum energy shift corresponding to the volume divergence. (This is not so for Y_2 as Robert Schrader will describe.) In addition one can make predictions regarding the domains of operators and the feasibility of estimates on the basis of the Rayleigh-Schrödinger series for the vacuum.

b) The Hamiltonian is the central object. One sneaks up on the physical Hamiltonian by approximating it with cutoff versions. While some of the cutoffs may be introduced for convenience (e.g., ultraviolet cutoff in $P(\phi)_2$), the cutoffs are in general necessary in order to make sense of a singular expression (e.g., box or space cutoff). With the cutoffs in force, cutoff dependent counter terms are added; if these counter terms are correctly chosen, then the theory converges as the cutoffs are removed.

One approach to the $P(\phi)_2$ Hamiltonian H that illustrates this procedure is to start with the space-(g), ultraviolet-(K) and box-(V) cutoff Hamiltonian $H_{K,V}(g)$. This reduces the problem to one with a finite number of degrees of freedom and one (c.f. Jaffe[28] can gain control over $H_{K,V}(g)$. No counter terms are required and as $K,V \to \infty$, $H_{K,V}(g) \to H(g)$ on Fock space(section 3.) This enables us to prove that $H(g)$ is a self-adjoint semibounded operator. Since we wish the spectrum of $H(g)$ to start at 0 we renormalize $H(g)$ by the only counter term of the theory, a constant E_g,

$$H(g) = H_0 + :P(\phi):(g) - E_g \quad . \tag{1.1}$$

As $g \to 1$, $E_g \to -\infty$, but $H(g) \to H$ in the weak sense of automorphisms of the algebra of observables, as described in the previous talk.

c) This brings me to my next point - the role of estimates. It is crucial for the success of the above cutoff strategy that we can derive estimates on the cutoff operators that are <u>independent of the cutoffs</u>. The various convergence statements are established in this way. In general one first proves estimates and then feeds them as input into functional analysis arguments. One could make a good case for the principle that every estimate has some use. In support of this principle, I have compiled a representative list of estimates and their consequences in the next section.

d) Roughly speaking the estimates serve to control the two sources of divergences in this model - the number and volume singularities. The first arises from the fact that every power of the field behaves like $N^{\frac{1}{2}}$ where N is the number operator; this is obvious from the definition of the annihilation operator,

$$(a(k)\Psi)^{(n-1)}(p_1,\ldots,p_{n-1}) = \sqrt{n}\,\Psi^{(n)}(k,p_1,\ldots,p_{n-1}). \quad (1.2)$$

The second arises from translation invariance and is reflected in the divergence of g dependent quantities, such as E_g, as $g \to 1$.

e) Much of the success with $P(\phi)_2$ has depended on exploiting the "Q-space" representation of the theory. Q-space can be thought of as infinite dimensional Euclidean space and is obtained by expanding the fields in harmonic oscillator variables with each oscillator expressed in the Schrödinger representation but with Gaussian measure[29]. Its virtue is that under the unitary equivalence $F \approx L^2(Q)$, ϕ and $H_I(g)$ go into multiplication operators, the Fock vacuum goes over into the function identically 1, and H_0 becomes the Hermite operator (in a continuum of variables). Q-space seems to be suited to making statements about the positivity of operators, whereas Fock space lends itself to estimates on the norms of operators.

f) As $g \to 1$ and particles are allowed to interact in all of space, we exploit this principle; although events in one region of space are not independent of events in a far away region, at least the coupling is exponentially small as a function of the distance between the two regions. This situation is of course due to the relativistic form of the fields. Thus let $\phi_{NW}(x) = 2^{-1/2}[\hat{a}(x) + \hat{a}^*(x)]$ be the

Newton-Wigner field corresponding to annihilation and creation at the point x; here $\hat{a}$ is the Fourier transform of (1.2). The relativistic field $\phi(x) = \mu^{-1/2}\,\phi_{NW}(x)$, where $\mu = (-\Delta + m_0^2)^{1/2}$. Now μ is not a local operator but it almost is. More precisely, if we write $\mu^{\tau}f(x) = \int k_{\tau}(x-y)\,f(y)\,dy$ then the kernel[30]

$$k_{\tau}(x) = 0(|x|^{-(\tau+1)/2}\,e^{-m_0|x|}) \tag{1.3}$$

as $x \to \infty$. Consequently if suppf.$\subset$ B, the smeared field $\phi(f)$ is affiliated with the observables $A(B)$, but the one particle state $\phi(f)\Omega_0$, for example, does not have support in B. However by (1.3) its wave function, $(2\mu)^{-1/2}\,f$, vanishes exponentially at large distances from B.

2. Some Estimates

2.1. N_{τ}- Estimates

Let

$$W = \int w(k_1,\ldots,k_n)\,a^*(k_1)\ldots a^*(k_j)\,a(k_{j+1})\ldots a(k_n)\,dk$$

be a Wick monomial whose kernel $w \in L^2(R^n)$. Because $a, a^* \sim N^{1/2}$, W is an unbounded operator. However, since $W \sim N^{n/2}$ the operator $(N+1)^{-a/2}\,W(N+1)^{-b/2}$ is bounded if $a + b \geq n$; in fact,

$$||(N+1)^{-a/2}\,W(N+1)^{-b/2}|| \leq \text{const.}\ ||w||_{L^2} \tag{2.1}$$

as an application of Schwarz' inequality shows.

It is possible to improve on (2.1) by replacing N by the number-energy operator N_{τ} which is the biquantization of the operator μ^{τ}, $N_{\tau} = \int a^*(k)\,\mu(k)^{\tau}\,a(k)\,dk$. Every factor of $(N_{\tau}+1)^{-1/2}$ on the right (up to a maximum of j) introduces an energy factor $\mu(k_i)^{-\tau/2}$ in $||\ \ ||_{L^2}$, and similarly on the right[23]. For example,

$$||(N_{\tau}+1)^{-j/2}\,W(N_{\tau}+1)^{-(n-j)/2}|| \leq \text{const.}\ ||\prod_{i=1}^{n}\mu(k_i)^{-\tau/2}W||_{L^2}\,. \tag{2.2}$$

Consequences a) The Wick power $:\phi^{2n}(g):$ is a well-defined operator on $D(N^n)$ if $g \in L^2$. For if we expand $:\phi^{2n}(g): = \sum_{j=0}^{2n} W_j$ as a sum of Wick monomials W_j with j creators and (2n-j) annihilators, the kernels

$$w_j = \binom{2n}{j} \hat{g}(-k_1-\ldots-k_j+k_{j+1}+\ldots+k_{2n}) \prod_i \mu(k_i)^{-1/2}$$

are in L^2. Note that this fact relies on the assumption that the number of space dimensions s = 1.

b) The time zero fields $\phi(x)$ and $\pi(x)$ are continuous in x as bilinear forms on a suitable domain[2]. For consider a typical term $W = \int e^{-ikx} \mu(k)^{-1/2} a^*(k)\, dk$ occurring in $\phi(x)$. By (2.2), if $\tau > 0$,

$$||(N_\tau+1)^{-1/2} W|| \leq \text{const.}\ ||\mu(k)^{-1/2-\tau} e^{ikx}||_{L^2} < \infty. \quad (2.3)$$

It follows from (2.3) and its adjoint relation that $\phi(x)$ is a well-defined bilinear form on $D(N_\tau^{1/2}) \times D(N_\tau^{1/2})$ and as such is a continuous function of x.

2.2. Q-space Estimates-Hypercontractivity

Considered as a function on Q-space, $V = H_1(g)$ enjoys the following two properties: for any $p < \infty$,

$$||V||_{L^p} < \infty \quad (2.4)$$

$$||e^{-V}||_{L^p} < \infty\ . \quad (2.5)$$

We are assuming here and in what follows that $g \geq 0$ and $g \in L^1 \cap L^2$. The estimate (2.4) says that V is "almost bounded"; it follows from the fact the Fock vacuum Ω_0 corresponds to the function 1 in Q-space so that

$$||V||_{L^{2r}} = (\Omega_0, V^{2r}\Omega_0)^{1/2r} < \infty\ ,$$

by repeated application of the result (1a). The estimate (2.5) says that V is "almost bounded below" and is the rigorous expression of the fact that the interaction is formally bounded below[31,32].

Since Q space has total measure 1 [29], $L^q \subset L^p$ if $p < q$, and we can think of L as consisting of the nicest functions. An important property of H_0 is that the semigroup e^{-tH_0} $(t \geq 0)$ is smoothing in the sense that for any $p > 1$ and $q < \infty$, e^{-TH_0} is a contraction from L^p to L^q for T sufficiently large:

$$||e^{-TH_0}\psi||_{L^q} \leq ||\psi||_{L^p} \, . \qquad (2.6)$$

Such a semigroup is said to be "hypercontractive"[16]. Furthermore we can deduce that e^{-tH_0} is "positivity preserving" on Q-space, i.e.

$$\psi \geq 0 \quad \text{implies} \quad e^{-tH_0}\psi \geq 0 \, . \qquad (2.7)$$

Consequences. The estimates (2.4) - (2.6) formed the basis for the original Nelson-Glimm proof of semiboundedness of H(g)[31,32]. Segal[13] and Simon-Hoegh-Krohn[16] have taken (2.4) - (2.6) as the starting point for a systematic investigation into the "perturbation of hypercontractive semigroups". Their abstract theory applied to $P(\phi)_2$ yields a self-adjointness proof for H(g) alternate to those given by Rosen[10,14] of which one is sketched in section 3.

Since e^{-tV} is obviously positivity preserving, one deduces from (2.7) by the Trotter product formula that so is $e^{-t(H_0+V)}$. This fact is an important ingredient in the proof that any ground state for H(g) can be chosen to be strictly positive (almost everywhere) in Q-space; since two strictly positive functions cannot be orthogonal, the ground state Ω_g must be unique.

2.3. Semi-boundedness of H(g)[31,32,13,16]

Perhaps the most important estimate is the stability condition

$$H_0 + H_I(g) \geq E_g \qquad (2.8)$$

where E_g is a (finite, negative) constant depending on g. Choosing E_g to be the largest possible constant, we renormalize H(g) as in (1.1) so that its spectrum starts at 0. Given (2.5) and (2.6), we can establish (2.8) as follows. Consider the operator $B = e^{-TH_0}e^{-2TV}e^{-TH_0}$. For suitable T,

e^{-TH_0} is a contraction from $L^2(Q)$ to $L^4(Q)$ by (2.6). Since e^{-2TV} is in L^4 by (2.5), it is a bounded operator from L^4 back to L^2 by Hölder's inequality. Finally e^{-TH_0} is a contraction on L^2 so that B is a bounded operator on L^2. Since $B \leq ||B||$, we obtain

$$e^{-2TV} < ||B||e^{+2TH_0}$$

and by the monotonicity of the logarithm,

$$-(\log ||B||)/2T \leq H_0 + V .$$

Consequences. Besides its physical significance, (2.8) lies at the heart of most of the known results for H(g), in particular, self-adjointness. As an immediate consequence we have the first order estimate: for any $a > 1$, there is a constant b_g such that

$$H_0 < a(H(g) + b_g) . \tag{2.9}$$

Note that inequalities like (2.9) hold in the sense of bilinear forms on the form domain of the right side, in this case, $D(H(g)^{\frac{1}{2}}) \times D(H(g)^{\frac{1}{2}})$. The estimate (2.9) is nothing more nor less than the semiboundedness estimate (2.8) applied to the Hamiltonian $H_0+a/(a-1)H_I(g)$. In like fashion we can deduce that if Q is a polynomial whose degree does not exceed that of P and if $|f|/g <$ const., then

$$:Q(\phi):(f) \leq a(H(g) + b) \tag{2.10}$$

for suitable constants a and b.

As an application of (2.9) we can prove that the field at time t

$$\phi(x,t) = e^{itH(g)}\phi(x,0)e^{-itH(g)} \tag{2.11}$$

is continuous in x and t as a bilinear form on $D(H(g)) \times D(H(g))$. For by (2.9), $e^{-itH(g)}$ takes $D(H(g))$ into $D(H_0^{\frac{1}{2}})$

and we have already seen that $\phi(x,0)$ is continuous in x on $D(H_0^{\frac{1}{2}}) \times D(H_0^{\frac{1}{2}})$. (Note that $H_0 = N_1$.)

2.4. Higher Order Estimates

The first order estimates (2.9) and (2.10) lead to higher order estimates[14]: for any $j > 0$

$$N^j \leq a(H(g) + b)^j \tag{2.12}$$

and for any $\varepsilon > 0$ there is an integer j such that

$$H_0^{3-\varepsilon} \leq a(H(g) + b)^j \tag{2.13}$$

where a and b are suitable constants. These estimates control the bare number and energy singularities of the theory in the sense that the expectation of N^j and $H_0^{3-\varepsilon}$ is finite in any state of finite total energy (with respect to $H(g)$).

According to perturbation theory (2.13) is best possible in the sense that we cannot take $\varepsilon \leq 0$. For if we could, then $(\Omega_g, H_0^3\Omega_g) = ||H_0^{3/2}\Omega_g||^2$ would be well-defined, whereas $H_0^{3/2}$ cannot be applied to the first order approximation $\Psi_1 = \Omega_0 - H_0^{-1}H_I(g)\Omega_0$ to Ω_g: If $H_I(g) = :\phi^{2n}:(g)$, then

$$||H_0^{3/2}\Psi_1||^2 = ||H_0^{\frac{1}{2}} H_I(g)\Omega_0||^2$$

$$= \sum_{j=1}^{2n} \int |\hat{g}(k_1+\ldots+k_{2n})|^2 \prod_{i\neq j} \mu(k_i)^{-1}\, dk = \infty .$$

Consequences a) The higher order estimates yield a proof of self-adjointness of $H(g)$ based on the resolvent convergence of the cutoff Hamiltonians $H_{K,V}(g)$ (section 3).

b) As above we can deduce differentiability properties of the field in x and t considered as a bilinear form $(\Psi,\phi(x,t)\Psi)$. Again the idea is that we are able to say that $\Psi(t) = e^{-itH(g)}\Psi$ is a nice vector if Ψ is. Thus the higher order estimates imply that if $\Psi \in D(H(g)^j)$ then $\Psi(t) \in D(H(g)^j) \subset D(H_0^{3/2-\varepsilon}) \cup D(N^j)$. Using this fact we can establish that on $C^\infty(H(g)) \times C^\infty(H(g))$, $\phi(x,t)$ is C^2 in x and t and that it satisfies the field equation

$$\phi_{tt}(x,t) - \phi_{xx}(x,t) + m_0{}^2\phi(x,t) + \vdots\, P'(\phi(x,t))\, \vdots = 0$$

where the nonlinear term

$$\vdots P'(\phi(x,t))\vdots = e^{itH(g)} :P'(\phi(x,0)): e^{-itH(g)} .$$

c) As a consequence of (2.12) and (2.13) we can make two assertions about the spectrum of $H(g)$. The first is that in $(-\infty, m_0)$ $H(g)$ has only point spectrum of finite multiplicity with no points of accumulation; this is proved by using the norm resolvent convergence $H_{K,V}(g) \to H(g)$ to transfer to $H(g)$ the fact that the spectrum of the boxed Hamiltonian $H_{K,V}(g)$ is discrete below m. The second assertion is that the interval $[m_0, \infty)$ lies in the continuous spectrum of $H(g)$ [12,15]. The method of proof consists of using (2.12) to establish the existence of the asymptotic limits of modified annihilation and creation operators

$$a_{\pm}(k) = \lim_{t\to\pm\infty} e^{itH(g)} e^{-itH_0} a(k) e^{itH_0} e^{-itH(g)} .$$

Since $[a_{\pm}(k), H(g)] = \mu(k) a_{\pm}(k)$, $H(g)$ acts like H_0 on the Fock space $\mathcal{F}_{\pm}$ generated by the $a_{\pm}^*$ from the vacuum Ω_g. The assertion follows since H_0 clearly has continuous spectrum $[m_0, \infty)$.

d) One can also prove higher order estimates with the local generator of Lorentz rotations replacing $H(g)$ [11,17]. The result is that the theory is Lorentz covariant in the algebraic sense that every Lorentz transformation $\{a,\Lambda\}$ is unitarily implemented on local algebras $A(B)$.

e) "Complex higher order estimates" are possible for $H_\beta(g) = H_0 + \beta{:}\phi^4(g){:}$, where the coupling constant β is in the complex plane cut along the negative real axis. This leads to summability results as described later by Barry Simon.

2.5. Linear Lower Bound

The semiboundedness of $H(g)$ can be sharpened to an estimate on the size of the vacuum shift [2,7,20]: for c independent of g [33]

$$E_g \geq -|\text{supp } g| \qquad (2.14)$$

where $|\text{supp } g|$ is the length of the interval where $g \neq 0$. Perturbation theory predicts (2.14) since the second order approximation to E_g

$$-(H_I(g)\Omega_0, H_0^{-1} H_I(g)\Omega_0) = 0(|\text{supp } g|).$$

The estimate (2.14) is reasonable in view of point f in the previous section. If different regions of space were completely independent, then the calculation of

$$E_g = \min_{||\Psi||=1} (\Psi, (H_0 + H_I(g))\Psi)$$

would decouple into a sum of the answers to similar problems over the different regions of space where $g > 0$.

Consequences Using (2.14) we obtain a refinement of the first order estimate (2.9),

$$H_0/|\text{supp } g| < H(g) + b \qquad (2.15)$$

where now b is independent of g. (2.15) says that as $g \to 1$ the bare energy per unit volume remains bounded. It is this property that leads to the Local Fockness of $P(\phi)_2$[2,3].

2.6. Local First Order Estimate

T. Spencer[34] has obtained the following improvement of (2.15). For $\tau < 1/2$ let $N_{\tau,B}$ be the biquantization of $\chi_B \mu^{\tau} \chi_B$ where χ_B is the characteristic function of the finite interval B. $N_{\tau,B}$ is the local (bare) number-energy operator for the region B. Then

$$N_{\tau,B} \leq c\,(H(g) + 1) \qquad (2.16)$$

where the constant c depends on the volume of B but is otherwise independent of g and B.

Consequences. It follows immediately from (2.16) that

$$\omega_g(N_{\tau,B}) \leq c \qquad (2.17)$$

where $\omega_g(A) = (\Omega_g, A\Omega_g)$ is the vacuum state for $H(g)$. Note that the "Locally Fock" estimates (2.15) - (2.17) are all predicted by perturbation theory since the expectation value $(\Psi_1, N_{\tau,B}\, \Psi_1)$ is bounded independently of g, where Ψ_1 is the first order approximation to Ω_g. The estimate (2.17) is the basic ingredient in the proof that as $g \to 1$ the states ω_g have a norm convergent subsequence on each local algebra (section 4).

2.7. Estimates on the Fields

Let f be a real function in $C_0^\infty(R^2)$. Then it is a simple consequence of (2.9) that

$$\phi(f), \phi(f)^2 \leq a(H(g)+b_g) \ . \tag{2.18}$$

A much deeper result[8] is that since f has compact support, polynomials in $\phi(f)$ can be dominated by $H(g)$ with constants a and b <u>independent of g</u>. These estimates transfer to the physical Hilbert space when $g \to 1$ for instance,

$$\phi(f) \leq a(H+b) \ . \tag{2.19}$$

Moreover, since $\phi(f_t) = i\,[\phi(f), H(f)]$ we have the estimate (2.19) for the commutator

$$i[\phi(f), H] < a(H+b) \ . \tag{2.20}$$

Consequences

a) Together (2.19) and (2.20) imply that $\phi(f)$ is a self-adjoint operator and that any core for $D(H^{1/2})$ is a core for $\phi(f)$[24,35].

b) Since $\phi(f)$ is self-adjoint one can define functions of $\phi(f)$, in particular, the algebra of observables A. The statement of locality can be rigorously formulated: if f and h have space-like separated supports, then on $D(H^{3/2})$

$$[\phi(f), \phi(g)] = 0 \ .$$

c) It is a simple consequence of (2.19) and (2.20) that

$$\phi(f): \quad D(H^{j/2}) \to D(H^{(j-1)/2})$$

for any positive integer j. Thus arbitrary products of fields are defined on $C^{\infty}(H)$ and the vacuum expectation values of the theory are tempered distributions[24].

d) On the basis of (2.19) and (2.20) Nelson[35] has proved that the time-ordered and retarded functions are well-defined tempered distributions.

2.8. Exponential decoupling

There are a number of estimates, all based on (1.3), which express the fact that the interference between distant regions of space is exponentially small[36]. We formulate here an estimate[3] whose content is this: Let A be an observable localized in the finite interval $B \subset R$ in the sense that $A \in \mathcal{A}(B)$. A is not localized in any finite interval in the sense of Newton-Wigner co-ordinates but the Newton-Wigner localization of A in distant regions is exponentially small.

More precisely, let $\mathcal{A}_o(B)$ be the algebra generated by finite linear combinations of operators $e^{i[\phi(f)+\pi(h)]}$ where $f, h \in C_o^{\infty}(B)$ and $\phi(f)$ and $\pi(h)$ are time zero fields. Expand $A \in \mathcal{A}_o(B)$ in a Taylor series of Wick monomials $A = \sum_{j,k=0}^{\infty} C_{jk}$, where

$$C_{jk} = \int c_{jk}(x_1,x_2,\ldots,x_{j+k})\, \hat{a}^*(x_1)\ldots\hat{a}^*(x_j)\, \hat{a}(x_{j+1})\ldots\hat{a}(x_{j+k})\,dx. \tag{2.21}$$

Then the kernels c_{jk} decrease exponentially at ∞ in the following sense. If $L = (\ell_1,\ldots,\ell_{j+k})$ where ℓ_i is an integer, we set

$$c_{jk}^L(x) = \mathrm{Sym}\ (c_{jk}(x) \prod_{i=1}^{j+k} \chi_{\ell_i}(x_i))$$

where $\chi_\ell(x)$ is the characteristic function of $[\ell,\ell+1]$ and the symmetrization takes place separately over the j creation and the k annihilation variables. Then there is a constant M such that

$$||c_{jk}^L|| \leq ||A||\ M^{j+k} \exp(-m_o \sum_{i=1}^{j+k} |\ell_i|) \tag{2.22}$$

where $||c^L_{jk}||$ is the norm of the operator from $F^{(k)}$ to $F^{(j)}$ with kernel c^L_{jk}.

Consequence Local norm compactness of the ω_g (section 4).

3. Removal of the Box and Ultraviolet Cutoffs[14]

Recall that the cutoff field is defined as

$$\phi_{K,V}(x) = (4\pi)^{-1/2} \int_{-K}^{K} \mu(k_V)^{-1/2} e^{ikx}(a^*(-k) + a(k))\, dk$$

where k_V is the point in the lattice $(2\pi/V)Z$ closest to k; and that $H_{K,V}(g) = H_{0,V}+H_{I,K,V}(g)$, where $H_{0,V} = \int a^*(k)\mu(k_V)a(k)$ and $H_{I,K,V}(g) = :P(\phi_{K,V}):(g)$.

Let $R_{K,V}(z) = (H_{K,V}(g)-z)^{-1}$ be the resolvent of $H_{K,V}(g)$. There are three steps involved in the removal of the K,V cutoffs: (i) as $K,V \to \infty$, the $P_{K,V}(z)$ converge in norm to an operator R(z) provided z = -b, a large negative real; (ii) P(z) is the resolvent of a self-adjoint operator T, R(z) $= (T-z)^{-1}$; (iii) T can be identified with H(g). These steps follow from higher order estimates of the form[14]

$$H_{0}^{2}, (N+1)^j \leq a(H_{K,V}(g)+b)^j \qquad (3.1)$$

where the constants a and b are independent of K and V. In the limit $K = V = \infty$ (3.1) transfers to the higher order estimates, stated for H(g) in section 2.

Proof of (i). Let K,V and L,W be two pairs of cutoff parameters with $K \leq L$ and $V \leq W$. In what sense is the operator $\delta H_0 = H_{0,V} - H_{0,W}$ small as $V \to \infty$? Now δH_0 is multiplication by $\sum_{i=1}^{n} \delta\mu_i$ on the n-particle space where $\delta\mu_i = \mu(k_{i_V}) - \mu(k_{i_W})$ $= 0(V^{-1})$. Thus $\delta H_0 \simeq N/V$ so that δH_0 is an unbounded operator; however if $i + j \geq 2$,

$$||(N+1)^{-i/2}\delta H_0 (N+1)^{-j/2}|| = 0(V^{-1}) \quad . \qquad (3.2)$$

Similarly δH_I is a sum of Wick nomomials of the form $\int \delta w(k)a^*(k_1)\ldots a(k_{2n})dk$ where the L^2 norm of the kernel δw goes to zero as $K,V \to \infty$. Again δW is an unbounded operator but by the estimate (2.1),

$$||(N+1)^{-i}\delta W(N+1)^{-j}|| \le \text{const. } ||\delta w||_{L^2} \to 0 \qquad (3.3)$$

provided $i + j \ge n$, where $2n$ is the degree of $P(\phi)$.

The proof of resolvent convergence now follows from the estimates (3.1)-(3.3) and the identity

$$R^{n-1}_{L,W} - R^{n-1}_{K,V} = \sum_{i=1}^{n-1} R^{n-i}_{L,W}(\delta H_0 + \delta H_I)R^{i}_{K,V} \; . \qquad (3.4)$$

Each term on the right of (3.4) converges to 0; for example,

$$\begin{aligned} &||R^{n-i}_{L,W}(-b)\delta H_I R^{i}_{K,V}(-b)|| \\ &\quad \le ||R^{n-i}_{L,W}(-b)(N+1)^{n-i}||\cdot||(N+1)^{-n+i}\delta H_I(N+1)^{-i}|| \\ &\qquad\qquad \cdot||(N+1)^{i}R^{i}_{K,V}(-b)|| \\ &\quad \le \text{const. } ||(N+1)^{-n+i}\delta H_I(N+1)^{-i}|| \end{aligned} \qquad (3.5)$$

by (3.1), where the constant is independent of the cutoffs. Therefore (3.5) converges to zero by (3.3) and similarly for the δH_0 terms by (3.2). We conclude that $R^{n-1}_{K,V}(-b)$ converges in norm and by the continuity of (n-1)st roots so does $R_{K,V}(-b)$.

Proof of (ii). Since $B(z)$ is a limit of resolvents it is obviously a pseudoresolvent[37] in the sense that

$$R(z_1) - R(z_2) = (z_1-z_2)R(z_1)R(z_2) \; .$$

But if $R(z)$ is to be the resolvent of an operator it must be invertible and so it remains to show that the null space $N(R(z)) = 0$[37]. The additional information we use is that on the dense domain $D_n = D(H_0) \cap D(N^n)$, $H_{K,V}(g) \underset{s}{\to} H(g)$; this strong convergence can be easily checked from the estimates (3.2) and (3.3). Now let $\Psi \in N(R(-b))$ for $b > 0$, and take Φ arbitrary in D_n. Then

$$\begin{aligned} (\Phi,\Psi) &= ((H_{K,V}(g)+b)\Phi,\ R_{K,V}(-b)\Psi) \\ &\to ((H(g)+b)\Phi,\ R(-b)\Psi) = 0 \end{aligned}$$

so that $\Psi = 0$. Therefore $R(-b)$ is invertible and

$$T = R(-b)^{-1} - b$$

is a densely defined closed symmetric operator with the negative real axis in its resolvent set; hence T is actually positive self-adjoint[38].

Proof of (iii). The operator T is clearly an extension of $H(g)$. For let $\Phi \in D_n$. Then $\Phi = R_{K,V}(-b)(H_{K,V}(g) + b)\Phi$ so that in the limit $K, V \to \infty$, $\Phi = R(-b)(H(g) + b)\Phi$. In other words, $\Phi \in D(T)$ and

$$(T+b)\Phi = (T+b)R(-b)(H(g)+b)\Phi$$

$$= (H(g) + b)\Phi .$$

This proves that

$$H(g) \upharpoonright D_n \subset T . \tag{3.6}$$

To prove the reverse inclusion we note that since the estimate (3.1) is cutoff independent it extends to the limiting operator T when the cutoffs are removed:

$$H_o^2 + (N+1)^j \leq a(T+b)^j \tag{3.7}$$

when $j \geq 2n$ [39]. This inequality implies that

$$C_n \equiv D(T^n) \subset D_n .$$

Therefore from (3.6)

$$T \upharpoonright C_n \subset H(g) \upharpoonright D_n .$$

But by the spectral theorem, C_n is a core for T; hence

$$T = (T \upharpoonright C_n)^- \subset (H(g) \upharpoonright D_n)^- .$$

But a self-adjoint operator has no proper symmetric extensions so that $T = (H(g)\restriction D_n)^-$. The natural domain for $H(g)$ is $\mathcal{D} = D(H_0) \cap (D(H_I(g)))$. What we have proved is that $H(g)$ is essentially self-adjoint on $\mathcal{D}$. It remains an open question whether $H(g)$ is actually self-adjoint (i.e. closed) on $\mathcal{D}$[40].

4. Renormalization of the Hilbert Space

As Arthur Wightman has outlined, the removal of the space cutoff ($g \to 1$) involves passing to a new Hilbert space $\mathcal{H}_\omega$ in which to represent the theory. $\mathcal{F}_{ren} = \mathcal{H}_\omega$ is produced by applying the GNS construction to a vacuum state ω obtained as a limit of the approximate vacua ω_g when $g \to 1$. Here $\omega_g(A) = (\Omega_g, A\Omega_g)$ for an observable $A \in \mathcal{A}$, where Ω_g is the unique ground state of $H(g)$ in $\mathcal{F}$.

It is instructive to examine this procedure in the case where $H(g) = H_0 + \frac{\lambda}{2}\int :\phi^2(x): g(x)dx$. This $(\phi^2)_{s+1}$ model is trivial in the sense that it can be solved exactly, where the number of space dimensions $s = 1, 2,$ or 3. The details are as follows[41,42]. Let A be a generator of $\mathcal{A}$, $A = e^{i[\phi(f) + \pi(h)]}$, where ϕ and π are time-zero fields and f and h are real C_0^∞ functions. Then by an explicit calculation

$$\omega_g(A) = \exp(-[\langle f, \mu_g^{-1} f\rangle + \langle h, \mu_g h\rangle]/4) \tag{4.1}$$

where

$\mu_g = [-\Delta + m_0^2 + \lambda g(x)]^{1/2}$, acting on the one particle space $L^2(R^s)$. We assume $m_0^2 + \lambda > 0$ so that μ_g is a well-defined positive operator with $\mu_0 = \mu$. If we compare the form (4.1) of ω_g with that of the free field vacuum ω_0 we see that the effect of the ϕ^2 interaction is simply to replace the single particle energy μ by μ_g.

Now let $g \to 1$. Clearly $\omega_g(A)$ converges to

$$\omega(A) = \exp(-[\langle f, \mu_1^{-1} f\rangle + \langle h, \mu_1 h\rangle]/4)$$

where $\mu_1 = [-\Delta + m_0^2 + \lambda]^{1/2}$, the energy operator for a particle with mass $m_1 = (m_0^2 + \lambda)^{1/2}$. Thus the Hilbert space $\mathcal{F}_{ren}$ that GNS gives us from ω is Fock space for a free field with mass m_1. The passage from F to F_{ren} in this model corresponds to shifting the mass from m_0 to m_1, a result which is formally obvious from the structure of the free Hamiltonian $H_0 = \frac{1}{2} \int : \pi^2 + (\nabla\phi)^2 + m_0^2 \phi^2 : dx$[43].

For $P(\phi)_2$ the removal of the space cutoff is hardly so simple[44]. In fact it is not known whether ω_g converges; we can assert only that there is a convergent subsequence $\{\omega_{g_n}\}$ of the ω_g. In this connection we note that by a general compactness argument there will always be at least one convergent (generalized) subsequence $\{\omega_{g_\nu}\}$. [Theorem: The unit sphere of the dual space of a Banach space is compact in the w*-topology.] So if we take ω as the limit of this subsequence, what is there left to prove about the convergence? The answer is that such a vacuum state ω is obtained too cheaply. It may be that ω is physically unacceptable. In particular, we can ask whether the representation π_ω obtained from ω permits the definition of fields and a Hamiltonian. The difficulty is that the operators which are naturally defined on H_ω, namely the observables in A, are all bounded operators; whereas the fields and Hamiltonian are not. Another way of posing the same question is to ask whether the unitary groups on F_{ren}, $V(t) = \pi_\omega(e^{it\phi(f)})$ and $U(t)$ = translation by time t, are strongly continuous in t. If so, then they are generated by unbounded self-adjoint operators $\phi_{ren}(f)$ and H_{ren} defined on F_{ren}.

The problem then is to prove that π_ω is continuous in an appropriate sense [locally ultraweakly continuous]. This will follow from a continuity statement about ω [namely, that ω is locally normal]. A state ρ is said to be normal if $\rho(A_\nu) \to \rho(A)$ whenever $A_\nu \uparrow A$, or equivalently, if ρ is given by a density matrix $\rho(A) = \mathrm{Tr}(DA)$. For example, the vector state $\omega_g(A) = (\Omega_g, A\Omega_g)$ is normal. It is easy to see that the normalcy of ω_g transfers to ω if ω_g converges to ω not just in the w*- topology but in norm; i.e., $\sup_A |\omega_g(A) - \omega(A)|/||A|| \to 0$. However norm convergence as states on A is too much to expect since ω is not given by a density matrix on Fock space. But we can show that on each local algebra $A(B)$ a subsequence ω_{g_n} converges in norm. This "local norm convergence" implies that ω is "locally normal", i.e. the restriction $\omega \upharpoonright A(B)$ to each $A(B)$ is normal, and this is enough for us to draw the desired conclusions about the fields and Hamiltonian.

I should now like to discuss the ideas involved in the proof of local norm convergence and the role that the estimates play. Fix the bounded open interval B. The intuitive picture is that when g is already 1 on a large set containing B, the effect of $H_I(g)$ in B is almost independent of the changes in g near infinity; consequently Ω_g and ω_g are practically settled in B. This viewpoint is supported by the estimate (2.17),

$$\omega_g(N_{\tau,B}) \leq c \ , \tag{4.2}$$

since the constant c can be chosen to be independent of g (and of translations of B). Let $F(B)$ be the Fock space constructed from the single particle space $L^2(B)$ with respect to Newton-Wigner co-ordinates, i.e. generated from Ω_o by the $\hat{a}^*(f)$ with $f \in L^2(B)$. Denote the algebra of bounded operators on $F(B)$ by $L(F(B))$. If B is bounded and $\tau > 0$, then $(N_{\tau,B}+1)^{-1}$ is a compact operator on $F(B)$. Hence we can deduce from (4.2) that the states $\omega_g \upharpoonright L(F(B))$ are norm compact, i.e., that a subsequence ω_{g_n} converges in norm on $L(F(B))$. [The relevant theorem is: Let A be a positive operator on a Hilbert space H such that A^{-1} is compact. Then the set of states $\{\omega \text{ normal } | \omega(A) \leq 1\}$ is norm compact[45].]

Unfortunately the norm compactness of $\{\omega_g \upharpoonright L(F(B))\}$ does not directly imply that of $\{\omega_g \upharpoonright A(B)\}$. For $\mu^{\pm 1/2}$ is not a local transformation and hence the two notions of localization "$A \in L(F(B))$" and "$A \in A(B)$" are not comparable: Operators in $A(B)$ are functions of $\phi(f)$ and $\pi(f)$ which produce wave functions involving $\mu^{\pm 1/2}f$ that have unbounded support. Hence operators in $A(B)$ cannot be regarded as operators on $F(I)$ no matter how large the interval I is. However the effect of an $A \in A(B)$ outside of a large interval I does become exponentially small.

The remainder of the proof thus depends on showing that any A in $A(B)$ can be suitably approximated by an

$A_I \in L(I)$) for some large but finite interval I. By a density argument [Kaplansky Density Theorem] it is sufficient to do this for A in $A_o(B)$ which we expand as $A = \Sigma\, C_{jk}$ as in Estimate 2.8. The approximating A_I is constructed by truncating the integrations in C_{jk} to lie within the interval I. Still, because of number singularities, it is not true that $||A-A_I||$ is small. However using the regularity of ω_g with respect to the number of particles, as expressed by (4.2), we can show that for arbitrary $\delta > 0$ there is an interval I such that

$$||\omega_g(A-A_I)|| \leq \delta||A|| \quad . \tag{4.3}$$

The norm compactness of $\omega_g \restriction A(B)$ then follows by applying a diagonal process to the $\omega_g \restriction L(F(I))$. Admittedly the proof of (4.3) is somewhat involved but it relies basically on the exponential decrease of the kernels appearing in $(A-A_I)$, as stated in (2.22).

As a further bonus we can deduce from local normalcy that, so far as each local algebra $A(B)$ is concerned, the representation on F_{ren} by π_ω is unitarily equivalent to the representation on F. This "Locally Fock" property of $P(\phi)_2$ vindicates the prediction of perturbation theory stated in Section 2.6.

REFERENCES

1. J. Glimm and A. Jaffe, Phys. Rev. 176(1968) 1945-1951.
2. J. Glimm and A. Jaffe, Ann. Math. 91(1970) 362-401.
3. J. Glimm and A. Jaffe, Acta Math. 125(1970) 203-267
4. J. Glimm and A. Jaffe, J. Math. Phys. 10(1969) 2213-2214.
5. J. Glimm and A. Jaffe, J. Math. Phys. 11(1970) 3335-3338.
6. J. Glimm and A. Jaffe, Comm. Math. Phys. 22(1971) 1-22.
7. J. Glimm and A. Jaffe, Comm. Math. Phys. 22(1971) 253-258.
8. J. Glimm and A. Jaffe, The $\lambda(\phi^4)_2$ quantum field theory without cutoffs IV: "Perturbations of the Hamiltonian" to appear.
9. I. Segal, Proc. Nat. Acad. Sci. 57(1967) 1178-1183.
10. L. Rosen, Comm. Math. Phys. 16(1970) 157-183.
11. J. Cannon and A. Jaffe, Comm. Math. Phys. 17(1970) 261-321.
12. R. Hoegh-Krohn, Comm. Math. Phys. 18(1970) 109-126. On the spectrum of the space cut-off $:P(\phi):$ Hamiltonian in two space-time dimensions, to appear.
13. I. Segal, Ann. Math. 92(1970) 462-481.
14. L. Rosen, Comm. Pure App. Math. 24(1971) 417-457.
15. Y. Kato and N. Mugibayashi, Prog. Theor. Phys. 45(1971) 628-639.
16. B. Simon and R. Hoegh-Krohn, J. Funct. Anal., to appear.
17. L. Rosen, J. Math. Anal. Applic., to appear.
18. D. Masson and W. McClary, Comm. Math. Phys., to appear.
19. J. Konrady, Comm. Math. Phys., to appear.
20. B. Simon, On the Glimm-Jaffe linear lower bound in $P(\phi)_2$ field theories, to appear.
21. B. Simon, Continuum embedded eigenvalues in a spatially cutoff $P(\phi)_2$ field theory, to appear.
22. A. Klein, Self-adjointness of the locally correct generator of Lorentz transformations for $P(\phi)_2$, to appear.
23. J. Glimm and A. Jaffe, "Quantum field theory models". In : 1970 Les Houches Lectures (C. Dewitt and R. Stora, ed.) Gordon and Breach, New York, 1972.
24. J. Glimm and A. Jaffe, Boson quantum field models, preprint.
25. B. Simon, Studying spatially cutoff $(\phi^{2n})_2$ Hamiltonians, preprint.

26. A. Jaffe, Comm. Math. Phys. 1(1965) 127-149).

27. J. Glimm, "Models for quantum field theory". In: Local Quantum Theory, ed. R. Jost, Academic Press, New York, 1969.

28. A. Jaffe, Princeton University Thesis, 1965. See also: A. Jaffe, O. Lanford, A. Wightman, Comm. Math. Phys. 15 (1969) 47-68.

29. The reason for the Gaussian measure can be seen by considering the cutoff field $\phi_{K,V}(x)$ in the box. Then there are a finite number M of discrete harmonic oscillator modes and $L^2(Q)$ is a tensor product $\bigotimes_{\ell=1}^{M} L^2(R)$. Since we are interested in the limit $M \to \infty$, we realize each oscillator on $L^2(R,\rho(q)dq)$ where $\rho(q) = (\mu/\pi)^{1/2}e^{-\mu q^2}$ is the zeroth Hermite function. Then each factor has measure 1 so that in the limit $M \to \infty$ there is a measure defined on Q-space. In fact Q-space has total measure 1. For details see[24].

30. G. Watson, A treatise on the theory of Bessel Functions, Cambridge University Press, 1944.

31. E. Nelson, "A quartic interaction in two dimensions". In: Mathematical Theory of Elementary Particles, ed. R. Goodman and I. Segal, M.I.T. Press, 1966.

32. J. Glimm, Comm. Math. Phys. 8(1968) 12-25.

33. When we assert that constants are independent of g in the estimates of this section, it is understood that $g \to 1$ in a smooth fashion. For example, in the estimate (2.14) it is sufficient that g be uniformly bounded.

34. T. Spencer, New York University Thesis, 1972.

35. E. Nelson, Time-ordered operator products of sharp-time quadratic forms, to appear.

36. Besides the estimate of Ref. 3 that we state here, similar estimates[7,20] play a role in the proof of (2.14). See also K. Osterwalder and R. Schrader, On the uniqueness of the energy density in the infinite volume limit for quantum field models, preprint.

37. T. Kato, Perturbation Theory for Linear Operators, Springer Verlag, New York, 1966. (page 428).

38. $T + b$ is symmetric and closed since it is the inverse of a (closed) self-adjoint operator; it is densely defined since the orthogonal complement of the range of $R(-b)$ is $N(R(-b)) = 0$. With the negative reals in its resolvent set, T has deficiency indices (0,0) and is thus self-adjoint.

39. In other words, when $\varepsilon = 1$ in (2.13) we may take $j = 2n$ 14.

40. The fact that $H(g)$ is closed on $\mathcal{D}$ is equivalent, by the closed graph theorem, to the quadratic inequality $H_o^2 \leq a(H(g)+b)^2$. In the particular case $P(\phi) = \phi^4$ this inequality has been proved[1] but it remains open for general $P(\phi)$. This is one of the few points where present technology distinguishes between ϕ^4 and ϕ^{2n}.

41. J. Eachus, Syracuse University Thesis, 1970.

42. L. Rosen, Renormalization of the Hilbert space in the mass shift model, preprint.
When $s = 3$ an infinite vacuum shift ($E_g = -\infty$) is necessary for $(\phi^2)_s$.

43. It is worth pointing out that two representations of the CCR's on Fock spaces associated with different masses are <u>not</u> unitarily equivalent.

44. See reference[3] for the complete details of the material in the remainder of this section.

45. This theorem[3] can be understood by relating it to the Arzela-Ascoli (respectively Rellich) compactness theorems which tell us that if a closed family of functions in $C(B)$ (respectively $L^2(B)$) is bounded and has uniformly bounded derivatives then the family is compact. The significance of (4.2) is that the wave functions making up ω_g have uniformly bounded L^2 derivatives of fractional order $\tau > 0$.

THE YUKAWA QUANTUM FIELD THEORY IN TWO SPACE-TIME DIMENSIONS*

Robert Schrader
Harvard University
Cambridge, Massachusetts

I. INTRODUCTION

In this talk we review recent progress made on a particular model in quantum field theory, the Yukawa model in two space-time dimensions. In four space-time dimensions this model is supposed to give a "realistic" description of the interaction between massive nucleons of spin 1/2 and massive scalar mesons. As for the theory of polynomial boson self interaction, which has been described to you by L. Rosen, the reason for choosing two space-time dimensions is in order to make the theory as easy as possible. Compared to the $P(\phi)_2$ theory this model poses new problems which are related to the fact that an infinite energy and boson mass renormalization is necessary. It is an open question whether the perturbation series converges, but nonperturbative methods have been established which permit a discussion. Nevertheless perturbation theory will again be a good guide in guessing the right form of the equations and estimates involved in the theory. To study the dynamics of the theory, the canonical way is to start with the Hamiltonian and express the interaction in terms of the free nucleon and meson field operators at time zero. This approach is obviously noncovariant, but the covariance will then hopefully be regained at a later stage. It turns out that the estimates needed in order to keep a quantitative control of the theory are those which are obtained by comparing the expressions in question with suitable powers of the Hamiltonian. That this is possible is intimately connected to the

* Work supported in part by Air Force Office of Scientific Research under contract F44620-70-C-0030.

fact that the Hamiltonian is positive as we will see. Our first task will therefore be to obtain a good knowledge of the Hamiltonian.

II. The locally correct Hamiltonian[3,5,6,7]

Let us first look at the Hamiltonian

$$H = H_0 + \lambda \int : \overline{\psi}(x)\ \psi(x):\ \phi(x)dx$$

where ψ and ϕ are the time zero fields of the free fermions and bosons with mass $m_f > 0$ and $m_b > 0$ respectively. H_0 is the free Hamiltonian operator in the Fockspace $\mathcal{F}$ associated with the free fields. As usual the double dot denotes Wickordering which amounts to a first renormalization of a trivial sort. λ is a (real) coupling constant. Though H is a densely defined bilinear form on Fockspace, it is not an operator. For later purposes we recall that a bilinear form is an expression defined to have expectation values on a suitable (dense) set of vectors, whereas an operator maps a suitable (dense) set of vectors into vectors. As a first remedy to obtain an operator we introduce a momentum cut-off κ on the fields and a space cut-off g in the interaction in a way familiar from standard textbooks.[1] Thus we obtain

$$H_{unren}(g,\kappa) = H_0 + \lambda \int : \overline{\psi}_\kappa(x)\psi_\kappa(x):\ \phi_\kappa(x)g(x)dx = H_0 + H_I(g,\kappa) \quad .$$

Having constructed a well defined operator, the intention is of course to keep control over the theory when both cut-offs are removed. It is reasonable to try to remove the momentum cut-off first, since this will be possible in Fockspace, whereas the removal of the space cut-off will necessarily be accompanied by a change in Hilbertspace due to Haag's theorem on the non-existence of the interaction picture. The important fact we now want to point out is that $H_{unren}(g,\kappa)$ will not stay bounded below when $\kappa \to \infty$.

To see this let us calculate the groundstate to first order in perturbation theory

$$|\Omega_1(g,\kappa)\rangle = |\Omega_0\rangle - H_0^{-1} H_I(g,\kappa)|\Omega_0\rangle \quad .$$

Ω_0 is the free vacuum.

Then the shift of the ground state energy in second order perturbation theory is

$$E_2(g,\kappa) = -\langle\Omega_1(g,\kappa)|H_0|\Omega_1(g,\kappa)\rangle = \langle\Omega_1(g,\kappa)|H_{unren}(g,\kappa)|\Omega_1(g,\kappa)\rangle$$

which is logarithmically divergent as $\kappa \to \infty$. Since $|\Omega_1(g,\kappa)\rangle$ converges as $\kappa \to \infty$, this relation shows that $H_{unren}(g,\kappa)$ is not uniformly bounded below. The boson mass shift in second order is also logarithmically divergent as $\kappa \to \infty$. We have

$$\delta_2 m_b^2(g,\kappa) = -\langle\delta(k)|\{H_I(g,\kappa)(H_0-m_b)^{-1}H_I(g,\kappa)-E_2(g,\kappa)\}|\delta(k)\rangle$$

where $|\delta(k)\rangle$ is the one-particle boson state with zero momentum. On the other hand, all higher order contributions to the vacuum energy and the boson and fermion mass stay finite. More generally all divergent expressions in perturbation theory may be traced back to these two divergences. This suggests the following form of the renormalized self adjoint Hamiltonian

$$H(g,\kappa) = H_0 + H_I\ (g,\kappa) + c(g,\kappa)$$

where the counterterm $c(g,\kappa)$ is given by

$$c(g,\kappa) = -\tfrac{1}{2}\ \delta_2 m_b^2\ (g,\kappa)\int\ :\phi\ (x)^2:g(x)^2 dx - E_2(g,\kappa)\ .$$

Note that $c(g,\kappa)$ is homogeneous of second order in the coupling constant λ. With this new renormalized Hamiltonian, perturbation theory stays finite in every order.

We are now in a position to state the main results involving the Hamiltonian operator.

(i) The time translation groups $\exp(i\ t\ H(g,\kappa))$ converge to a unitary time translation group $\exp(i\ t\ H(g))$. The limit is independent of the particular construction of the momentum cut-off.

(ii) $H(g)$ is a self adjoint operator, which after an additional finite energy renormalization $E(g)$ has zero as the lowest point in the spectrum.

(iii) $H(g)$ gives locally the correct dynamics. In particular, influence propagates not faster than light.

A few remarks concerning the first two results are in order. Statement (i) is exactly what we expect from a reasonable theory where a removal of the high-momentum cut-off is possible: The dynamics described by $H(g,\kappa)$ should smoothly go over to a limiting dynamics as $\kappa \to \infty$. As for the second result, positivity of the Hamiltonian is a feature of all dynamical theories in which stability is expected. The g dependence of $|\delta E(g)|$ may be estimated by a constant times the volume described by g, a fact which again is predicted by perturbation theory. We will come back to statement (iii) below. Finally, we want to stress the fact that the above results are valid for all values of the coupling constant. For given g, however, the groundstate could possibly be finitely degenerate and is only known to be unique if λ is chosen sufficiently small. The proof of these statements makes heavy use of the so-called N_τ estimates, by which Wick-ordered monomials are estimated by powers of the operators

$$N_\tau = \sum_{\varepsilon=0,\pm 1} \int \mu^\tau(k,\varepsilon) b^*(k,\varepsilon) b(k,\varepsilon)\, dk \quad .$$

Here $b(k,\varepsilon)$ is the annihilation operator for a boson ($\varepsilon = 0$), fermion ($\varepsilon = +1$) or antifermion ($\varepsilon = -1$) of momentum k and relativistic energy $\mu(k,\varepsilon)$. Thus for the boson for example $\mu(k,0) = (k^2 + m_b^2)^{\frac{1}{2}}$. N_τ is the number operator for $\tau = 0$ and the free Hamiltonian for $\tau = 1$. The linear and quadratic estimates

$$\left.\begin{aligned} N_{2\tau} &\leqslant \text{const } (H(g,\kappa) + \text{const}) \\ (N_\tau)^2 &\leqslant \text{const } (H(g,\kappa) + \text{const})^2 \end{aligned}\right\} \quad \tau < \frac{1}{2}$$

which are uniform in κ then make it possible to compare certain Wick ordered monomials directly with $H(g,\kappa)$. These estimates in particular show that $H(g,\kappa)$ is bounded below uniformly in κ. This will permit us to establish that $H(g)$ is bounded below. The idea for proving the linear estimate is taken from the theory of dressing transformations. Formal perturbation theory suggests there is a unitary operator T, the wave operator, such that

$$H(g,\kappa) = T\, H_o T^* = \sum_{\varepsilon=0,\pm 1} \int \mu(k,\varepsilon)\hat{b}^*(k,\mu)\ \hat{b}(k,\varepsilon)dk$$

where

$$\hat{b}(k,\varepsilon) = T\ b(k,\varepsilon)\ T^*$$

are the dressed annihilation operators. Since H_o is positive the positivity of $H(g,\kappa)$ would follow. Expanding $\hat{b}(k,\varepsilon)$ as a power series in λ, the divergences occur only for the first order terms. But the first order ansatz

$$T = 1 + X,\ X = -X^*$$

$$H_o + H_I = (1 + X)\ H_o(1 - X) = H_o + [X,\ H_o]$$

gives $X = -\Gamma H_I$, where $[H_o,\ \Gamma H_I] = H_I$. Note that the counter-term is of second order in λ. Hence to first order

$$\hat{b}(k,\varepsilon) = b(k,\varepsilon) + [b(k,\varepsilon),\ \Gamma\ H_I(g,\kappa)] \quad .$$

Some minor modifications give a new $b(k,\varepsilon)$ such that the positive operator

$$\hat{H}(g,\kappa) = \sum_{\varepsilon=0,\pm 1} \int \mu(k,\varepsilon)\ \hat{b}^*(k,\varepsilon)\hat{b}(\ ,\varepsilon)\ dk$$

is almost $H(g,\kappa) - \frac{1}{2} N_{2\tau}$. The error may be estimated uniformly in κ. This in particular proves the first order estimate

$$N_{2\tau} \leqslant \text{const}(H(g,\kappa) + \text{const}) \quad .$$

The convergence of the unitary groups $\exp(i\ t\ H(g,\kappa))$ follows from the convergence of the resolvents $R_\kappa(z) = (H(g,\kappa) - z)^{-1}$ to the resolvent of a self adjoint operator, which by definition will be $H(g)$. This is essentially established by showing that R_κ is a Cauchy sequence such that

$$R_\kappa(z) - R_{\kappa'}(z) = -\ R_\kappa(z)\ (H(g,\kappa) - H(g,\kappa'))\ R_{\kappa'}(z) =$$
$$= R_\kappa(z)\ \delta(H_I + c)\ R_{\kappa'}(z)$$

goes to zero as $\kappa,\kappa' \to \infty$. The cancellation of the singularities arising from $H_I(g,\kappa)$ with the infinities in the counter term $c(g,\kappa)$ is then obtained by using the so-called pull through formula which describes how the creation operators $b(k,\varepsilon)$ may be commuted through the resolvent $R_{\kappa'}(z)$:

The relation

$$b(k,\varepsilon)H_o = H_o\, b(k,\varepsilon) + \mu(k,\varepsilon)b(k,\varepsilon)$$

immediately gives

$$b(k,\varepsilon)(H(g,\kappa')-z) = (H(g,\kappa') + \mu(k,\varepsilon)-z)b(k,\varepsilon) +$$

$$+ [b(k,\varepsilon),\ H_I(g,\kappa') + c(g,\kappa')]$$

where [,] denotes the commutator. Reexpressed in terms of the resolvents, this relation reads

$$b(k,\varepsilon)\ R_{\kappa'}(z) = R_{\kappa'}(z-\mu(k,\varepsilon))b(k,\varepsilon) - R_{\kappa'}(z- \ (k,\varepsilon))$$

$$[b(k,\varepsilon),\ H_I(g,\kappa') + c'(g,\kappa')]\ R_{\kappa'}(z)$$

which is the pull through formula. To show that $R_\kappa(z)$ is indeed a Cauchy sequence, we split δH_I into three parts: The fermion pair creation part $\delta H_I{}^C$, the boson emission and absorption part $\delta H_I{}^D$ and the fermion annihilation part $\delta H_I{}^A$. If we choose $\kappa < \kappa'$, then the two terms

$$R_\kappa\ \delta H_I{}^C\ R_{\kappa'},\ R_\kappa\ \delta H_I{}^D\ R_{\kappa'}$$

separately go to zero. To estimate the remaining part, we pull the fermion annihilation operators in $\delta H_I{}^A$ through $R_{\kappa'}$. This gives a sum of terms, each of which individually goes to zero with the exception of a term of the form

$$\int R_\kappa(z)\phi(k)\ R_{\kappa'}(z - \mu(p,1) - \mu(p',-1))$$

$$\{b(p,1),[b(p',-1),H_I(g,\kappa')]\}R_{\kappa'}(z)a_{\kappa,\kappa'}(k,p,p')dk\ dp\ dp'$$

where {,} denotes the anticommutator. Note that the commutator expression is linear in the boson field ϕ. We may replace $R_{\kappa'}(z - \mu(p,1) - \mu(p',-1))$ by $(\mu(p,1) + \mu(p',-1))^{-1}$ giving an error which may be controlled. Thus we are left with an expression of the form

$$\int R_{\kappa}(z)\ \phi(k)\ \phi(k')\ a_{\kappa,\kappa'}\ (k,k')\ R_{\kappa'}(z)\ dk\ dk' \quad .$$

This expression does not go to zero when κ and κ' go to infinity. We have thus extracted the singular part of $R_{\kappa}\ \delta H_I\ R_{\kappa'}$. However, this expression cancels with the counterterm: The contraction arising from the Wick ordering cancels the divergence of the energy counter term and the Wick ordered term cancels the divergence of the boson mass renormalization counterterm, thus establishing the convergence of $R_{\kappa}(z)$.

We turn to a discussion of the finite propagation speed. If we assume for a moment that we may treat $H(g)$ as a sum

$$H(g) = H_o + H_I(g) + c(g)$$

the finite propagation speed follows as in the $P(\phi)_2$ theory using Trotters product formula, since both $H_I(g)$ and $c(g)$ are local. This argument, however, is not correct and it is necessary to deal with the cut-off Hamiltonians $H(g,\kappa)$ which do not describe a dynamics with finite propagation speed since a momentum cut-off destroys locality. It may be remedied by introducing an additional time dependent perturbation $\delta H_I(g,\kappa,t)$. This perturbation restores the finite propagation speed for all finite κ. On the other hand for $\kappa\to\infty$ the contribution from $\delta H_I(g,\kappa,t)$ becomes negligible and the dynamics described by $H(g)$ is again recovered. This whole procedure may be contrasted with the situation for the $P(\phi)_2$ theory, where the introduction of a momentum cut-off is sometimes a convenient but not a necessary procedure.

III. Fields, field equations and currents[2,3,8,9,10]

To exploit the results obtained so far it is convenient first to look at the theory in the Heisenberg picture, where the dynamics is expressed in terms of the fields. We start with the smeared out time zero fields as Cauchy data:

$$\phi(f,0) = \int f(x)\ \phi(x)\ dx$$

$$\pi(f,0) = \int f(x)\ \pi(x)\ dx$$

$$\psi(f,0) = \int f(x)\ \psi(x)\ dx \quad .$$

The sharp time fields

$$\phi_g(f,t) = \exp(i\ H(g)t)\phi(f,0)\ \exp(-\ i\ H(g)t)$$

$$\pi_g(f,t) = \exp(i\ H(g)t)\pi(f,0)\ \exp(-\ i\ H(g)t)$$

$$\text{resp.}\quad \psi_g(f,t) = \exp(i\ H(g)t)\psi(f,0)\ \exp(-\ i\ H(g)t)$$

are selfadjoint (resp. bounded). Due to the finite propagation speed, they are independent of g if g = 1 on the set which has distance less than $|t|$ to the support of f. These fields satisfy the usual equal time commutation or anticommutation relations. Furthermore we may construct fields which are smeared out in space-time and which are g-independent:

$$\phi(f) = \ f(x,t)\ \exp(i\ t\ H(g))\phi(x)\ \exp(-\ i\ t\ H(g))dx\ dt$$

$$\text{resp.}\ \psi(f) = \int f(x,t)\ \exp(i\ t\ H(g))\psi(x)\ \exp(-\ i\ t\ H(g))dx\ dt\ .$$

They are selfadjoint (resp. bounded) if f is real and smooth with compact support. These field operators commute or anticommute, if the test functions have space-like separated support. The dynamics is described by the field equations, which have the form

$$\left(\frac{\partial^2}{\partial t^2} - \frac{\partial^2}{\partial x^2} + m_b^2\right)\ \phi(x,t) + j(x,t) = 0$$

$$\left(\gamma_o \frac{\partial}{\partial t} + \gamma_1 \frac{\partial}{\partial x} + m_f\right)\ \psi(x,t) + J(x,t) = 0$$

in the sense of operator valued distributions. They are hyperbolic equations reflecting the fact that H(g) describes a dynamics with finite propagation speed. The terms j, J are renormalized currents. Formally they are

$$j = \lambda :\ \overline{\psi}\ \psi :\ \ -\ \delta_2\ m_b^2\ \phi = -\ i\ [H_I + c,\ \pi]$$

$$J = -\ \lambda\ \phi\ \psi \qquad\qquad = i\ \gamma_o[H_I + c,\ \psi] \quad .$$

More precisely these equations are the limit of the field equations for

$$\phi_\kappa(x,t) = \exp(i\ t\ H(g,\kappa))\phi(x)\ \exp(-\ i\ t\ H(g,\kappa))$$

$$\psi_\kappa(x,t) = \exp(i\ t\ H(g,\kappa))\psi(x)\ \exp(-\ i\ t\ H(g,\kappa))$$

which exhibit the dynamics given by $H(g,\kappa)$. The renormalized scalar current is

$$j_\kappa(x) = -\ i[H_I(g,\kappa) + c(g,\kappa),\ \pi(x)]$$

which in the limit $\kappa \to \infty$ is a bilinear form continuous in x. Also the space-time smeared out scalar current

$$j_\kappa(f) = \int f(x,t)\ \exp(i\ t\ H(g,\kappa))j_\kappa(x)\ \exp(-\ i\ t\ H(g,\kappa))\ \ dt$$

$$= \int f(x,t)\ j_\kappa(x,t)\ dxdt$$

is a selfadjoint operator in the limit $\kappa \to \infty$ and independent of g. On the other hand, perturbation theory predicts that the sharp time renormalized scalar current, which is smeared out in space only, is not an operator.

We note that the field equations are quite insensitive to a finite mass renormalization δm_b^2 (resp. δm_f) in $H(g)$; $j(x,t)$ (resp. $J(x,t)$ is then replaced by $j(x,t) + \delta m_b^2\ \phi(x,t)$ (resp. $J(x,t) + \delta m_f\ \psi(x,t)$) which is again well defined. Stated differently this means that in this model it is possible to obtain field equations without a detailed knowledge of the particle spectrum of the interaction theory.

We remark that the space-time smeared out vector and pseudovector currents are well defined operators

$$j_\mu(f) = \int f(x,t)\ \exp(i\ t\ H(g)):\overline{\psi}(x)\gamma_\mu\psi(x):\exp(-\ i\ t\ H(g))dxdt$$

$$j_{\mu,5}(f) = \int f(x,t)\ \exp(i\ t\ H(g)):\overline{\psi}(x)\gamma_\mu\gamma_5\psi(x):\exp(-i\ tH(g))dxdt\ .$$

The sharp time vector and pseudovector currents which are smeared out in space also exist as operators. The space-time smeared out pseudoscalar current presumably does not exist.

The whole analysis may be carried out for a pseudoscalar coupling where a pseudoscalar meson replaces the scalar meson; also SU(3) symmetry may be incorporated giving a non-trivial dynamics with an algebra of currents. The Jacobi identity is not valid due to the occurrence of finite Schwinger terms. Mathematically this is related to the fact that the product of two currents only gives a densely defined bilinear form, whereas the product of three currents is not defined.

IV. The Infinite Volume Limit[4,10]

We have seen that the Heisenberg picture permitted us to obtain a dynamic theory in Fockspace in the limit $g \to 1$. However, this description is a theory without a Hamiltonian. So what we look for is a representation space for the algebra $\mathcal{A}$ of the space-time smeared out fields in which the g-independent field equations are obtained from a Hamiltonian operator. In other words, this Hilbertspace should provide us with two equivalent descriptions: the Heisenberg picture and the Schrödinger picture. This aim can indeed be achieved, though not as much is known as in the $P(\phi)_2$ theory:

There exists a Hilbertspace $\mathcal{F}_{ren}$, a continuous representation U of the space-time translation group, a representation of the algebra $\mathcal{A}$ and a groundstate Ω in $\mathcal{F}_{ren}$ such that

(i) $$U(a,\tau) = \exp(i\, a\, P - i\, \tau\, H)$$

with $H > 0\ ;\ [H,P] = 0$

(ii) $$H\,\Omega = P\,\Omega = 0$$

(iii) $$U^{-1}(a,\tau)\,\Pi(A)U(a,\tau) = \Pi(\sigma_{a,\tau}(A))\ ;\ A \in \mathcal{A}\ .$$

Here $\sigma_{a,\tau}$ is the space-time automorphism on $\mathcal{A}$ given locally by conjugation with

$$V(a)\, \exp(-\, i\, t\, H(g))$$

V(a) being the standard unitary representation of the translation group in Fockspace $\mathcal{F}$. Using the representation Π, the field equations carry over into the new space $\mathcal{F}_{ren}$.

The crucial estimate necessary to obtain these results is of the form

$$(\Omega_g, N_\tau \Omega_g) \leqslant \text{const measure (supp g)}$$

where Ω_g is any groundstate for H(g). Taking a limit point of the sequence of functionals $(\Omega_g \cdot \Omega_g)$, $\mathcal{F}_{ren}$, U, Π and Ω are then obtained through the Gelfand-Naimark-Segal construction. This representation is locally Fock.[4] It is not yet known whether the joint spectrum of H and P lies in the forward cone $H^2 - P^2 \geqslant 0$. In particular nothing is known about the particle spectrum. We note that here perturbation theory fails to make predictions. It is also not known whether the vacuum lies in the zero sector of the nuclear number operator, which is predicted by perturbation theory at least for small coupling constants λ. Nothing is known about uniqueness. The existence of Wightman functions and Lorentz covariance have yet to be established.

REFERENCES

1. See e.g. Bjørken, J. D. and S. Drell, Relativistic quantum fields, McGraw Hill, N.Y. (1965).
2. Dimock, J. Estimates, renormalized currents and field equations for the Yukawa_2 field theory, to appear in Ann. Phys.
3. Glimm, J. The Yukawa coupling of quantum fields in two dimensions I, II Comm. Math. Phys. 5 (1967) 343-386 and 6 (1967) 61-76.
4. Glimm, J. and A. Jaffe, The $\lambda(\phi^4)_2$ quantum field theory without cutoffs III; The Physical vacuum, Acta Math. 125, (1971) 203-268.
5. Glimm, J. and A. Jaffe, Selfadjointness of the Yukawa_2 Hamiltonian, Ann, Phys. 60 (1970) 321-383.
6. Glimm, J. and A. Jaffe, The Yukawa_2 quantum field theory without cutoffs, Jour. Funct. Anal. 7 (1971) 323-357.
7. Glimm, J. and A. Jaffe, Les Houches Lectures 1970, C. deWitt and R. Stora, eds., Gordon and Breach, 1971, New York.
8. McBryan, O., Harvard University Thesis, to appear (1972).
9. Schrader, R., A remark on Yukawa plus boson selfinteraction in two space-time dimensions, Comm. Math. Phys. 21 (1971) 164-170.
10. Glimm, J, Yukawa quantum field theory in two space-time dimensions without cutoffs, to appear in Ann. Phys.

PERTURBATION THEORY AND COUPLING CONSTANT ANALYTICITY IN TWO-DIMENSIONAL FIELD THEORIES

Barry Simon*
Princeton University
Princeton, New Jersey

Introduction

One obvious question to ask if one has a family of honest field theories obtained from Lagrangian models is the relation of the theory to the Feynman perturbation series. One is interested in this question for two reasons. First, one would like to understand why perturbation theory is such a good guide; put differently, one would like to show that perturbation theory "determines" the theory in some way. Secondly, one hopes to prove rigorously that some (or all) of the theories are non-trivial. If one could show a Feynman series is asymptotic for a truncated four-point Green's function, one would know some theories are nontrivial in that their S-matrix is nonzero (modulo the proof of a mass gap and the existence of one particle states).

This talk will differ in many ways from the last two talks. One does not yet have control over the analytic properties in the coupling constant in the infinite volume limit, although one can hope the development of the localization techniques discussed by Prof. Jaffe will eventually lead to control on the infinite volume limit. Thus, this talk will contain more conjectural material than the previous talks, and the proven results we do discuss are mostly over a year old. In addition, much of the talk will be devoted to discussing general properties of perturbation series rather than questions of field theory.

---

*A Sloan Fellow.

1. Conventional Wisdom

Once Feynman and Dyson had systematically developed perturbation theory it was natural to ask if the series converged for sufficiently small coupling. Let me remind you of Dyson's argument[1] predicting divergence of the series in Q.E.D. It has two elements:

(i) Consider a world with $e^2 < 0$. In such a world, electrons attract one another and repel positrons. If we ignore relativistic corrections, it is easy to see that under those conditions the groundstate energy of an n-electron system grows as n^2. Thus, for n large, an n-electron, n-positron state with the n-electrons close to one another but far from the positrons will have energy lower than $-2nmc^2$. Thus, "the vacuum" is not the lowest charge zero state. We conclude that there is no theory for $e^2 < 0$.

(ii) Thus, Dyson says, the perturbation series (in e^2) does not converge for $e^2 < 0$ since it has nothing to converge to. Since series converge in circles, the series must diverge for all e^2.

(Parenthetically, we mention that arguments of type (ii) can be misleading, for a series could converge and have nothing to do with the non-existent quantity. For example, let $H_0 = -\Delta - 1/r$ and let $V = 1/r$. If $\lambda > 1$, $H_0 + \lambda V$ has no discrete ground state, so one might suppose the ground state energy $E(\lambda)$ of $H_0 + \lambda V$ should diverge if $\lambda > 1$. But $E(\lambda) = -1/4(\lambda-1)^2$!)

Dyson's arguments suggest that the best analyticity for Green's functions that one might conjecture is for e^2 in a cut plane: $e^2 \not\in (-\infty,0]$. And one might well expect the possibility of singularities at values of e where bound states first occur, if not in Q.E.D., then perhaps in Yukawa.

Proofs of the divergence of actual perturbation series go back to Hurst, Thirring and Peterman[2] who studied the propagator in $(\phi^3)_4$ - this theory is super-renormalizable, which helped in the analysis. Jaffe[3] proved all the Green's functions have divergent series for $(\phi^4)_2$. All these proofs involve counting the number of diagrams and obtaining bounds on the contribution of each diagram. As remarked by

Frank[4], these divergences should not be surprising: the perturbation series for the ground state energy of $\rho^2 + x^2 + \beta x^4$ is divergent, essentially by Dyson's argument (Bender and Wu[5] proved a bound $a_n > CD^n\Gamma(n/2)$ for the coefficients in this series).

Alas the results for fermions are not so clear cut. There are relative minus signs between diagrams due to the Pauli principle and so divergent lower bounds are hard to come by.

The earliest results on fermion theories suggest convergence. One studies "regularized" perturbation series - that is, one puts in space-time and ultraviolet cutoffs although in a much cruder way than that described to you by Rosen and Schrader: One writes the Gell-Mann-Low series in x-space and by fiat makes the xt integrals over a finite region of space and time and assumes that there is some sort of ultraviolet cutoff which makes the Feynman propagator bounded in x-space. Caianiello[6] used Hadamard's formula for determinants to prove the Feynman series for a regularized and unrenormalized theory of Q.E.D., Yukawa or Fermi interactions had a finite radius of convergence. Several authors[7] then showed both the numerator and denominator of the Gell-Mann-Low series were entire functions.

These convergence results were all for unrenormalized theories. Renormalization, while necessary for the removal of the ultraviolet cutoff, tends to partially suppress certain diagrams and thereby partially destroys the cancellation mechanism. Simon[8] studied the effect of renormalization in Y_2 and found that when regularized the series still had a finite radius of convergence but he was unable (i) to recover the entirety of numerator and denominator in the Gell-Mann-Low series, and (ii) to obtain a circle of convergence that didn't shrink to 0 as either cutoff was removed. This shrinkage of the proven circle of convergence may indicate that the renormalized series diverges. A difference in the behavior of the renormalized and unrenormalized series has been noted in a variety of simplified Fermion models[9].

What morals should we draw from all this sparring with the perturbation series? If we are willing to be permissive and take the inability to prove certain things as indications that they may be false, one should expect:

(1) In pure boson theories, perturbation theory diverges even when ultraviolet and space cutoffs are put in. Various objects are analytic in the coupling constant, in cut planes, or at least cut circles near 0.

(2) In theories with fermions, perturbation theory diverges but not when cutoffs are put in. Singularities in the cutoff theories arise from renormalizations and approach the origin in the coupling constant plane as the cutoff goes away. Various objects are analytic in a cut plane, in the square of the coupling constant, or at least in cut circles near 0.

2. Asymptotic and Strong Asymptotic Conditions

We thus see that one expects Feynman series to diverge. One can thus ask about the relation between the series and the objects one hopes to construct in a constructive field theory program.

In this section, we will discuss "what to do when your perturbation theory diverges". This discussion will explain why we have been so concerned with analyticity properties in the last section.

Suppose one has a function, $f(\beta)$, for β real and positive or for β in a cut plane. And suppose one has a formal power series $\Sigma a_n \beta^n$. What are the various senses in which $\Sigma a_n \beta^n$ is "the perturbation series for $f(\beta)$"? There are two usual "classical" answers:

(1) $\Sigma a_n \beta^n$ converges to $f(\beta)$ at least for β small

(2) $\Sigma a_n \beta^n$ is asymptotic to f, that is for each N:

$$\lim_{\beta \downarrow 0} \left| f(\beta) - \sum_{n=0}^{N} a_n \beta^n \right| / \beta^N = 0$$

where $\lim_{\beta \downarrow 0}$ is in the sense of β going to zero along the real axis or in some sector.

Since we already expect (1) is too strong a condition to hope for (since power series converge in circles), let us concentrate on (2). There are two important facets of asymptotic series:

(a) A function has at most one asymptotic series; for if $f(\beta) - \sum_{n=0}^{N} a_n\beta^n = 0(\beta^N)$, then $\lim_{\beta\downarrow 0} (f(\beta) - \sum_{n=0}^{N-1} a_n\beta^n)/\beta^N = a_N$ which determines the a_n inductively from f.

(b) However, any series is asymptotic to an infinity of different functions. This can be illustrated by noting that as $\beta \to 0$ along the positive real axis (or in suitable sectors) then $\exp(-\beta^{\alpha})/\beta^N$ goes to zero for any $N > 0$, $\alpha < 0$.

This means that while one can hope to prove that Feynman series are asymptotic to actual answers, one should hope for something more. That is, one should seek a relation between series and functions which is weaker than (1) but stronger than (2), and which has a uniqueness result associated with it. Uniqueness results are of course associated with a condition that a function associated with the zero series be zero. Thus we are seeking a condition stronger than merely saying that f has zero asymptotic series but weaker than saying f is analytic with a zero Taylor series but which implies f is zero. Such a result is the famous theorem of Carleman [10]:

Theorem Let f be analytic in a region, $R = \{z| \ |\arg z| < \pi/2; 0 < |z| < B\}$ for some B. Suppose for all $z \in R$ and n:

$$|f(z)| \leq a_n |z|^n$$

where

$$\sum_{n=1}^{\infty} a_n^{-1/n} = \infty \quad .$$

Then

$$f \equiv 0 \quad .$$

Notice that the statement that $|f(z)| \leq a_n|z|^n$ is essentially a statement that f has 0 asymptotic series. The condition that $\sum_{n=1}^{\infty} a_n^{-1/n}$ diverges is a statement that $a_n^{-1/n}$ does not fall off too quickly or equivalently that a_n not grow too fast.

We note that Carleman's theorem under the stronger hypothesis where R is replaced by $\{z \mid |\arg z| < \pi/2 + \varepsilon;\ 0 < |z| < B\}$ and where $|a_n| < AB^n\, n!$ replaces $\sum_{n=1}^{\infty} a_n^{-1/n} = \infty$ is much easier to prove[10], following from Stirling's formula and the Phragmen-Lindelöf principle. This weaker theorem (note that $\Sigma(n!)^{-1/n}$ is barely divergent so it isn't much weaker) suffices for most purposes and for a variety of reasons we single out:

Definition We say $\Sigma\, a_n z^n$ is a strong asymptotic series for a function f(z) if and only if:

(i) f is analytic in a region

$$R_{\varepsilon,B} = \{z \mid |\arg z| < \pi/2 + \varepsilon;\ 0 < |z| < B\}$$

(ii) For some A and B and all N = 0,1,... and $z \in R$:

$$\left|f(z) - \sum_{n=0}^{N} a_n z^n\right| \leqslant AB^{N+1}(N+1)!\,|z|^{N+1} .$$

The idea which we wish to push is that for a wide variety of field theories, the renormalized Feynman series is a strong asymptotic series for Green's functions (and vacuum energies per unit volume). We will shortly discuss this idea in the $P(\phi)_2$ and Y_2 theories. Before doing this, we make a long series of remarks:

(1) At the risk of overstating it, we note that Carleman's theorem implies that at most one function, f, can have $\Sigma a_n z^n$ as strong asymptotic series for if both f and g have $\Sigma a_n z^n$ as strong asymptotic series, then $|f(z) - g(z)| < 2\tilde{A}(\tilde{B})^n n!\,|z|^n$ for all n and suitable $\tilde{A}$, $\tilde{B}$. It is this uniqueness criterion that makes the strong asymptotic condition more attractive than an asymptotic condition. It suggests that field theories are uniquely determined by their asymptotic series.

(2) If f is uniquely determined by a strong asymptotic condition, one would expect that there is some more or less explicit method for determining f from the a_n. There is, in

fact, a method known as Borel summability which does recover f. The essence of this method is the formula $n! = \int_0^\infty x^n e^{-x}\,dx$ so that formally $\Sigma a_n z^n = \int_0^\infty e^{-x}[\Sigma(a_n/n!)(zx)^n]dx$. It is a theorem of Watson (the one physicists associate with Whittaker and with Bessel functions)[11] that if f obeys a strong asymptotic condition then (a) The function $g(z) = \Sigma a_n z^n/n!$, called the Borel transform of f, which converges in a circle of radius B^{-1}, has an analytic continuation into the sector $\{z \mid |\arg z| < \varepsilon\}$ (b) If $|z| < B$ and $|\arg z| < \varepsilon$, then the integral $\int_0^\infty e^{-x}|g(zx)|dx$ is finite (c) If $|z| < B$ and $|\arg z| < \varepsilon$, then

$$f(z) = \int_0^\infty e^{-x} g(xz)\,dx \quad .$$

(3) There are a variety of other summability methods which have been proven to be applicable to divergent perturbation series. The series are generally simpler but of a similar nature to the $P(\phi)_2$ series and the methods are usually computationally superior to the Borel method. Included are the Padé[12], a modified Euler method[13], and a mixed Padé-Borel method[14].

(4) A variety of summability methods have been shown to be applicable to a quantized electron in a static, homogeneous external unquantized field[15].

(5) A strong asymptotic condition implies a bound $|a_n| < AB^n\, n!$ For $(\phi^{2m})_2$ theories there are strong indications[16] that $|a_n|$ behaves as $AB^n[n(m-1)]!$ as $n \to \infty$. Thus strong asymptotic conditions in the sense defined above probably do not hold for $P(\phi)_2$ if $\deg P > 4$.

(6) There is a simple extension of the strong asymptotic condition allowing worse growth. f is required to be analytic in $\{z \mid |\arg z| < k\,\pi/2 + \varepsilon\}$ and the remainder need only be bounded by $AB^{N+1}[k(N+1)]!\,|z|^{N+1}$. This extended notion of strong asymptotic condition could hold for $P(\phi)_2$ theories with $\deg P > 4$. This would require some control on various objects when the coupling constant β is continued off the cut plane onto a second sheet. That such control might be obtained is indicated by the fact that it has been obtained

for anharmonic oscillators $p^2 + x^2 + \beta x^{2n}$ [17].

(7) Most importantly, it is an open question as to when a sequence a_n obeying $|a_n| \leq AB^n\, n!$ is the strong asymptotic series of some function. It seems like a hard question to answer. For example, $a_n = (-1)^n\, n!$ is the strong asymptotic series of a function while $a_n = n!$ is not. This means that summability methods cannot be used as an existence tool. Only after an object has been shown to exist can one use these methods to establish a link with the perturbation series. However, the uniqueness result might be useful in the constructive program. For example, if one could prove all Green's functions summable for $|\beta| < B$, $|\arg \beta| < \varepsilon$ for some fixed ε and B, then the Lorentz invariance of the Feynman graphs would imply Lorentz invariance of the Green's function when $|\beta| < B$.

3. $(\phi^4)_2$

The original results on summability methods and divergent perturbation series were obtained for $p^2 + x^2 + \beta x^4$ [18] which is a sort of "zero-space dimensional" field theory. Thus one would expect results on strong asymptotic conditions for $(\phi^4)_2$. We emphasize once again that one can only prove this after one has proven the existence of the objects of interest and after one has some control over these objects. For this reason, we have nothing to say about $(\phi^4)_2$ in the infinite volume limit[19] or about the Green's functions[20], but:

Theorem[21] Let g be a positive function of R in $L^1 \cap L^2$ and with a first derivative in L^2.

$$H(\beta) = H_0 + \beta \int g(x) \;:\phi^4(x):\; dx \quad .$$

Then the ground state energy, $E(\beta)$, and the space-smeared equal-time VEV's

$$W(\beta) = \int \langle \Omega_\beta \phi(x_1) \ldots \phi(x_n) \Omega_\beta \rangle\, f_1(x_1) \ldots f_n(x_n)\, d^n x$$

have strong asymptotic series.

Remarks 1. We know that these series are not convergent[2,3].

2. The series for $E(\beta)$ is given by Feynman diagrams; it is also the usual Rayleigh-Schrödinger series.

While I do not want to describe all the details of the proof of this theorem, I would like to describe how some of the ideas discussed by Rosen fit in. I will restrict myself to $E(\beta)$. To treat $W(\beta)$ one needs to use higher order estimates for complex coupling.

<u>Step 1</u> Construct "nice" operators $H(\beta)$ for any β in the cut plane[22], $\{z|z\notin(-\infty,0)\}$.

<u>Step 2</u> Prove that as $|\beta| \to 0$, $(H(\beta) - \lambda)^{-1}$ converges in norm to $(H_0 - \lambda)^{-1}$ as long as $\lambda \notin [0,\infty)$ and arg β stays in some sector, $|\arg \beta| < \pi - \varepsilon$[23].

<u>Step 3</u> Use the result of Step 2 to conclude that $H(\beta)$ for $|\beta|$ small has only one eigenvalue $E(\beta)$ near 0. This is a general perturbation theory result[24].

<u>Step 4</u> Note the formula, that for β in a sector, $|\arg \beta| < \pi - \varepsilon$ and $|\beta|$ small[25]

$$E(\beta) = E_0 + \beta \frac{\langle\Omega_0, H_I(g)P(\beta)\Omega_0\rangle}{\langle\Omega_0, P(\beta)\Omega_0\rangle}$$

where $P(\beta) = (-2\pi i)^{-1} \int_{|\lambda|=\delta} (H(-\beta)-\lambda)^{-1} d\lambda$.

<u>Step 5</u> Note that quotients of functions with strong asymptotic conditions obey a strong asymptotic condition (if the denominator is not 0 at $\beta = 0$). To obtain bounds on an integral $\int_{|\lambda|=\delta}$ we need only prove uniform bounds in λ with $|\lambda| = \delta$.

<u>Step 6</u> Expand $(H(\beta) - \lambda)^{-1}$ in a geometric series with remainder

$$(H(\beta) - \lambda)^{-1} = \sum_{n=0}^{N} (-\beta)^n (H_0-\lambda)^{-1}[H_I(H_0-\lambda)^{-1}]^n$$

$$+ (-\beta)^{N+1}(H(\beta)-\lambda)^{-1}[H_I(H_0-\lambda)^{-1}]^{N+1} .$$

<u>Step 7</u> From Step 2 (and a compactness argument) note that $||(H(\beta) -\lambda)^{-1}||$ is uniformly bounded as $|\beta| \to 0$ with $|\arg \beta| < - \varepsilon$ and $|\lambda| = \delta$. Thus conclude we need only prove

$$||[H_I(H_0-\lambda)^{-1}]^{N+1}\Omega_0|| \leq AB^{N+1}(N+1)!$$

Step 8 Prove this last inequality using N_τ estimates.

4. Y_2

There are no theorems about coupling constant analyticity in Y_2 without ultraviolet cutoff even if the space cutoff is included. Let me describe what I think happens.

Consider first the cutoff theory $H_0 + \beta H_{I,\kappa} = H_\kappa(\beta)$. It is not too hard to prove that the resolvent of $H_\kappa(\beta)$ is entire in β[26]. This is the analogue of the entirety results for perturbation theory[7]

Before removing the cutoff for β real we know we must renormalize and so look at $H_0 + \beta H_{I,\kappa} + \beta^2(C_{I,\kappa}) = H_\kappa^{ren}(\beta)$. What do we expect the counter terms to do to the analyticity of the resolvent? The energy shift is harmless but the mass renormalization should not be. As an analogue consider $p^2 + x^2 + \alpha^2 x^2$. This has a ground state energy $\sqrt{1 + \alpha^2}$ which has singularities at $\alpha^2 = -1$. $H_\kappa^{ren}(\beta)$ has mass terms of the form:

$$1/2 \int m_0^2 :\phi^2(x): dx + 1/2\ \beta^2 \delta m_\kappa^2 \int g^2(x)\ :\phi^2(x): dx\ .$$

Thus:

Conjecture 1 $H_\kappa^{ren}(\beta)$ has a resolvent analytic in a plane cut along the imaginary axis from $\pm\, im_0/\delta m_\kappa$ to $\pm\, i\infty$. There are singularities at the end points of the cuts.

Thus we conjecture that as a consequence of the renormalization, there are introduced singularities which as $\kappa \to \infty$ approach $\beta = 0$. But we expect that no other singularities should occur.

Conjecture 2 As $\kappa \to \infty$, the resolvents $H_\kappa^{ren}(\beta)$ converge in norm to the resolvents of operators $H^{ren}(\beta)$ if $\text{Re}\ \beta > 0$.

Because of the convergence along the real axis, conjecture 2 will follow from the Vitali convergence theorem, if one can prove that the $R_\kappa(\beta)$ are uniformly bounded. This might follow from a generalized Glimm lower bound[27].

The next key step in a proof of strong asymptotic conditions would be:

Conjecture 3 As $|\beta| \to 0$ with $|\arg\ \beta| < \pi/2 - \varepsilon$, the resolvent at $H(\beta)$ goes in norm to the resolvent of H_0.

The proof of conjecture 3 might follow from some sort of complex version of the Dimock quadratic estimate[28]. Because $E(\beta)$ is only a function of β^2, we have analyticity in a cut β^2 plane (à la Dyson) and can hope to prove:

<u>Conjecture 4</u> $E(\beta)$ has a strong asymptotic series in β^2.

REFERENCES

1. F. Dyson, Phys. Rev. 85, 631 (1952).
2. C. Hurst, Proc. Camb. Phil. Soc. 48, 625 (1952), W. Thirring, Helv. Phys. Acta 26, 33 (1953), A. Peterman, Helv. Phys. Acta 26, 291 (1953).
3. A. Jaffe, Commun. Math. Phys. 1, 127-149 (1965).
4. W. Frank, Ann. Phys. 29, 175 (1964).
5. C. Bender and T.T. Wu, Phys. Rev. 184, 1231 (1969). This paper is quite basic to understanding what is going on in the anharmonic oscillator. While it is nonrigorous, its computations provide useful intuition.
6. E. Caianiello, Nuovo Cimento 3, 223 (1956).
7. D. Yennie and S. Gartenhaus, Nuovo Cimento 9, 59 (1958); A. Buccafurri and E. Caianiello, Nuovo Cimento 8, 170 (1958).
8. B. Simon, Nuovo Cimento 59A, 199 (1969).
9. F. Guerra and M. Mariano, Nuovo Cimento 42A, 285 (1966).
10. Carleman's theorem is proven in T. Carleman, Les Fonctions Quasianalytiques, Gauthier-Villars, Paris, 1926. The proof of the simpler theorem is discussed in G. Hardy, Divergent Series, Oxford Univ. Press 1949 and in M. Reed and B. Simon, Methods of Modern Mathematical Physics, Vol. II, Academic Press, 1973.
11. G. Watson, Phil. Trans. Roy. Soc. 211, 279-313 (1912); see also Hardy, ref. 10.
12. The general method is described in G. Baker, Adv. Teho Phys. 1, 1 (1966). The application to Rayleigh-Schrödinger series is J.J. Loeffel, A. Martin, B. Simon, and A.S. Wightman, Phys. Lett. 30B, 656 (1969).
13. The general method is discussed in G. Hardy, ref. 10. The application to perturbation series is in J. Gunson and D. Ng. (Univ. of Birmingham preprint).
14. S. Graffi, V. Grecchi and G. Turchetti, Nuovo Cimento 4B, 313 (1971).
15. This is an exactly soluable model solved in J. Schwinger, Phys. Rev. 82, 664 (1951). The relation to summability is found in V. Ogieveski, S.A.N.S.S.R. 109, 919 (1956)

and S. Graffi, J. Math. Phys., to appear.

16. This is the behavior of the main term in a Rayleigh Schrödinger series, i.e. $(-1)^n \sum_{i_1=0 \ldots i_{n-1} \neq 0}$

$$\langle 0|V|i_1\rangle \ldots \langle i_{n-1}|V|0\rangle \, (E_{i_1}-E_0)^{-1} \ldots (E_{i_n}-E_0)^{-1}.$$

For the anharmonic oscillator there is a numerical analysis by Bender and Wu (ref. 5) and a set of reasonable arguments by C. Bender and T.T. Wu, Phys. Rev. Lett. 27, 461 (1971) which predict [n(m-1)]! behavior.

17. B. Simon, Ann. Phys. 58, 79 (1970); S. Graffi, V. Grecchi, and B. Simon, Phys. Lett. 32B, 631 (1970).

18. The Padé method was proven to work for one dimensional oscillators by Loeffel et al. (ref. 12) and the Borel method by Graffi et al. (ref. 17).

19. It is our feeling that the localization techniques that Jaffe will discuss will eventually produce enough control on the infinite volume limit to prove strong asymptotic conditions in that case.

20. Similarly, we think the ideas in E. Nelson, "Time-ordered operator products of sharp-time quadratic forms" Princeton Preprint, Dec. 1971 should lead to control over the Green's functions.

21. The result for E(β) is from B. Simon, Phys. Rev. Lett. 25, 1583-1586 (1970). The result for W(β) is from L. Rosen and B. Simon Trans A.M.S., to appear.

22. The operator H(β) are "sectorial" in the sense of M. Reed and B. Simon, Methods of Modern Mathematical Physics, Vol. I, pp. 281-282, Academic Press, 1972. This is a simple consequence of the linear lower bound as noted by B. Simon and R. Hoegh-Krohn, J. Func. Anal., to appear.

23. I know of at least four proofs of this fact: (i) The original proof in Simon, Phys. Rev. Lett. (ref. 21) used the linear lower bound (ii) A proof using the $(\phi^4)_2$ quadratic estimate was suggested by L. Rosen and appears in B. Simon, Adv. Math., to appear. (iii) A proof using "the method of hyperconstractive semigroups

appears in Rosen-Simon (ref. 21) (iv) A proof due to B. Gidas, Univ. of Michigan preprint, 1971 which relies on techniques of P. Federbush.

24. This and many other very useful perturbation theory results appear in T. Kato, Perturbation Theory for Linear Operators, Springer, 1966.

25. $P(\beta)$ is the projection onto the eigenvector with eigenvalue $E(\beta)$. The formula for $E(\beta)$ is very useful. One can compute the higher order of the Rayleigh-Schrödinger series from it by using a geometric series on $(H_0 + \beta H_I - \lambda)^{-1}$ in the formula for $P(\beta)$ and one can compute the error easily using a geometric series with remainder.

26. By this we mean not that $(H(\beta) - \lambda)^{-1}$ is entire for any particular λ, but that for any β_0 there is some λ_0 with $(H(\beta) - \lambda_0)^{-1}$ analytic for β near β_0. The entirety follows from an estimate $H_{I,\kappa}^2 \leq \varepsilon H_0^2 + b^2$ for ε arbitrarily small and with b ε-dependent.

27. J. Glimm, Commun. Math. Phys. 5, 343-386 (1967); 6, 61-76 (1967).

28. J. Dimock, Harvard Preprint, 1971; a similar but divergent bound is found in J. Glimm and A. Jaffe, Ann. Phys. 60, 321-383 (1970).

Postscript: Other Super-renormalizable Theories

Prof. Wightman has asked me to conclude with a few words about other S.R. theories which are more singular than $P(\phi)_2$ and Y_2. There are three of them: $(\phi^4)_3$, Y_3 and $(\phi^3)_4$. As one would guess, even after renormalization, $(\phi^3)_4$ leads to a Hamiltonian which is not bounded from below[P1].

While self-adjointness or even semiboundedness for spatially cutoff $(\phi^4)_3$ has not been proved, the serious infinite cancellations have been controlled by Glimm[P2]. It turns out that even with a space cutoff, one must use a non-Fock representation of the CCR[P3].

Let me say a brief word about why Fock space is not enough and about a new "number infinity" which is not apparent in Feynman perturbation theory since it does not appear as a counter term in H. The three infinities that the student of Q.E.D. is familiar with are mass renormalization, field strength renormalizations and vertex renormalizations. We have already seen mass renormalization and a vacuum energy divergence in Y_2. Field strength renormalization is only necessary if coupling constants are dimensionless (or has positive dimension of length) and SR theories are marked by coupling constants with dimensions of a power of inverse length. Thus this number infinity is another infinity which appears in constructive field theory but which doesn't appear in the Feynman series. To see where this non-Fock property comes from, consider lowest order Rayleigh-Schrödinger theory. The lowest order vacuum corrections is $\Omega_1 = H_0^{-1}H_I\Omega_0$ where H_0^{-1} is 0 on $\mathcal{F}_0$ and is $(\mu(k_1) + \ldots + \mu(k_n))^{-1}$ on $\mathcal{F}_n$. The correction to the energy is $E_2 = (\Omega_0, H_I H_0^{-1} H_I \Omega_0)$. E_2 is clearly given by a less convergent integral than $||\Omega_1||^2 = (\Omega_0, H_I H_0^{-2} H_I \Omega_0)$. For $P(\phi)_2$ both E_2 and $||\Omega_1||^2$ are finite. For Y_2, $E_2 = \infty$, but $||\Omega_1||^2 < \infty$. We interpret this as telling us that an infinite energy renormalization is needed but that the new theory and its vacuum can be found in Fock space. For $(\phi^4)_3$, $||\Omega_1||^2 = \infty$. This we interpret as telling us that we cannot build the theory in Fock space.

Glimm[P2] constructs H = "$H_0 + H_I$ + counter terms" by

using a dressing transformation T. Formally, dressed particles should have energy given by H_0 (if we take the proper mass renormalization) so we expect $HT = TH_0$. Glimm uses an approximate dressing transformation, T, with $HT = TH_0$ + error. If one puts in an ultraviolet cutoff, one finds an approximate dressing transformation, T_σ. That $||\Omega_1|| = \infty$ is mirrored by $\lim_{\sigma\to\infty} ||T_\sigma \Omega_0|| = \infty$. The infinity is a "number infinity" in that "dressed" particles contain infinitely many bare particles. The key fact is that for nice vectors Ψ, $\lim_{\sigma\to\infty} ||T_\sigma \psi||/||T_\sigma \Omega_0|| < \infty$. One thus introduces a "renormalization of inner product" using a multiplicative factor of $||T_\sigma \Omega_0||^{-2}$. This renormalization called "wave function renormalization" by Glimm[P4] is the new renormalization; the renormalized inner product leads to a new Hilbert space and a non-Fock representation of the CCR.

REFERENCES TO POSTSCRIPT

P1. K. Osterwalder, Fort. der Phys., to appear.

P2. J. Glimm, Commun. Math. Phys. 10, 1-47 (1968).

P3. K. Hepp in Systèmes a un nombre infini de degrés de liberté, C.N.R.S. publ. No. 181, Paris, 1970.
J. Fabray, Commun. Math. Phys. 19, 1 (1968).
K. Osterwalder and J.P. Ekmann, Helv. Phys. Acta., to appear J.P. Eckmann, Brandeis University, Preprint.

P4. It is unfortunate that some books call field strength renormalization "wave function renormalizations". Glimm's "wave function renormalization" is very different from field strength renormalization.

PANEL DISCUSSION

JAFFE: I was thinking what would happen in this discussion session in view of possibly being asked to discuss my talk before I gave it. So to avoid that, I wrote down one or two things that might be reasonable to talk about here. First I would like to elicit some comments from the other people in the panel. Now that we have good control over theories in one space, one time dimension and some preliminary estimates in two space, one time dimensions, I would like to ask the members of the panel firstly how they feel the subject can come closer to what other people in elementary particle physics are doing and secondly how to bring out the physical ideas, perhaps to the point where the mathematical ideas can be somewhat suppressed. Now that might sound like heresy but I think we are headed in the direction where the proofs are becoming more and more complicated. We can look back 20 or 30 years at people who have worked on early renormalization theory and they gave up at a point where the things they were trying to establish became so technically complex they couldn't see their way through the problems. I wonder what the other members of the panel think we should do to avoid that happening to us.

SIMON: I think if you give up,Arthur, we all have to give up.

WIGHTMAN: That summarizes my view too.

SIMON: Perhaps I'd like to make one other comment. That is that if things do get too complicated in higher dimensions we are beginning to get enough control over the two dimensional theories so that they can still serve as a laboratory for certain questions. What we have done so far in constructive field theory is just setting the background. But, I think, as the last remarks by the chairman indicate, we are beginning to get to the point where it will be of interest in particle physics. So, if in fact, we come to a screeching halt, i.e. we hit four dimensions and bounce off, then I still think that there is interesting work we can do in two dimensions, in answering questions like dynamical instability and questions about scattering theory. I hope we can do at least one three-dimensional theory, so that there is angular momentum and Regge around. However, I don't think it would necessarily be a complete catastrophe or that we'd have to give up and go home, if we found there was no idea for four dimensional theories at all.

JAFFE: I think also we shouldn't loose contact with the fact that we should be doing some physics, and perhaps keep our minds on doing approximate calculations, or thinking hard about calculations that haven't been done before. It would be nice to just compute some numbers that had physical relevance or at least physical relevance for a model.

WIGHTMAN: There is one aspect of that which interests me and for which the original version of the Goldstone Criterion offers little guidance: What is the significance of other relative minima of $P_1(\xi) = \frac{1}{2} m_o^2 \xi^2 + P(\xi)$ in $P(\phi)_2$? It is natural to extend Goldstone's ideas to conjecture that there are metastable states with average values for the field equal to the values of those relative minima and quasi-particle excitations of a mass approximately equal to the curvature at the minima. This leads immediately to the more general question: What can we really say about composite particles in this kind of theory?

SCHRADER: Well, I would like to make one general remark. The work which has been done until now has been mostly guided by perturbation theory. We have in most cases asked what does perturbation theory say, and how can we make this rigorous (in cases for example where we definitely know that perturbation series diverge). We are still at the level where we have not exploited perturbation theory to its end. Other questions in which perturbation theory fails to make predictions (as in the theory of composite particles) have presumably still to wait a long time. The first thing I would say concerning physical applications is to understand why the predictions of perturbation theory in quantum electrodynamics are so good, and more generally why have the predictions coming from perturbation theory been such a good guide?

SIMON: There is a perturbation theory that might tell us something about composite particles; it's a lack of my education that I am ignorant of many of the details of this Bethe-

Salpeter equation. I would think that it would be a useful thing to look at it in connection with $P(\phi)_2$ theories.

ROSEN: I think most of the questions that have been raised here tend to cast a negative aura over the discussions. I am going to throw in another two questions to add to that and I think these two questions summarize the reaction that I find I get when I try to describe constructive field theory to an outsider. These are (i) What's going to happen when the theories are no longer super renormalizable? and (ii) Why does such a round-about procedure have to be followed in order to arrive at the physical representation. After the cut-off approach has been kicked around for a few years, will there be some insight that enables us to go more directly to the physical representation? I wonder if any of the panelists can comment on these questions.

OSTERWALDER: I think there has been a very interesting approach to that question in a very much simplified situation by Michael Reed. He constructed a model where he can, on the one hand, get to the physical representation just by the procedure that has been described by Rosen and Schrader, (namely by a GNS construction) and on the other hand he can guess what the physical representation of the commutation relations should be. Obviously he makes that guess from knowing the result of the GNS construction, but anyway his approach suggests that one might start from a representation which is not Fock, at the very beginning. Of course I realize that in a more realistic model it is very hard to make any other guess to begin with than just a Fock representation.

JAFFE: I'd like to remind you that the infinite volume divergence is non-superrenormalizable since it occurs in every order of perturbation theory. We have some more information about that divergence than was known, as I am going to talk about this afternoon. These techniques are also useful in the study of ultraviolet divergences in more singular models.

KASTRUP: I have some questions which may sound somewhat uneducated in this field for experts:

1. What does one know about the asymptotic states and the S-matrix?
2. What is the qualitative relevance of the results in one

space dimension for the higher dimensions? There are simple mathematical examples which indicate large qualitative differences in different dimensions, e.g. of all unit spheres S^m only S, S^3 and S^7 are parallisable.

3. What does one know about the short distances properties of the models investigated?

SIMON: About the asymptotic state situation: In the actual physical Hilbert space we don't even know at this point whether the vacuum is unique. So we don't know there is a mass gap or one particle states. On the other hand the three things go together, the Goldstone picture that Professor Wightman talked about does say in fact that for certain theories we expect there won't be a mass gap and that there won't be a one particle state. The thing that I think is perhaps bad about all the results so far is that they hold for every polynomial. Some day soon we're going to have to get a first result that says if the polynomial has such and such a property then such and such holds. At that point we will probably get a mass gap and one-particle states and then good old axiomatic field theory in the sense of Haag-Ruelle scattering theory will take over and we will be on our horses with an S-matrix. But we are not there yet. Someone reminded me that we do know however that when there is a space cut-off, asymptotic states do exist. These asymptotic states of course are essentially the bare particles. What happens is that if you let a particle go it runs off to infinity where there is no interaction because of the space cut-off. Hoegh-Krohn and Kato-Mugiboyoshi in $P(\phi)_2$ and J. Dimock in Y_2 have constructed these asymptotic states for the space cut-off. So one has a little bit of information; but I think these states are unrelated to physical scattering but only to mathematical properties of the cut-off Hamiltonians.

JAFFE: I'd like to comment on Kastrup's question about 2-dimensions. The main real reason for studying two dimensions is that there are fewer divergences in two dimensions than in four. The main difficulty in all these problems comes from the ultraviolet divergences, and they don't seem to be associated with the geometry of space-time. We feel that once we understand the renormalization program then we will be able to solve more and more complicated theories. In four dimensions all the model local field theories one would study are renormalizable but non-superrenormalizable. The simply renormalizable theories are easier to study because as was mentioned in the previous talks the counter terms can be written down explicitly , at least in a finite volume, by using perturbation theory. So the study of lower dimensional models is really a practice ground to understand renormalization theory, that is, more and more singular divergences. In order to attack problems you have to build up a great deal of technique because we know the problems are extremely difficult since they have been around for a long time. About the short distance behavior a few things are known but not very much. For instance how the matrix elements of fields behave at short distances. In the $P(\phi)_2$ theory you can use various forms of the pull-through expansion that Schrader talked about to isolate at least for the spatial cut-off the singularity of the two point function or more generally the n-point vacuum expectation values. You find there is a relativistic-type singularity as you would expect, and using the type of estimates that Spencer has proved one should be able to extend these results to the infinite volume limit and be able to make statements about growth of the n-point functions. Now that Nelson has proved that the retarded functions exist, we can ask about growth of the retarded function at short distances or in other words high energy behavior. So maybe there will be some abstract bounds which come from these investigations by isolating the most singular terms motivated by the perturbation expansion. Most of the work has just begun in this direction. Mr. Oliver O'Brian has studied the short distance behavior in currents in the SU(3) Yukawa model but he so far has mainly studied equal-time currents. The results, at other than equal times, are incomplete. Jon Dimock has shown that the scalar Y_2 current is a bilinear form, but not an operator.

OSTERWALDER: I have a remark which is related to the last question. There has been some attention paid to the question of a Wilson expansion of operator products at the same point, and you could ask whether there are interesting answers to that problem in one of these two dimensional models that have been studied. Unfortunately the Wilson expansion is trivial for all superrenormalizable models, as all fields have canonical dimension.

DURR: Confronted with the tremendous mathematical difficulties in proving the mathematical consistency of local quantum field theories, one may ask the question whether one cannot make the task easier by weakening some of the conditions. In particular, one may think about relaxing the condition on locality or the condition on the positive definiteness of the metric of the state space. In doing so one may be guided by the experience we gained in quantum electrodynamics which indicates that by insisting on positive definiteness of the metric in Hilbert space we cannot require the theory to be manifestly local. In particular we observe that only observables and not the vector potential commute for space-like distances. Also the "nonlocal" Coulomb interaction shows up, which however does not violate Einstein causality. If, on the other hand, we try to identify causality with manifest locality of the theory we are forced to enlarge the state space to the Gupta-Bleuler state space with indefinite metric. Hence it may appear that quantum electrodynamics is actually teaching us that interaction is not always interpretable by virtual exchange of locally coupled entities which correspond to physical particles asymptotically but is a more general phenomenon which, in fact, may include nonlocal, causal interactions of the general type of Coulomb interactions without being necessarily tied up to the long-range feature or the infrared problem. I do see the general problem connected with arbitrariness one will get into if one relaxes some of the conditions of axiomatic quantum field theory. But I would like to get your opinion on this.

WIGHTMAN: I would like to make several points about the situation you described. First of all, the introduction of zero mass particles into the models considered in constructive field theory brings with it a series of technical infrared problems. So far people have avoided these problems by assuming strictly positive masses; they have troubles enough without worrying about the infrared. Consequently, relatively few results are known. Now that some serious progress has been made for massive particles, it seems natural to ask to what extent the results can be generalized to models with massless particles. Second, you have emphasized that in your view it is both desirable and promising to open up the formalism to the possibility of indefinite metric and non-locality. In response, let me say that one of the main points of constructive field theory is to show that the standard models really have solutions with positive definiteness and locality strictly satisfied. We have as yet no evidence that one must abandon these principles in order to obtain solutions. Third, the fact that for massive particles in two-dimensional space time, there is no indication of the necessity of introducing such a generalized formalism does not mean that it will not happen in the presence of massless particles (it will!), nor in four-dimensional space time. I think it's an interesting possibility well worth study.

TODOROV : I wonder whether the assumption that everything is defined for equal times should not be considered as a step backward from the axiomatic formulation of quantum field theory. It seems to me that insisting on canonical hamiltonian formulation is the origin of some of the difficulties in extending the present results in $P(\phi)_2$ to four dimensions. For instance, the results of Wilson, Symanzik, etc. on small distance behaviour in presence of anomalous dimensions suggest that the product of two fields may be more singular than required in canonical quantization. Would you comment on that?

WIGHTMAN: I think it is a fine idea to rewrite the results of the present theory directly in terms of Green's functions, to find the appropriate spaces of distributions, to locate appropriate analogues of the estimates that are currently proved using the canonial formalism, and to do the whole theory over again entirely in terms of Green's functions. Such a formulation might be more easily generalizable to four dimensions. It's a possibility that almost everyone on this panel has in the back of his mind. But may I remind you that every previous attempt to prove existence theorems in field theory using the Green's function formalism has resulted in a ship wreck.

JAFFE: I will mention that in principle the question of

unequal times occurs only when you remove the cut-off. So it's possible to start with a theory satisfying the canonical commutation relations. You expect in a limit that for the theories we are discussing at least the canonical commutations will be modified by an infinite field strength renormalization: that by first averaging the field operators on the cut-off theory over time you can then pass to the limit in the time averaged theory taking the appropriate field strength renormalization which would be only one more renormalization dealt with in the manner of the other renormalizations.

SIMON: I think it would perhaps be wrong not to end on a positive note since most of the discussion has been so negative. I would like to say that the problems ahead do look hard, but, looking backwards, one does see a very large number of impressive results that have been obtained by Glimm and Jaffe and those of us that followed in their footsteps. I would point out that four years ago most of those problems looked hard and insoluble also.

CONSTRUCTIVE FIELD THEORY, PHASE II

Arthur Jaffe*
Harvard University
Cambridge, Massachusetts

1. INTRODUCTION

It now appears possible to study a realm of problems in CFT which, two years ago, appeared an order of magnitude too difficult to tackle. The finishing touches have just been put on the first paper of this sort[1], and in this talk I will describe some of the basic ideas; these ideas are simple, although the details of the paper are complicated. The main question that I will discuss today has been around as long as field theory: Namely how does an interacting field model react to perturbations by an external source? Such perturbations of $P(\phi)_2$, or in fact any reasonable local perturbations of the $P(\phi)_2$ model, have been shown to be regular by James Glimm and myself[1]. As a consequence, (i) further properties of the original model have been established[1], (ii) some new breakthroughs are being made[2] and (iii) these methods appear promising for use in the study of more singular models than $P(\phi)_2$.

Before delving into these new results, I will describe the external source problem. Let H denote a renormalized $P(\phi)_2$ Hamiltonian without cutoffs. With a perturbation due to an external source h, the modified Hamiltonian is

$$H(h) = H + \phi(h) \tag{1}$$

which has vacuum energy $\delta E(h)$, namely

$$H(h) - \delta E(h) \geq 0, \ \{H(h) - \delta E(h)\}\Omega(h) = 0 \quad . \tag{2}$$

*Supported in part by the Air Force Office of Scientific Research, Contract AF 44620-70-C0030 and the National Science Foundation, Grant GP-31239X.

Let us assume that our source is <u>bounded</u>, $|h| \leq 1$ and <u>localized</u> $h(x) = 0$ for $|x| > 1$. Such a source produces a finite energy shift.

<u>Theorem[1]</u>. There exists a constant M, independent of h, such that

$$|\delta E(h)| \leq M. \tag{3}$$

What are some consequences of this regularity?

(i) For the $P(\phi)_2$ theory, the vacuum expectation values $\langle\Omega, \phi(x_1,t_1)\ldots\phi(x_n,t_n)\Omega\rangle$ exist. They determine, via the reconstruction theorem, unique (i.e. essentially self adjoint) field operaoors $\phi(f)$, (see ref. 1). The close connection between these two results is made clear by the formal identity

$$\langle\phi(x_1,t_1)\ldots\phi(x_n,t_n)\rangle_\Omega$$

$$= (-1)^n \frac{\delta}{\delta f(x_1,t_1)} \cdots \frac{\delta}{\delta f(x_n,t_n)} \langle e^{-(H+\phi(f))}\rangle_\Omega \Big|_{f=0} .$$

(ii) Using a generalized form of the theorem, Osterwalder and Schrader[2] have established the first result about convergence for a quantum field model as $V \to \infty$. (Convergence, rather than the existence of convergent subsequences.) Their paper is especially interesting because it brings the study of the $V \to \infty$ limit in quantum field theory into contact with the circle of ideas that have been so successful in analyzing the thermodynamic limit in statistical mechanics[3]. (The first such connection was made by Symanzik[4], but he did not deal directly with the energy density.) In statistical mechanics the ground state energy is monotonic and subadditive; in quantum field theory these properties are approximate, rather than exact. The difference arises in field theory because of the kinematic factors $\mu^{-\frac{1}{2}}$ appearing in the definition of the field, and which are required to ensure Lorentz covariance.

It should be possible to establish further connections between the properties of local perturbations of the vacuum and convergence as $V \to \infty$. In fact, one expects the convergence of all the vacuum expectation values.

2. <u>Why Phase II?</u>

In the past five years, considerable progress has been

made in establishing properties of various models. Some of this progress is illustrated in Figure 1, where I have plotted a few results, labelled by the years in which they were established. For further general discussion of $P(\phi)_2$ or Yukawa_2, see Rosen's and Schrader's talks in these proceedings. For a comprehensive survey with details about $P(\phi)_2$ or Yukawa_2, see ref. 5-6 . These results are a combination of setting forth a general mathematical framework, on the one hand, with a proof of concrete estimates in particular models, on the other hand. The older estimates generally take the following form: One must show that some renormalization constant differs by a finite amount from an explicitly given approximation. For instance, in the $P(\phi)_2(g)$ model, the vacuum energy $E(g)$ differs by a finite amount from zero,

$$|E(g)| \leq \text{const},$$

or in the Yukawa_2 model with a spatial cutoff, $E(g)$ differs by a finite amount from the (divergent) second order constant $E^{(2)}(g)$. This type of estimate is possible when dealing with superrenormalizable phenomena, for in that case the explicitly given constant can be computed by perturbation theory.

The estimate of the above theorem is qualitatively different, because it is not a superrenormalizable situation. The theorem deals with a finite shift $\delta E(h)$ in the vacuum energy E for $H_0 + H_I$. Here E diverges (as the volume tends to infinity) in every order of perturbation theory. It is not given explicitly, but is implicitly defined by the conditions $H \geq 0$, $H\Omega = 0$. Thus E is not superrenormalizable.

The remarkable feature of the theorem is that the proof comes sufficiently close to the physical situation to deal with a finite correction to the implicitly defined renormalization constant E. It is because the new method to deal with $P(\phi)_2$ comes so close to the physics, that it has the potential to generalize to more singular models. The previous methods were too crude, either to estimate δE or to deal in detail with higher dimensions. The new methods come close to the physics by yielding an expansion into Feynman graphs which is closely connected with the perturbation theory that every physicist learns.

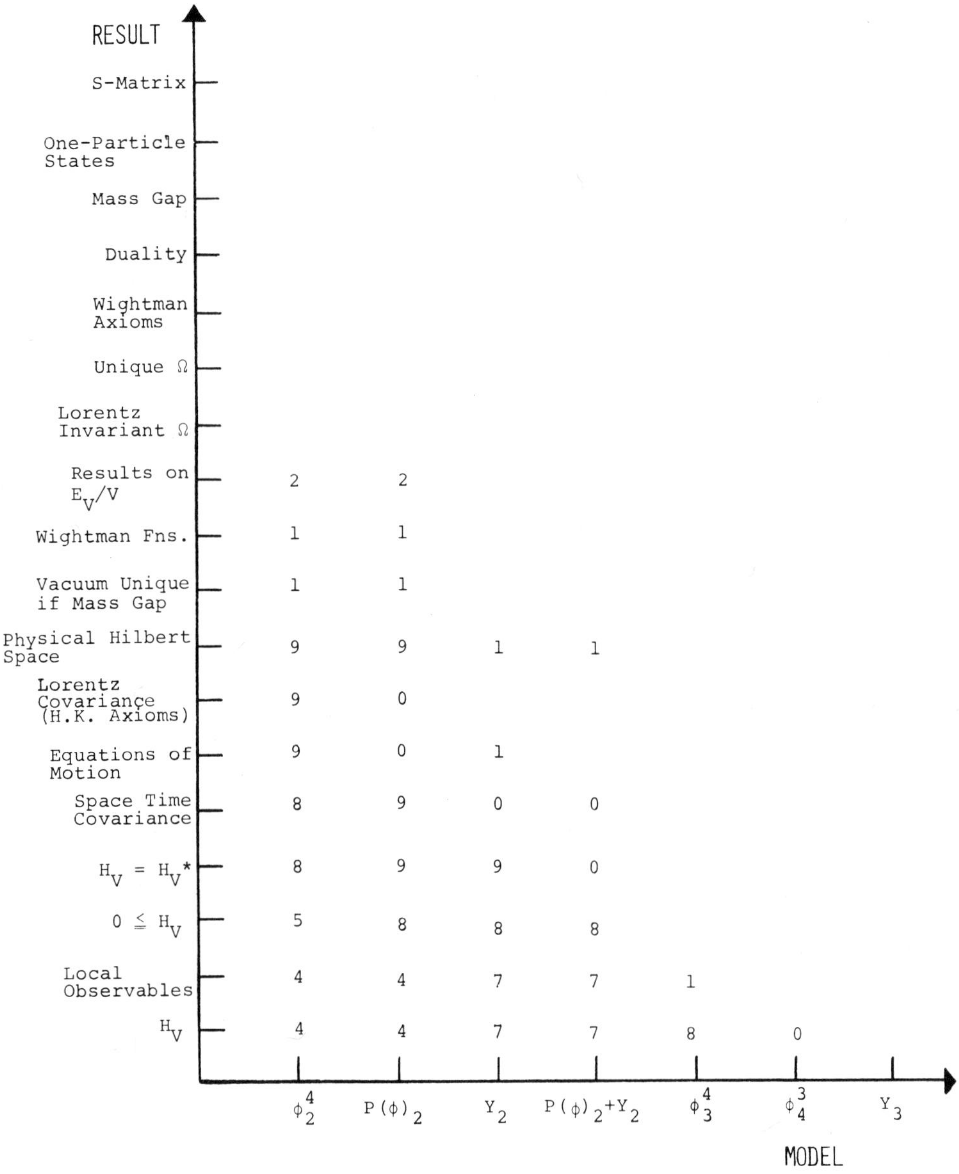

FIGURE 1

Figure 1. Progress in Constructive Field Theory, illustrated by year 196(4) - 197(2) that problems have been solved.

3. Expansion in Graphs (Formal Ideas)

We start with the exact formula

$$-(E + \delta E) = \lim_{T\to\infty} \frac{1}{T} \log \langle e^{-T(H_0+H_I+\phi(h))} \rangle_{\Omega_0} , \tag{4}$$

where Ω_0 is the no particle vector (Fock vacuum). Of course, we wish to isolate the contribution to (4) from the interaction, and we do not wish to expand E in a power series in the coupling constant, since such an expansion would diverge. One convenient representation (though by no means the only method) is to use the Feynman-Kac formula

$$\langle e^{-T(H_0+H_I+ \ (h)} \rangle_{\Omega_0}$$

$$= \quad dq \exp(- \textstyle\int_0^T \{H_I + \phi(h)\}(q(\sigma))d\sigma) \quad , \tag{5}$$

to express the Fock space expectation value as a path space integral. Here $q(\sigma)$ denotes a path in q space, parameterized by the time $\sigma \in [0,T]$. The integration over dq indicates integration over paths. The measure dq on path space is constructed from the kernel of $\exp(-TH_0)$, in the same manner that Wiener measure is related to $\exp(T\Delta)$. The Feynman-Kac formula is a standard tool in mathematical physics; it is useful not only to study quantum field theory, but also in potential scattering and in statistical mechanics, see for instance ref. 7.

Let us now estimate (5). Essentially, the main idea is to expand $\exp(-\int\phi d\sigma)$ in a power series, and then to bound the resulting terms. Eventually we obtain a bound

$$\int dq\, e^{-\int_0^T \{ H_I + \phi(h)\}(q(\sigma)d\sigma} \le e^{TM-TE} . \tag{6}$$

Thus our chain of inequalities (4)-(6) has E on both sides, yielding $\delta E \ge -M$. By reversing the sign of h we obtain $|\delta E| \le M$.

The constant term in the expansion of $\exp(-\int\phi \, d\sigma)$ gives the integral

$$\int dq\, e^{-\int_0^T H_I(q(\sigma))d\sigma} \le e^{-TE} .$$

For this term there is no difficulty in obtaining E on both sides of the chain of inequalities (4-6). Thus we need only consider terms with one or more powers of

$$\int_0^T \phi \, d\sigma .$$

The standard methods to estimate such terms would proceed by using a Schwarz (or Hölder) inequality to split the effect of H_I from the effect of

$$\int_0^T \phi \, d\sigma .$$

Unfortunately, this method raises $\exp(-\int_0^T H_I d\sigma)$ to some power and hence changes the value of the coupling constant. Doing so, however, produces an infinite shift in E, and therefore yields no information about the finite correction δE.

In order to avoid this difficulty, we use an expansion into graphs. We define the expansion inductively, in order to isolate the major contribution from the powers of $\phi(h)$. The intuitive idea of the bound is to use the localization of the perturbation. An expansion of $E + \delta E$ by standard perturbation theory yields a sum of two types of graphs, according to whether source (i.e. $\phi(h)$) vertices occur or not. The graphs with no $\phi(h)$ vertices formally sum to E, and each such graph is proportional to the volume. Graphs with at least one $\phi(h)$ vertex are bounded uniformly in the volume, and formally sum to δE. Since the perturbation series diverges, we use an asymptotic bound for (5) in the neighborhood of the localization of the source; we then show that the effect of the source on distant regions is exponentially small. Thus the source produces a finite energy shift in the region where it is located, plus a small correction due to the correlations between distant space-time regions, as required by Lorentz covariance.

4. A Bound on Graphs.

Consider a perturbation graph G composed of vertices and

legs. Let each vertex v be localized in some space-time square of area 1. We obtain such a vertex by considering the contribution to $\int_0^T H_I d\sigma$ in (5) from

$$S_{I,i} = \int_{i_0}^{i_0+1} d\sigma \int_{i_1}^{i_1+1} dx\, H_I(x; q(\sigma)) \; . \tag{7}$$

We write $\int_0^T H_I d\sigma$ as a sum over localized vertices (7). Let v_ν denote the kernel of the νth vertex, and $d_{r,\nu}$ the (Euclidean) contraction distance of the rth line at vertex ν.

Theorem[1]. For $\varepsilon > 0$, there exists a constant γ such that the value $I(G)$ of the graph G satisfies

$$|I(G)| < \prod_{\text{vertices}} (\gamma \|v_\nu\|_2) \prod_{\text{lines}} (e^{-(m-\varepsilon)d_{r,\nu}}) \; , \tag{8}$$

where m is the mass of the particles.

This bound exhibits the space-time localization of the particles, and leads to convergence of the expansion. Once a $\phi(h)$ vertex is introduced, an exponentially small factor occurs whenever a leg contracts by a large Euclidian distance.

We can use this estimate to take into account localization in momentum space. In particular, we will bound the low momentum part of $\exp(-S_{I,i})$ by removing the Wick ordering. In the remainder, at least one particle at each vertex has a momentum larger than the cutoff $\kappa(v_\nu)$. We obtain for such vertices $\|v_\nu\|_2 < 0\ (\kappa(v_\nu)^{-\frac{1}{2}+\varepsilon})$. Thus the bound (8) gives us control over simultaneous localization in space-time and in momentum space.

5. The Inductive Construction[1]

These ideas are made precise by an inductive bound to prove (6). At each step in the induction, our bound is a sum of terms

$$\int ds \int dq\, R(s) e^{-\int H_I(q(\sigma))d\sigma} \; , \tag{9}$$

where $R(s) = R(s_1,\ldots,s_n)$ is a polynomial on path space. Our expansion replaces (9) by a sum of similar terms, and this sum bounds (9) above. Each polynomial R(s) in (9) is represented by a graph, so our expansion has a convenient description in terms of graphs. As we are expanding a time ordered exponential, the graphs that occur have the usual structure for graphs in the expansion of an S-matrix. The vertices of the graphs are H_I or $\phi(h)$ vertices, and the legs at each vertex are contracted to form lines (two contracted legs) as described below. In some terms, R(s) is a constant function on path space. These constant polynomials are represented by graphs with no uncontracted legs (i.e. they are fully contracted) and are called vacuum graphs. The induction terminates for any vacuum graph; otherwise it continues through a repetition of four elementary substeps:

(a) The graph expansion

(b) Squaring

(c) The path space construction

(d) Counting the graphs.

A full cycle of the four steps (a-d) makes up one inductive step; these inductive steps continue indefinitely, reducing (5) to a sum of vacuum graphs.

(a) The graph expansion. During this step, the legs of vertices contract. We express H_I or ϕ in terms of creation operators $a^*(k)$ or annihilation operators $a(k')$. We move $a^*(k)$ to the left and $a(k')$ to the right in (5) until they are eliminated by $a(k')\Omega_0 = 0$. Each use of the commutation relations $[a(k'),a^*(k)] = \delta(k-k')$ is represented by the contraction of the legs labelled by momenta k and k' to form a line with momentum k.

In addition, it is necessary to commute $a^*(k)$ or $a(k')$ past the $\exp(-s(H_0 + H_I))$. For this purpose, we use the "pull through" identity.

$$a(k)e^{sH} = e^{-sH}e^{-s\mu}a(k) - \int_0^s ds'\, e^{-s'\mu}\, [a(k),\, H_I]\, e^{(s-s')H} \qquad (10)$$

The second term in (10) produces a "new" vertex and is called a "contraction to the exponent." The graph expansion continues until all old legs contract, or yield zero when applied to Ω_0.

The legs of new vertices are not pulled through until the following inductive steps are taken.

(b) Squaring R(s) is a technical step, to ensure that R is a positive polynomial on path space.

(c) The path space construction. The main idea is to use the fact that $S_{I,i}$ is formally positive, and to bound localized portions of the exponent $\exp(-S_{I,i})$ by a constant plus a sum of polynomials on path space. This is a standard bound in estimating boson interactions. The low momentum part of $-S_{I,i}$ is bounded by the Wick ordering constant, and the high momentum part yields the polynomial, and contributes a convergence factor $O(\kappa^{-\frac{1}{2}+\varepsilon})$ per vertex (see section 4). The new feature in our use of the path space construction is the way that we keep track of the localization of the vertices. The exponent is removed only in the neighborhood of the region where R(s) is localized. We continue the path space construction for each graph until each vertex is surrounded by a large region with no vertices or exponent. As a result, any further contraction of R(s) to the exponent introduces a small factor $\exp(-(m-\varepsilon)d)$, where d is the (large) distance from R(s) to the exponent. This small factor dominates the large number of new vertices.

(d) Of course, it is necessary to count the number of graphs to ensure convergence of the expansion, and it is convenient to do this counting in each inductive step.

In order to get a picture of the expansion, let us take the simplest case, where $\int R(s)ds$ is a localization $\int \phi_i ds$ of $\int \phi ds$, and the latter is the first term in the expansion of $\exp(-\int \phi ds)$. In this term, the perturbation is initially localized near the space time point i. We then apply the path space construction to each graph, to remove the exponent in squares of area 1 that are localized near i. We continue until we obtain a circle of radius r which contains all vertices, and which is surrounded by a circle of radius r + d containing neither exponent nor vertices except in the smaller circle. The path space construction then ends and we pass to the next inductive step. In the following graph expansion, the vertices of the previous inductive step are "old" vertices, and they contract, both to existing vertices and to the exponent. As a result, the exponent is removed in a larger circle, etc. At each stage the convergence factors $0(\kappa^{-\delta})$ for path space vertices, or $0(\exp(-md))$ for contraction to the exponent, dominate the number of new terms and ensure convergence.

6. Phase Space Localization[8]

The convergence of the inductive bound depends on simultaneous localization in space-time and in momentum space, i.e. phase space localization. In each allowed localization region we obtain a convergent expansion to bound $\exp(-tH)$, and we show that different regions of localization are approximately independent. Such simultaneous localization is limited by the uncertainty principle (dispersion of the Fourier transform): It must satisfy

$$(\delta x)(\delta p) \geq \text{const.} \tag{11}$$

in each variable. In the $P(\phi)_2$ case, we used localization in intervals of length greater than a fixed size, so the possibility of the desired phase space localization was assured. In more singular models, these ideas can be applied in principle. We wish to make an asymptotic expansion of $\exp(-tH)$ in space time region that is sufficiently small to yield bounded energy shifts in the unrenormalized theory. For a superrenormalizable interaction, the worst divergence is the vacuum energy. On dimensional grounds, a superrenormalizable vacuum energy grows no faster than $0(V\kappa^{s})$ in s space dimensions; on the other hand, in a renormalizable theory it grows as $0(V\kappa^{s+1})$. Here κ denotes an ultraviolet cutoff. For a fixed total volume V, the minimum localization consistent with (11) is κ^{-1} for each space variable. Allowing a time localization satisfying $(\delta t)(\delta E) \geqslant$ const, the minimum allowed space-time volume is $0(\kappa^{-s-1})$. Thus these ideas of localizing the interaction appear to apply in principle to any superrenormalizable model. In the case of renormalizable theories, the formal ideas are marginal, so these methods must be supplemented by other ideas to deal with such problems.

REFERENCES

1. James Glimm and Arthur Jaffe, The $\lambda\phi_2^4$ quantum field theory without cutoffs IV, Perturbations of the Hamiltonian, Harvard preprint.
2. Konrad Osterwalder and Robert Schrader, On the uniqueness of the energy density in the infinite volume limit for quantum field models, Harvard preprint.
3. David Ruelle, "Statistical Mechanics", Benjamin, New York 1969.
4. Kurt Symanzik, Euclidean quantum field theory, 1968 Varenna Lectures, "Local Quantum Theory", R. Jost, Editor, Academic Press, New York, 1969.
5. James Glimm and Arthur Jaffe, Quantum Field Theory Models, in 1970 Les Houches Lectures, C. DeWitt and R. Stora, Editors, Gordon and Breach Science Publishers, New York, 1972.
6. James Glimm and Arthur Jaffe, Boson Quantum Field Models, in 1971 London Lectures, R. Streater, Editor.
7. Jean Ginibre, Some applications of functional integration in statistical mechanics, 1970 Les Houches Lectures, C. DeWitt and R. Stora, Editors, Gordon and Breach Science Publishers, New York, 1972.
8. James Glimm and Arthur Jaffe, work in progress.

DISCUSSION

SIMON: If I understand the new results correctly, they suggest an interesting possibility which would explain the sense in which H_{ren} solves $P(\phi)_2$. One would hope that for $g \in \mathcal{S}$,

$$e(g) = \langle \Omega_{phys} , (\int g(x) \; T_{oo}(x) dx) \; \Omega_{phys} \rangle$$

exists, where T_{oo} is the interacting energy density. Translation invariance would imply

$$e(g) = e_o \int g(x) \; dx$$

for a constant e_o which is presumably $\lim\limits_{V \to \infty} E_V/V$.

Then H_{ren} might be a $P(\phi)_2$ Hamiltonian in the sense that

$$H_{ren} = \int [\; T_{oo}(x) - e_o] \; dx \quad .$$

as an integral of quadratic forms.

JAFFE: Well, I think these methods allow for the first time a computation to be performed in the infinite volume limit.

THEORY OF WEAK AND ELECTROMAGNETIC INTERACTIONS

Steven Weinberg
Massachusetts Institute of Technology
Cambridge, Massachusetts

The problem I am going to discuss is an old one, and I don't think it needs a great deal of convincing to show you it's an important one. The problem is how to construct field theories of the weak interaction which make sense, and in particular, how to construct theories in which divergences either do not appear at all, or can be eliminated in a physically reasonable way.

There have been a number of approaches to this problem in the past. One of the tricks that has been applied particularly in recent years by Gell-Mann, Goldberger, Kroll, and Low[1], and Glashow, Iliopoulos, and Maiani[2] goes by the name of the unitarity cutoff. The idea here can be seen by just looking at a simple test problem, one that was in fact used as a whipping boy by the first named group. Let's consider specifically neutrino-antineutrino production of a pair of intermediate vector bosons.

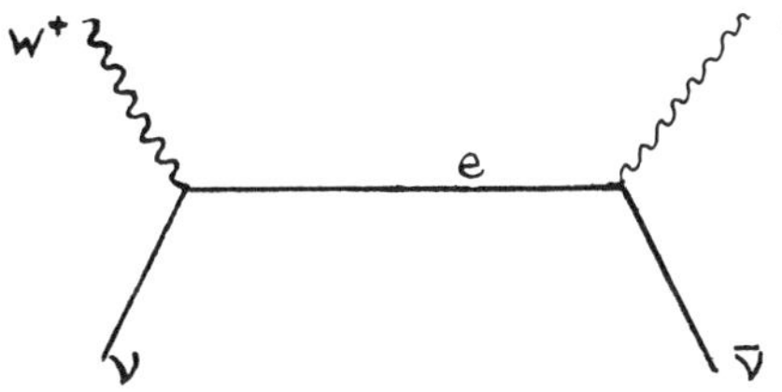

When you calculate the Feynman diagram you find a scattering amplitude of the form

$$\frac{e^{2i\delta}-1}{E} \sim GE \quad ; \; G = \text{weak coupling constant.}$$

Now it's clear that unitarity will be violated drastically when $E = 0(1/\sqrt{G})$. At that point, $e^{2i\delta}-1$ becomes comparable to 1, but it won't have the right phase to yield a unitary scattering amplitude. The above mentioned authors say, and again this is an old idea, that there must be a cutoff that comes in precisely because at these energies the perturbation series is no longer reliable and the weak interactions are not weak anymore. You can easily see that the so-called unitarity cutoff Λ comes in at an energy where $\Lambda \sim \frac{1}{\sqrt{G}}$.

The trouble with this idea, that perturbation theory breaks down thereby providing a natural cutoff, is that it's then hard to understand why perturbation theory works at all. The perturbation series is an expansion in G, and all the integrals diverge; and when you cut them off with the unitarity cutoff you see that you are then really expanding in powers of $G\Lambda^2 \sim 1$. If you are expanding in a parameter of order 1, why does anything work?

The work of the groups I mentioned was to show that it is possible to bury this difficulty, i.e. to show that the higher order effects could in fact be relegated to areas where the experiments haven't yet been done, and with a great deal of cleverness this was achieved.

There are of course other approaches to this problem which introduce cutoffs which come in at lower energies. A unitarity cutoff entering in at ~ 300 GeV is a catastrophic thing for perturbation theory, but if the cutoff comes in at 1 GeV or 10 GeV then the perturbation series will be nicely behaved expansions in a small parameter. Here I should particularly mention the recent work of T.D. Lee and Wick on the indefinite metric, and the work of Salam and others, which I believe was discussed here last year, and which consists of picking subseries of the perturbation series and summing them to all orders, though perhaps Salam wouldn't look at it quite like that.

In 1967 I was also thinking about this problem, and came up with a different suggestion which I described in a brief letter to Phys. Rev. Letters[3]. This idea I allowed to become buried for the last four years until it was recently resurrected. This idea begins with a question: Why does an intermediate vector boson theory of the weak interactions not provide a simple renormalizable theory just like quantum electrodynamics? After all, since in electrodynamics a spin one field interacts with leptons, why isn't this kind of theory with an intermediate vector boson just as good as QED? Well, the answer is very well known and is that the propagator of any kind of spin one object, whether it's a photon or a massive particle, will always have the form

$$\frac{g_{\mu\nu} + \alpha q_\mu q_\nu}{q^2 + m^2} .$$

If all you had was $\frac{q_{\mu\nu}}{q^2 + m^2}$ the theory would be renormalizable; the $\frac{1}{q^2 + m^2}$ would be sufficient to ensure convergence of the integrals. But you are going to have things like $\frac{q_\mu\, q_\nu}{q^2 + m^2}$ which don't vanish at all for large q^2. In fact, the theory is then just as bad as the conventional Fermi interaction theory. In electrodynamics, this problem doesn't really matter because, as is very well known, the $q_\mu\, q_\nu$ terms can simply be thrown away. The argument is that they depend on the gauge, but since the theory is gauge invariant you can choose a gauge in which they are absent, or, you can choose a gauge in which the function $\alpha \sim \frac{1}{q^2}$. There are various choices of gauge which have their various merits. The corresponding theory of a spin one massive particle, except for one very special simple case which is not relevant here, is not gauge invariant in any sense (one thinks) and therefore the $q_\mu\, q_\nu$ term cannot be thrown away and the theory is not renormalizable.

The suggestion I made in 1967 was based on an idea proposed some years earlier, particularly by Peter Higgs[4] and Tom Kibble[5] that in general a theory of vector mesons with mass could in fact be gauge invariant if the mass was not derived from a term in the Lagrangian which is put in by hand, but rather arose through a mechanism of spontaneous symmetry breaking.

Just to show you the idea, suppose I have a Lagrangian which describes a gauge invariant theory of scalar mesons represented by fields ϕ_i. Of course, for gauge invariance I shall need gauge fields which I call $(A_N)_\mu$: the corresponding generators t_N are represented on the gauge field by matrices $(t_N)_{ij}$, and there is a coupling constant g.

$$L = -\frac{1}{2}(\partial_\mu \phi_i - g\,(t_N)_{ij}\,A_{N\mu}\,\phi_j)^2 + \ldots \quad .$$

The first term written above is just the familiar kinematic term made gauge invariant for the scalar mesons alone. You should understand that there are additional terms for the gauge fields and for any interactions involving these fields with other fields. Suppose that the symmetry is broken by the scalar field having a nonvanishing expectation value

$$\lambda_i = \langle\phi_i\rangle_0 \quad .$$

Now it's pretty clear that there will be a term in the Lagrangian in which the A field occurs quadratically with a coefficient which is the mass matrix

$$m^2_{NM} = g^2\,(\lambda^T\,t^T_N\,t_M\,\lambda) \quad .$$

Thus, for every independent linear combination of generators for which $t\lambda$ does not vanish, in other words for every symmetry in the original gauge group which is broken by the vacuum, there will be a nonvanishing mass for some vector meson.

Now, looking at this you might say that there is a well known theorem stating that such nonvanishing vacuum expectation values which spontaneously break symmetries always entail the existence of mass zero scalar particles which are simply not observed in nature and which would rule out the possibility of any such theory. The Higgs, Kibble idea was to show that in fact such theories do not have mass zero scalar particles. The old Goldstone theorem which you might, at first, think implies such massless scalar bosons in fact, tells you to take the V.E.V., multiply by the generator as represented on the scalar multiplet and then multiply into the scalar field multiplet. These fields formed by this particular scalar product are the fields which according to the Goldstone theorem should have zero mass. What Higgs and Kibble realized was in effect that since the theory is completely gauge invariant, we can simply choose a gauge in which all these fields are zero. The gauge condition reads

$$\phi_i (t_N)_{ij} \lambda_j = 0 .$$

Therefore, although it is true that the mass matrix of the scalar mesons does have zero eigenvalues, the fields corresponding to these eigenvectors happen to vanish, which does not of course entail the existence of a mass zero scalar particle.

It may be a little unfamiliar, and in fact I found it so, to think in terms of a gauge being a restriction on the scalar multiplet. Usually, from our experience in electrodynamics, we think of a gauge condition as a condition on the vector fields: the Landau gauge is a condition for which the vector field is divergenceless, and so on. But after all, a gauge transformation affects all the fields in the theory, not just the vector fields, and one can simply choose a gauge not by saying anything about the vector field but by saying something instead about the scalar fields. This then is the gauge that one chooses to show that there are no Goldstone bosons in the theory.

This gauge has another advantage by the way. If you square out the kinematic term in the above expression for the Lagrangian, you see that the cross term is eliminated, and the term $(\partial_\mu \phi)(\partial^\mu \phi)$ stands alone; there is no term like $(\partial_\mu \phi)A^\mu$ which would confuse the physical interpretation of the field. That is a purely technical remark.

For those of you who may be interested, there is a certain amount of jargon which has sprung up from this. The original theory where the gauge invariance is manifest and all the fields look like they have zero mass is called the Stage I Lagrangian. If you let the scalar field have some V.E.V., but you do not choose this special gauge, you have the Stage II Lagrangian. The above form of the theory in the gauge we have chosen is called the Stage III Lagrangian theory. This is Ben Lee's terminology. I call the three stages A, C and B, but his names are better.

Now, what is the application of this to the weak

interactions? Well, one doesn't know a priori what the fundamental gauge group which governs the weak interaction is. One has to do a good deal of guess work, and so one guesses starting out in the simplest possible way. The simplest thing you can do is to forget the muons, and the hadrons, and just consider the electron neutrino together with the electron as being the simplest fields which certainly have something to do with each other, and try to make a gauge group out of them. Here one is immediately led to an old kind of symmetry which was suggested somewhere in the depths of the past by Schwinger[6] and Glashow[7], and later by Salam and Ward[8], and resurrected for this purpose by me in the 1967 paper.

Since we know that the electron and neutrino couple via the weak interactions only through their left-handed components, you form a little doublet L which is made up of just the left-handed part of the electron and electron-neutrino fields:

$$L = \left(\frac{1+\gamma_5}{2}\right) \begin{pmatrix} \nu_e \\ e^- \end{pmatrix} .$$

You must still worry about the right handed part of the electron field because after all that enters into electromagnetic interactions. Thus you introduce a singlet R, which is just the right handed part of the electron field,

$$R = \left(\frac{1-\gamma_5}{2}\right) e^- .$$

A very important guide in this business is to note that you don't want to introduce any symmetries which are not broken unless you are willing to face the possibility that the vector meson associated with that symmetry will wind up with zero mass. Recall that the only vector mesons which acquire a mass through the Higgs - Kibble mechanism are the ones that correspond to broken symmetries. If you introduce a symmetry which is in fact not spontaneously broken, the gauge field will retain its zero mass, and we know only one zero mass gauge field, namely the photon. So a constraint on this kind of model building is not to include things like lepton number in the gauge group because otherwise you would have a zero mass gauge field coupled to lepton number, and that would be a pity. It follows that the largest group allowed is SU(2) × U(1), in which the left-handed components behave like a little doublet, and the right hand components like a little singlet. The way you arrange for the symmetry to be spontaneously broken is to introduce a scalar doublet

$$\phi = \begin{pmatrix} \phi^+ \\ \phi^o \end{pmatrix}$$

with ϕ^o being the singlet, which is going to have nonvanishing V.E.V.

Then, corresponding to L, R and ϕ, we have the following "isospin - hypercharge" assignments:

	L	R	ϕ
T	$\frac{1}{2}$	0	$\frac{1}{2}$
Y	$\frac{1}{2}$	1	$-\frac{1}{2}$; $Q = T_3 - Y$.

Then, you simply write down the most general gauge invariant, simply renormalizable Lagrangian you can imagine according to this scheme.

$$\begin{aligned} L = &- \frac{1}{4} (\partial_\mu \vec{A}_\nu - \partial_\nu \vec{A}_\mu + g \vec{A}_\mu \times \vec{A}_\nu)^2 \\ &- \frac{1}{4} (\partial_\mu B_\nu - \partial_\nu B_\mu)^2 - \bar{R} \gamma^\mu (\partial_\mu - ig' B_\mu) R \\ &- \bar{L} \gamma^\mu (\partial_\mu - ig\vec{t}.\vec{A}_\mu - \frac{i}{2} g' B_\mu) L \\ &- | \partial_\mu \phi - ig \vec{A}_\mu . \vec{t}\phi + \frac{i}{2} g' B_\mu \phi |^2 \\ &- G_\phi (\bar{L} \phi R + \bar{R} \phi^\dagger L) + \mu_\phi^2 \phi^\dagger \phi + h (\phi^\dagger \phi)^2 . \end{aligned}$$

Here A_μ is the gauge field coupled to isospin, B_μ being the gauge field coupled to hypercharge. Now we take this Lagrangian, and allow it to become ugly by letting ϕ have a nonvanishing V.E.V. If we do this in what is called Stage III, we must first make a gauge transformation as I indicated before which will eliminate what would be the Goldstone boson field ϕ^+, and which will also eliminate what would be another Goldstone boson field, namely the imaginary part of ϕ^0. Thus ϕ becomes a purely neutral real field which you allow to have a V.E.V. A number of things come out of this immediately.

First you find that the gauge fields become redefined, that the mass matrix is not automatically diagonal and has to be diagonalized, and that the gauge fields which diagonalize the mass matrix to lowest order are a positively charged W^μ, together with of course its anti-field, and two neutral fields Z^μ and A^μ:

$$W^\mu = \frac{1}{\sqrt{2}} [A_1^\mu + i\, A_2^\mu]$$

$$Z^\mu = \frac{1}{\sqrt{g^2+g'^2}} [g\, A_3^\mu + g'\, B^\mu]$$

$$A^\mu = \frac{1}{\sqrt{g^2+g'^2}} [-g'\, A_3^\mu + g\, B^\mu] \ .$$

The masses associated with these fields are

$$m_W = \frac{\lambda g}{2}\ , \quad m_Z = \frac{\lambda}{2}\sqrt{g^2+g'^2}\ , \quad m_A = 0\ ;\ \lambda = \langle\phi^0\rangle_o\ .$$

Note that there are two coupling constants g, g'. This is because the gauge algebra we are dealing with is $SU(2) \times U(1)$ which is not simple, but merely semi-simple. For each simple component of a semi-simple group you will always have an independent coupling constant, i.e. the gauge invariance does not fix the ratio $\frac{g'}{g}$.

Now you notice immediately that the mass of the neutral field is predicted to be larger than that of the charged field. We know something about the mass of the charged field because we know experimentally what the strength of the weak interaction is. The usual weak interaction is mediated by the exchange of a very low momentum W-boson, so that its propagator is essentially $\frac{1}{m_W^2}$. The effective Fermi coupling constant which describes weak interactions at low energy turns out to be just

$$\frac{G}{\sqrt{2}} = \frac{g^2}{8m_W^2} = \frac{1}{2\lambda^2}$$

and since we know experimentally what G is, we thus know λ. Another thing that you see is that the electromagnetic field A^μ is a well-defined combination of B^μ and A_3^μ. Previously, in the Stage I Lagrangian B^μ entered with a coupling constant g' whereas A_3^μ entered in with coupling g. Both terms in A^μ enter proportionately to the combination

$$e = \frac{gg'}{\sqrt{g^2+g'^2}}$$

which is just the electric charge, another experimentally known parameter. Unfortunately, without knowing the ratio $\frac{g'}{g}$, you cannot then immediately solve for the masses. However you can easily see that both g and g' must be larger than e, so that

$$g \geqslant e \ , \ \sqrt{g^2 + g'^2} \geqslant 2e \ .$$

It then follows that

$$m_W > 37.3 \text{ GeV} \ , \text{ and } m_Z > 74.6 \text{ GeV} \ .$$

The first of these numbers was actually obtained as an equality some time ago by T.D. Lee[9], and also by Schechter and Ueda[10] in theories which have little to do with the present one; it's just that the above combination of coupling constants arises naturally in any attempt to unify weak and electromagnetic interactions. I would like to stress however that it would be most unnatural to find a charged boson with a mass of 37.3 GeV because that would only be possible if the mass of the neutral one was infinite and that would be unpalatable.

The interaction in this theory is quite involved because of the complications introduced by the spontaneous symmetry breaking. One finds

$$\begin{aligned}
L' = {} & \frac{ig}{\sqrt{g^2+g'^2}} \, [gZ^\nu - g'A^\nu][W^\mu(\partial_\mu W_\nu^\dagger - \partial_\nu W_\mu^\dagger) \\
& - W^{\mu\dagger}(\partial_\mu W_\nu - \partial_\nu W_\mu) + \partial^\mu (W_\mu W^{\mu\dagger} - W_\nu W_\mu^\dagger)] \\
& - \frac{g^2}{g^2+g'^2} W_\mu W_\nu^\dagger (g Z_\rho - g'A_\rho)(g Z_\sigma - g'A_\sigma)(\eta^{\mu\nu}\eta^{\rho\sigma} - \eta^{\mu\rho}\eta^{\nu\sigma}) \\
& + \frac{g^2}{2}(|W_\mu W^\mu|^2 - (W_\mu W^{\mu\dagger})^2) + P(\phi) - \frac{m_e}{\lambda}\phi \bar{e} e \\
& - \frac{1}{8}(\phi^2 + 2\lambda\phi)[(g^2+g'^2)Z_\mu Z^\mu + 2g^2 W_\mu W^{\mu\dagger}] + \frac{igg'}{\sqrt{g^2+g'^2}} \bar{e}\gamma^\mu e A_\mu \\
& + \frac{i}{\sqrt{g^2+g'^2}} \bar{e}\gamma^\mu \left[\left(\frac{1-\gamma_5}{2}\right)g'^2 + \left(\frac{1+\gamma_5}{4}\right)(g'^2 - g^2)\right] e \, Z_\mu \\
& + \frac{i}{2}\sqrt{g^2+g'^2}\, \bar{\nu}\gamma^\mu \left(\frac{1+\gamma_5}{2}\right)\nu Z_\mu + \frac{ig}{\sqrt{2}} \bar{\nu}\gamma^\mu \left(\frac{1+\gamma_5}{2}\right) e W_\mu^\dagger \\
& + \frac{ig}{\sqrt{2}} \bar{e}\gamma^\mu \left(\frac{1+\gamma_5}{2}\right)\nu W_\mu - 6i\,\delta^4(0) \ln(1 + \phi/\lambda) \; .
\end{aligned}$$

The last term is a peculiarity. Strictly speaking, the interaction Lagrangian is everything I have written here except for this last term. However, without this term, you find that calculation of the Feynman diagrams does not give the right answers, for a peculiar technical reason. In most field theories one is familiar with such as Q.E.D. and simple scalar meson field theories there is a procedure for generating the Feynman rules which is known to work. The fact that this procedure is known to work goes by the name of "Matthews' theorem", but it is a folk-theorem rather than a real one. (Matthews himself stated and proved this theorem only within the limited context where it <u>is</u> valid.) Matthews' theorem says if you use the interaction part of the Lagrangian as simply equal to, up to a minus sign, the interaction Hamiltonian and you forget all the noncovariant parts of the propagators, then you get the right answers. This present theory is sufficiently complicated that Matthews' theorem isn't right, and in fact you get the right answer here if you use the covariant parts of the propagators but add to the interaction Lagrangian the above singular term in order to generate the appropriate Feynman rules. The logarithm is understood in the sense of a power series. The whole theory is ugly enough, but this nonpolynomial interaction term is forced on you by the canonical commutation rules. It is not that there is anything mystical about the derivation of this term from the C.C.R., but that's what emerges.

Now this is the theory in the form I derived it in 1967, although at that time I did not bother deriving the last term. I looked at it for 4 years and it was clear it would not be renormalizable except for very delicate cancellations which I had no good reason to believe would occur. What I should have done was to do some sample calculations, but for obscure reasons I failed to do them. The whole thing was resurrected this summer by a graduate student in Utrecht, a student of Veltmans' named 't Hooft[11]. What 't Hooft did was not to work with what I called the Stage III Lagrangian but to go back to a different gauge. 't Hooft worked in a gauge in which the renormalizability was more or less manifest, but in which the absence of Goldstone bosons was not so clear. In the previously described gauge, it is obvious that there are no unphysical singularities, no fields with indefinite metric, no Goldstone bosons bothering you. It is obvious that you will have a unitary S-matrix in the sense of perturbation theory. What is not obvious here is that the theory is renormalizable. What 't Hooft did was to invent a gauge in which the renormalizability of the theory was more or less obvious, but in which the unitarity of the theory was not at all clear because his theory contained scalar mesons of indefinite metric, and so was very ugly from that point of view, but was very pretty otherwise. He then showed that these unphysical singularities don't in fact appear in the S-matrix, even though they seem to appear in the Lagrangian. This is not too surprising since the Stage III gauge exists, and the two theories are related by gauge transformations.

Subsequently, Ben Lee[12] did a thorough job showing that this program really works, at least in a simple model. He looked at an abelian version of the above theory, but which nevertheless has all the essential complicated features even down to the nonpolynomial last term. The virtue of Lee's model is of course that it just doesn't have the sheer algebraic complexity of the non-abelian interaction. He showed in fact that the theory really is renormalizable including major, but subtle facts such as that the theory has fewer counter terms than are required to do renormalization in the usual way. There are fewer free parameters than you usually have, so what you have to show is that the divergences satisfy certain simple relations so that the limited number of counter terms you have available can absorb all those divergences. By working with Ward identities he showed that this was in fact the case.

Inspired by 't Hooft and Lee I went back to start what I should have done all along, which was just to look at some simple examples to see how this theory could manage to solve the problems that had confronted the weak interactions. Since I have been talking to Francis Low, the first thing I did was to try the old thing that Gell-Mann, Goldberger, Kroll and Low had worked at a long time ago. The diagram

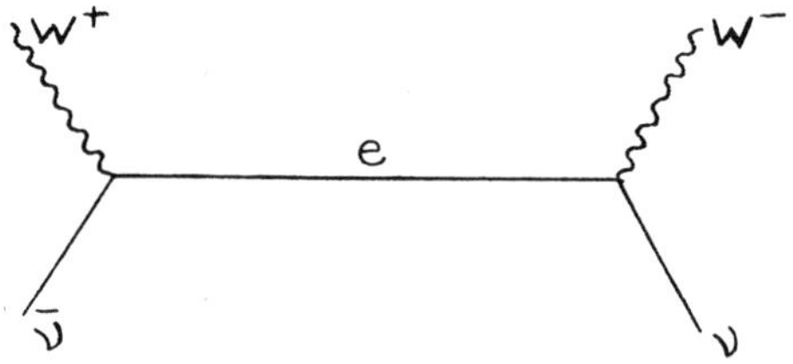

still exists remember, and gives a scattering amplitude f $\propto$ E, implying an unpleasantly large unitarity cutoff in the theory. However, remember that there is a neutral intermediate vector boson Z which gives the additional Feynman diagram

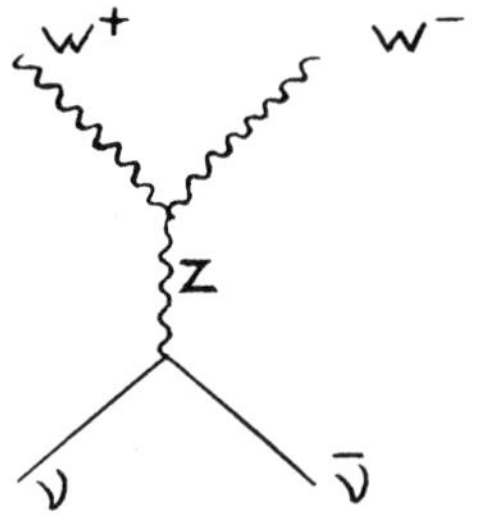

The scattering amplitude for this diagram again grows like E, and if you add the two up, you find the terms $\propto$ E cancel.

The sum in fact grows like 1/E which is just what you need in order to stay away from the unitarity bound. Indeed what happens is that this cancellation begins to occur when the invariant-mass^2 approaches the Z-mass^2, and that is essentially near threshold, so from threshold on, the amplitude begins to fall off.

Well, I then went on to look at a more challenging example, the first loop calculation. This is again a purely academic problem, but it is the simplest one where you can exercise the theory. Consider neutrino-neutrino scattering. In any conventional theory, as described in some lecture notes by Francis Low, there will be the following diagram:

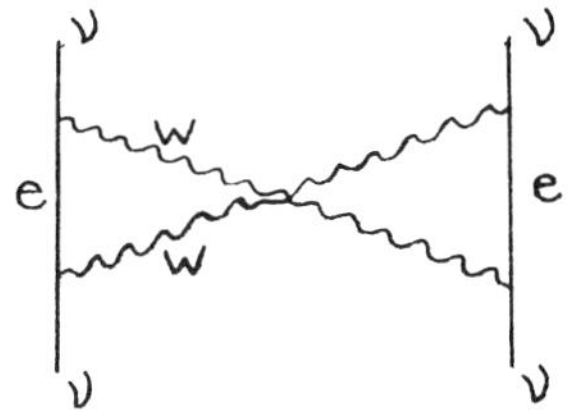

This diagram is quadratically divergent, which as Low pointed out, simply reflects the fact that if you cut the diagram in the t-channel you see that the process $\nu + \nu \rightarrow W + W$ blows up at high energy. Now, however, there are additional diagrams.

Having had the experience of the Z mesons saving you when you look at the trees, you naturally look at the two diagrams where Z's are exchanged:

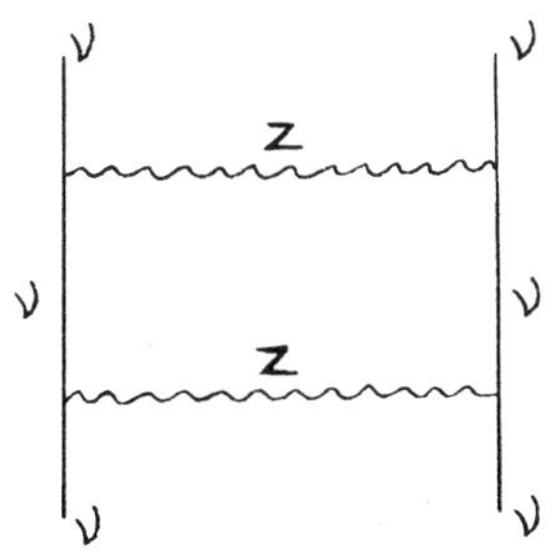

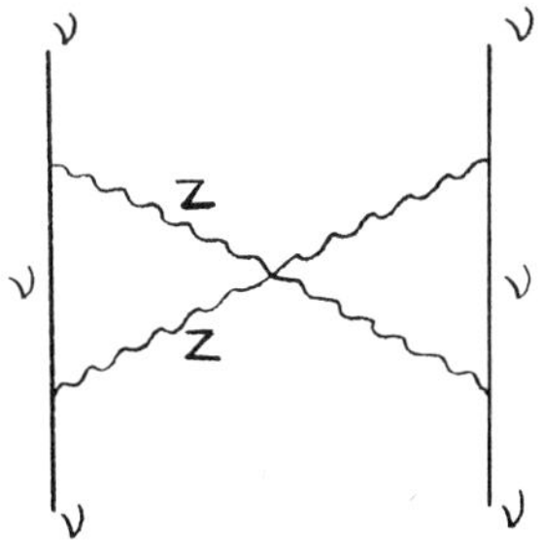

These are also quadratically divergent, but unfortunately these divergences cancel each other when the two diagrams are summed and they don't help you at all with the original quadratically divergent graph. I cried over this for a few weeks, and it seemed fairly obvious that the additional radiative correction diagrams

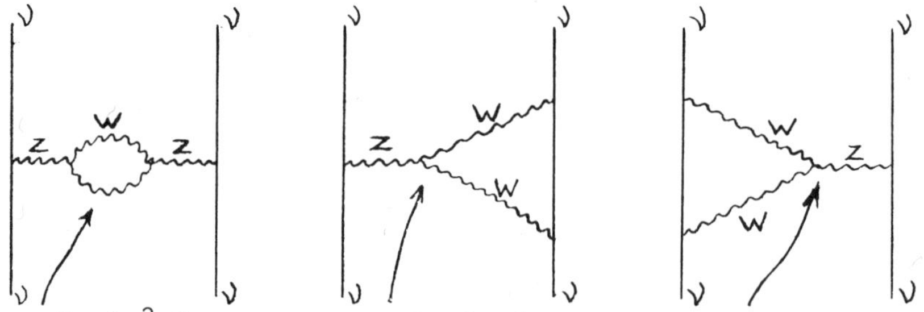

$(q^2+m_Z^2)^2\Lambda^2$ $\qquad$ $(q^2+m_Z^2)\Lambda^2$ $\qquad$ $(q^2+m_Z^2)\Lambda^2$

wouldn't help because, after all, the Z propagators have denominators $\frac{1}{q^2+m_z^2}$. Thus, even though the bubble diagrams could be horribly divergent, how could it cancel these propagators? It just did not look as if it had the right structure. But in fact, the bubble is so divergent that even though you have pulled out four factors of momentum the diagram nevertheless remains quadratically divergent. Likewise the other two diagrams are quadratically divergent, so that when everything is finally summed the quadratic divergences cancel.

I then looked at another example, one in which the photon comes in for the first time. This is electron-positron scattering, and again you find that $f \propto E$.

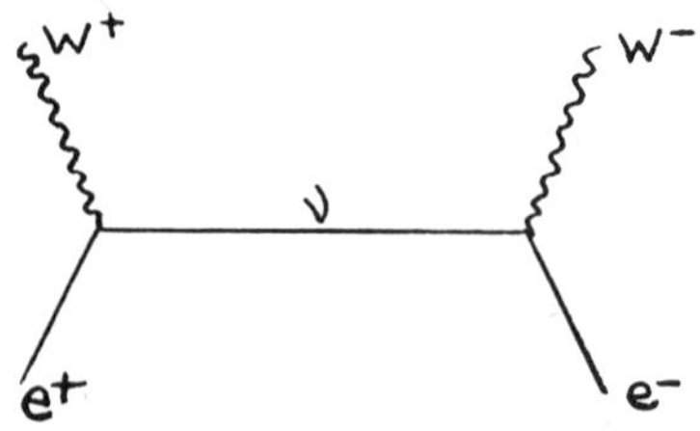

Again you have an additional diagram

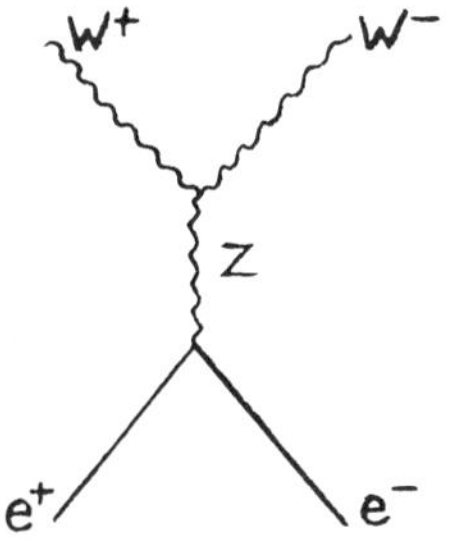

and again $f \propto E$. However this divergence does not cancel when the two contributions are added. You then remember that the theory is also intrinsically a theory of weak and electromagnetic interactions and so there is the further diagram

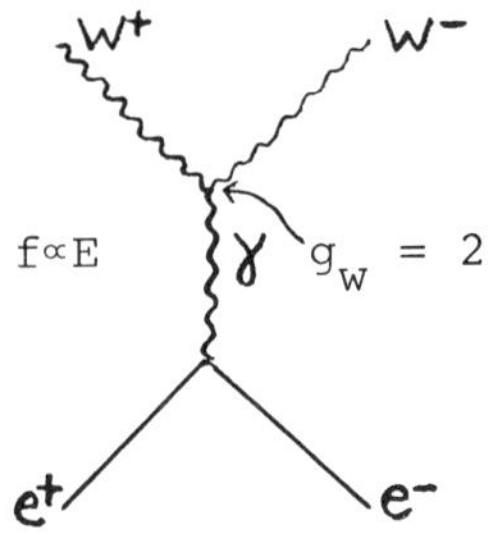

This is a process which a number of people have looked at, and it's not supposed to violate unitarity; but that is because most of the people that have looked at it have assumed that the W has a normal gyromagnetic ratio of 1. In fact, in this theory, the W necessarily has a gyromagnetic ratio of 2 and this fact causes the above diagram to behave badly at high energy, a fact pointed out by Glashow[7] many years ago in a different context. Again the quadratic divergence vanishes when the three graphs are added together.

Now if you look at this situation carefully, it is somewhat less convincing than it might at first appear. This is because we have only looked at quadratic divergences, and at this level the cancellations are rather elementary and they do not require the full Higgs-Kibble mechanism to be operating in all its glory. The simplest way to say that, is just that so far the scalar meson which is a real characteristic of the theory in that it defines how the symmetry is (spontaneously) broken, does not appear at this stage. I am not saying that

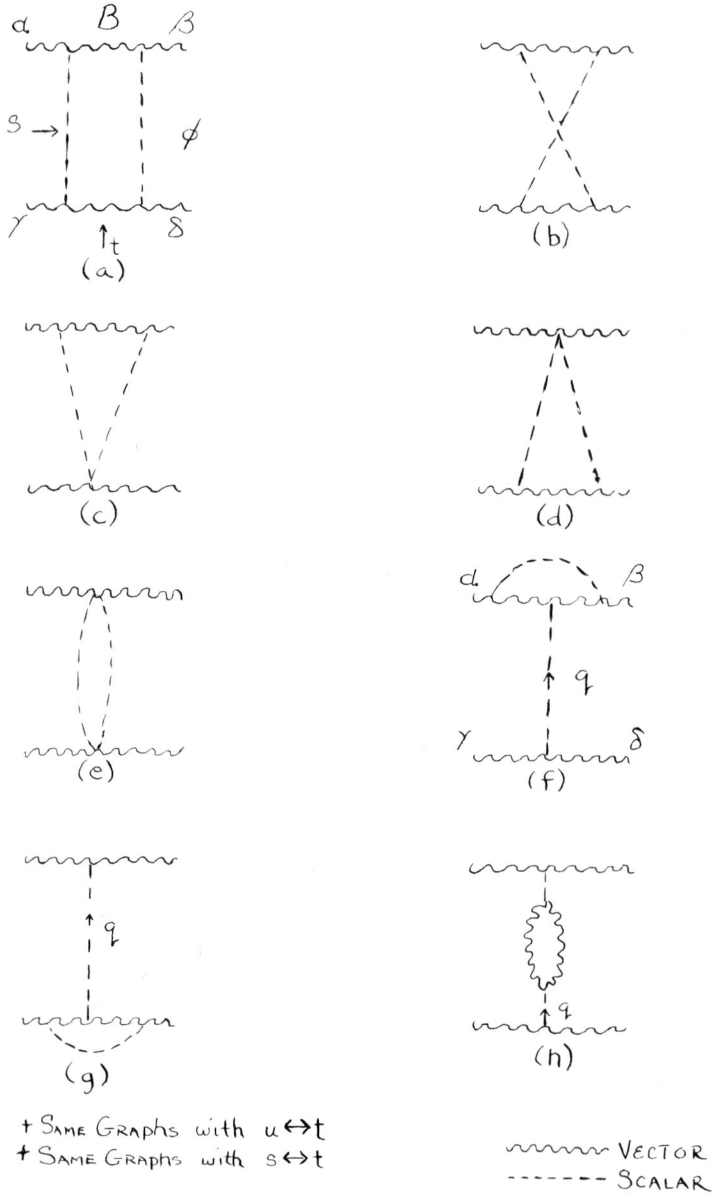

Figure 1: Diagrams for vector-vector scattering [14].

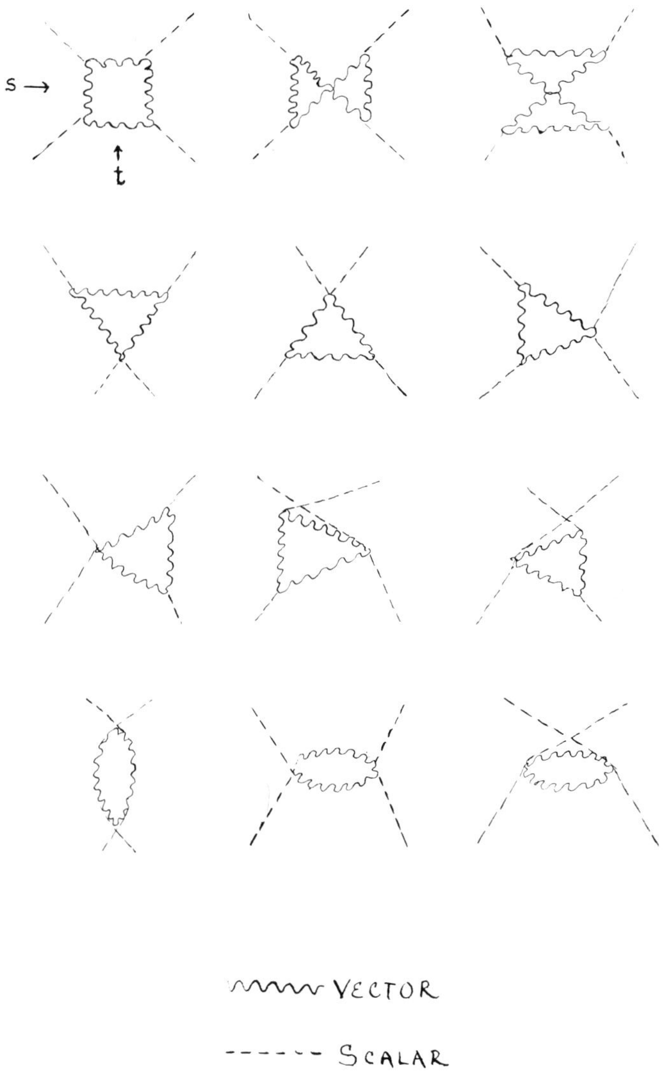

Figure 2 : Diagrams for scalar-scalar scattering [14].

I neglected it, but simply that it does not contribute to any of these divergences. So, it's clear that in order to really exercise the theory, it is necessary to go beyond the simple tests described above and in a recent letter of mine in Phys. Rev. Letters[13]. One has to look at the logarithmic divergences. This has now been done in Ben Lee's simpler abelian version of the theory by Appelquist and Quinn[14]. They have calculated vector-vector scattering (Fig. 1), scalar-scalar scattering (Fig. 2), lepton-lepton scattering (Figs. 3 and 4). In each case, the divergences are found to cancel.

I should perhaps point out a cute thing concerning the scalar-scalar scattering. If you add up the contributions of the diagrams shown in Fig. 2, you discover that there is not a complete cancellation, but there is a divergent term left over. This term however is exactly cancelled by the term of order ϕ^4 from the power series expansion of the term $\ln(1 + \phi/\lambda)$. Similarly for lepton-lepton scattering you find that not only does the scalar meson not hurt but in fact it is needed to cancel a remaining logarithmic divergence.

The next step clearly is to do the analogous calculations for the non-abelian theory. I do not think there are any important differences between the two theories (except tedium perhaps) which would cause you to doubt that a similar result will be also obtained for the non-abelian version.

There is a much more important problem with renormalizability in this kind of theory which has been pointed out by a number of people, and in particular has been written about in a recent preliminary preprint by Roman Jackiw and David Gross[15]. This is the problem of what are called the Adler, Bell, Jackiw triangle anomalies. If you look at renormalizability from the 't Hooft, Lee point of view you see that it is essential that the theory satisfy certain Ward identities, otherwise the infinities are not subject to the same algebraic relations that the counter terms are, meaning that there is no way the counter terms can absorb all the divergences. Also without the Ward identities there is no way of seeing whether the unphysical singularities which appear in the non-unitary gauge will cancel. However it's well known from recent experience in current algebra that if

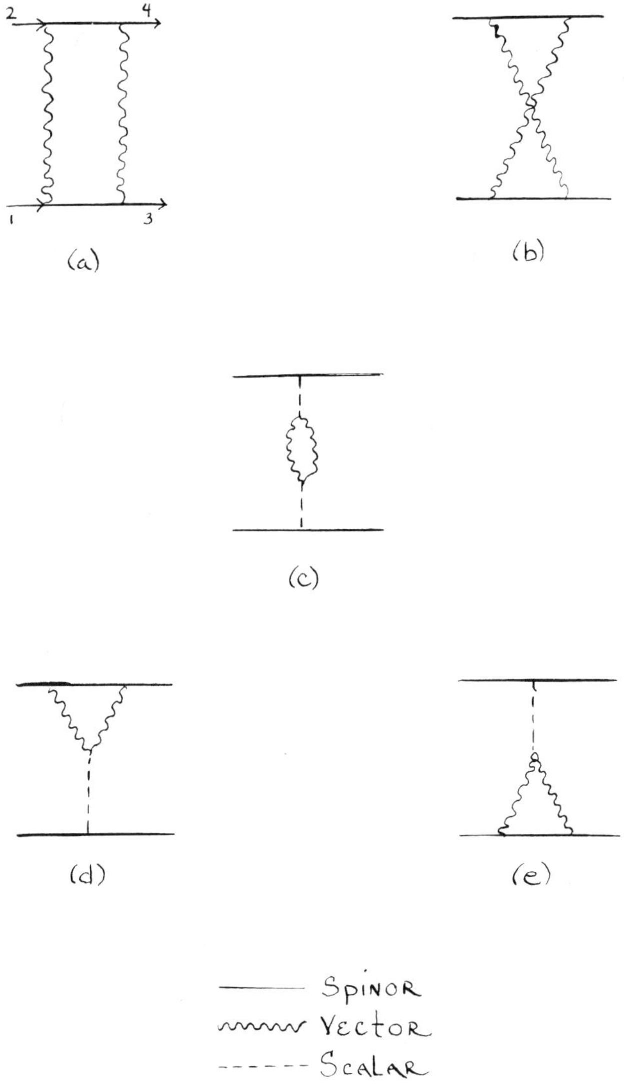

Figure 3: Diagrams for lepton-lepton scattering [14].

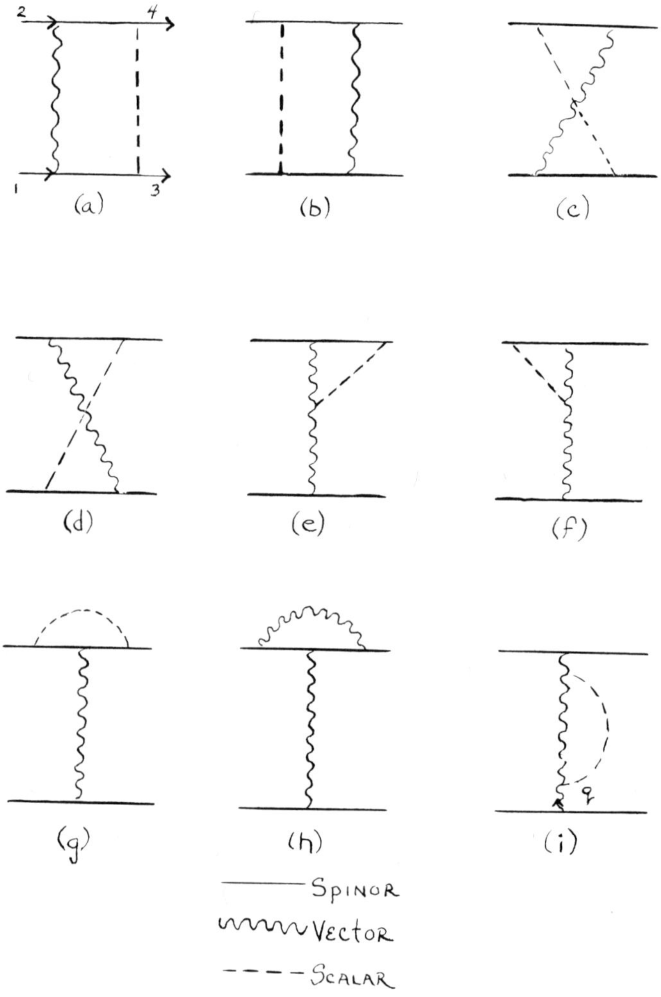

Figure 4: Further diagrams for lepton-lepton scattering.[14].

you have fermions going round in little triangles, and having three currents coming together on a fermion triangle, then the naive Ward identities are sometimes simply not true. This is a subtlety in the way currents are constructed as bilinear combinations of fields. This disease is a disease of all the theories proposed so far, both of my theory of weak and electromagnetic interactions, and of the simpler abelian version of Ben Lee. However this disease is curable, and in fact may be more of a pleasure than a liability. For instance, if you take a theory like Lee's abelian model you have a $1 + \gamma_5$ at each vertex, and add another fermion which happens to interact with the same gauge field with the same coupling but a $1 - \gamma_5$ at each vertex, then the triangles will simply cancel and the disease will go away. This is actually great because the theory doesn't tell you what the gauge group is and doesn't tell you how many leptons and fermions there are in the universe. It leaves you a good deal of freedom, and the condition that the triangle anomalies should cancel is another constraint on the theory which can be successfully used in model building. In particular, let me mention one example of what I mean. If you have a ν and e^-, which usually couples with a $1 + \gamma_5$, you also have another doublet, viz. the proton-neutron which also couples with $1 + \gamma_5$. Now these doublets don't have the same coupling constants to the gauge field so you really can't compare them. However, if you take the antineutron, antiproton doublet, it couples with $1 - \gamma_5$. The two doublets then cancel one another. That is, the triangle anomaly caused by baryons running around in triangles is exactly cancelled by the lepton triangle anomaly.

This does not provide us with a complete solution; for instance we have left out the muon-type leptons and the strange hadrons. In fact it must be said that there is as yet no satisfactory theory incorporating all particles including the strange hadrons, except some extremely artificial ones. The strange hadrons really pose the major difficulty which is reviewed in a recent preprint by Peter Freund[16]. There just doesn't seem to be a natural theory which incorporates hadronic SU(3) and leptonic "whatever it is".

Now let me turn briefly to experimental tests of the theory. Of course the obvious experimental task is to look for the vector boson, but that will take a while. For things that can be done in the next year or two, one can do neutrino scattering and look for the effects of the intermediate vector boson Z. This has been analyzed by 't Hooft[17] and more completely by Chen and Lee[18]. Fig. 5 shows the data from a reactor experiment of Reines[19] on anti-neutrinos scattering on electrons. Being a reactor experiment, one has of course electron-type anti-neutrinos. The process $\bar{\nu}_e + e \to \bar{\nu}_e + e$ has not of course been definitely observed so all you can say is that the point $x = \frac{g'^2}{g^2+g'^2}$ lies somewhere within the elongated Reines ellipse (see Fig. 5). Feynman and Gell-Mann claim that the point is located inside the ellipse at the point ⊕ indicated on the figure. For reasons which I shall come to, I suggest that the right point is $x = \frac{1}{4}$, which is equivalent to postulating $\frac{g'}{g} = \frac{1}{\sqrt{3}}$. Certainly in this theory x may lie anywhere on a line (cf. Fig. 5) but its precise position is not known since the ratio $\frac{g'}{g}$ is not determined.

Similar experiments were inadvertently done at CERN on muon-type neutrino scattering by electrons. These experiments are lousy for measuring electron-neutrino[20] scattering on electrons, but as pointed out by Albright, they are very good for the analogous scattering of muon-type neutrinos off electrons. Again the allowed region is an ellipse (cf. Fig. 6) which includes the line $0 < x < 1.0$. Again I would suggest that $x \sim 0.25$. (Figures 5 and 6 are cribbed from the paper of Chen and Lee.)

Concerning radiative corrections, Jackiw and I[21] have calculated the weak interaction contribution to the anomalous magnetic moment of the muon:

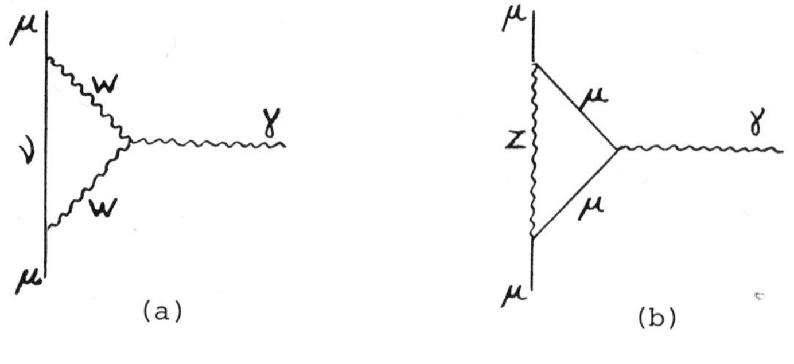

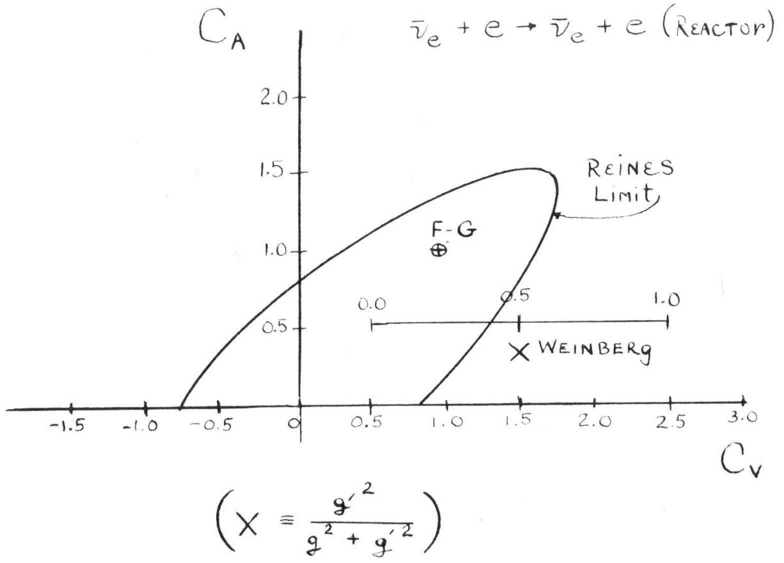

Figure 5: Experimental upper limits and theoretical predictions for $\bar{\nu}_e + e^- \to \bar{\nu}_e + e^-$ [18].

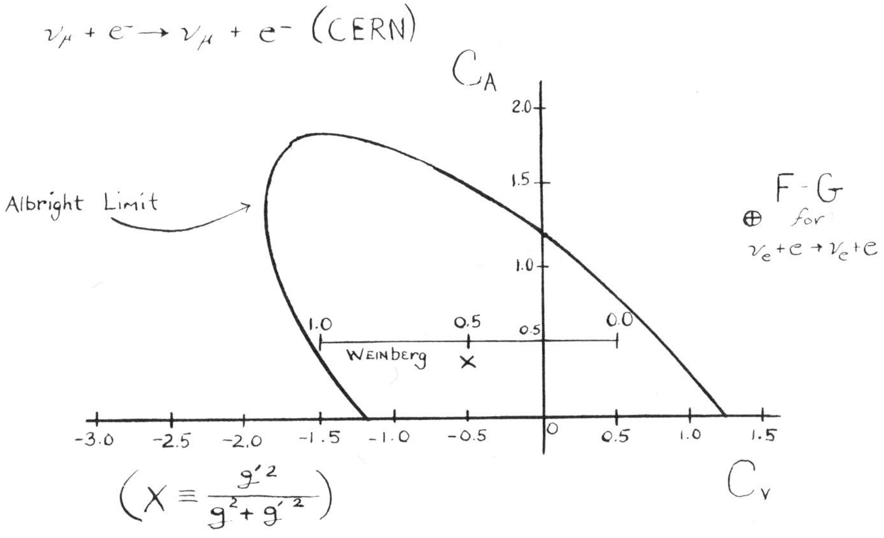

Figure 6: Experimental upper limits and theoretical predictions for $\nu_\mu + e^- \to \nu_\mu + e^-$ [18].

The diagram (a) has been calculated many times by various people and, as is well known, it is logarithmically divergent unless W has a gyromagnetic ratio of 2. In fact in this theory W has just such a gyromagnetic ratio, so that the diagram (a) is finite. The other diagram (b) which also enters is again finite, so that the total contribution gives a finite result depending on the parameter x:

$$\Delta g_\mu = \frac{G\, m_\mu^2}{\pi^2\sqrt{2}} \left[\frac{5}{12} + \frac{4}{3}\left(x - \frac{1}{4}\right)^2\right] .$$

Depending on the value of x, the change in the gyromagnetic ratio is

$$0.4 \times 10^{-8} < \Delta g_\mu < 1.0 \times 10^{-8}$$

which is too small to be detected. The scalar mesons could have any mass, unfortunately, and therefore you can't make any definite predictions. If their mass < 20 MeV they would produce anomalies in muonic atoms similar to the anomalies which have in fact been reported by Dixit et al,[22] but I don't know whether in fact those anomalies will survive. There is no particular reason the scalar meson should be so light anyway.

Now I haven't said anything yet about the interaction with the hadrons. In this theory, as I have said, you just close your eyes to the existence of strange particles. You can predict of course neutrino-nucleon processes. For example (see Fig. 7) you can predict the ratio of elastic scattering $\nu + p \rightarrow \nu + p$ to the usual inelastic process $\nu + n \rightarrow \mu^- + p$. The one standard deviation limits set by a CERN experiment are[23] shown in Fig. 7 as dotted lines, and the bold-face curve is the theoretical prediction.[24] Again I would suggest that $x \sim 0.25$. Clearly unless $x > 0.5$ the experiment does not put any constraint on the theory.

The problem of how to incorporate the strange particles is, as I have stressed, unsolved. There is one possible solution which is to be found in an earlier paper by Glashow, Iliopoulos, and Maiani,[2] which involves the introduction of a fourth quark which is a way of eliminating the strangeness

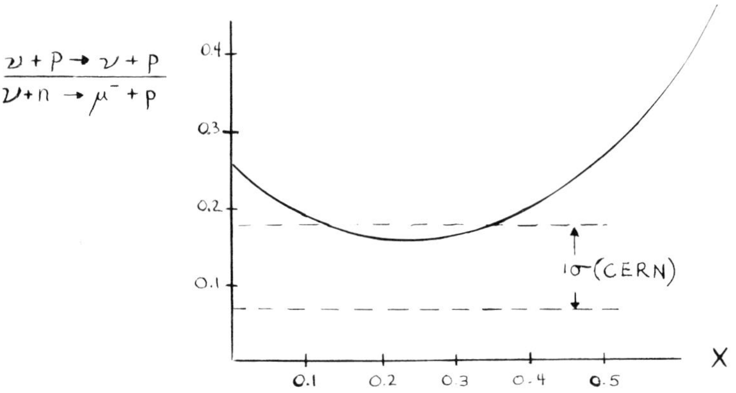

Figure 7: Experimental limits and theoretical predictions for $\dfrac{\nu_\mu + p \to \nu_\mu + p}{\nu_\mu + n \to \mu^- + p}$

changing neutral currents. However I find this theory unnatural.

Apart from the problem of how to deal with the strange particles, there is another problem which is just the plain ugliness of the above theory. The theory introduces, from the beginning, a distinction between left and right-handed leptons, it introduces a scalar particle just for the sole purpose of breaking the symmetry, and it introduces two coupling constants so that there is a free dimensionless parameter in the theory. I find all of these things rather repulsive. There is a theory[25] that I would like to promote which solves the first and third of these difficulties, but still has the rather ad hoc introduction of scalar fields. This is based on the old Konopinski-Mahmoud[26] idea of joining the leptons into a triplet, (μ^+, ν, e^-) and saying that the right-handed part of the ν is what we usually call the $\bar{\nu}_\mu$, and the left-handed part of the ν is identified with ν_e. Then the gauge group is not SU(2) × U(1), but is SU(3) × SU(3). With this gauge group, if you assume that parity is conserved, there is only one independent coupling constant. You do have a problem with this kind of theory in that there are 16 gauge fields, many of which would induce unobserved effects. However these effects could be suppressed if the spontaneous symmetry breaking takes a certain pattern which I just don't have time to go into here.

Anyway I had better conclude by saying that this kind of theory leads you to a model which looks in all essential respects just like the one I have already described, except for the important feature that $\frac{g'}{g}$ is fixed, and the value for $\frac{g'}{g}$ automatically comes out to be $\frac{1}{\sqrt{3}}$. This then makes a definite prediction for the mass of the intermediate vector bosons, namely, the charged boson will have a mass of 74.6 GeV, and the neutral one a mass of 86.2 GeV. I certainly look forward to discovering whether in fact this is true. However, I want to stress that the key idea, of achieving renormalizable theories of the weak and electromagnetic interactions through exact though spontaneously broken gauge symmetries, does not lead uniquely to any one model, and so does not stand or fall according to the results of any particular experiments.

REFERENCES

1. M. Gell-Mann, M.L. Goldberger, N.M. Kroll, and F.E. Low, Phys. Rev. 179, 1518 (1969).
2. S.L. Glashow, J. Iliopoulos, and L. Maiani, Phys. Rev. D2, 1285 (1970).
3. S. Weinberg, Phys. Rev. Letters 19, 1264 (1967).
4. P.W. Higgs, Phys. Letters 12, 132 (1964), Phys. Rev. Letters 13, 508 (1964), and Phys. Rev. 145, 1156 (1966). Also see F. Englert and R. Bront, Phys. Rev. Letters 13, 321 (1964); G.S. Guralnik, C.R. Hagen, and T.W.B. Kibble, Phys. Rev. Letters 13, 585 (1964).
5. T.W.B. Kibble, Phys. Rev. 155, 1554 (1967).
6. J. Schwinger, Ann. Phys. (N.Y.) 2, 407 (1957).
7. S.L. Glashow, Nucl. Phys. 22, 579 (1961).
8. A. Salam and J. Ward, Phys. Letters 13, 168 (1964); also see A. Salam in Elementary Particle Physics, edited by N. Svartholm (Almquist and Wiksells) Stockholm, 1968), p. 367.
9. T.D. Lee, Phys. Rev. Letters D3, 801 (1971).
10. J. Schechter and Y. Ueda, Phys. Rev. D2, 736 (1970).
11. G. 't Hooft, Nuclear Physics B35, 167 (1971).
12. B.W. Lee, to be published.
13. S. Weinberg, Phys. Rev. Letters 27, 1688 (1971).
14. T. Appelquist and H. Quinn, to be published.
15. R. Jackiw and D. Gross, to be published.
16. P.G.O. Freund, to be published.
17. G. 't Hooft, Physics Letters 37B, 197 (1971).
18. H.H. Chen and B.W. Lee, to be published.
19. F. Reines, invited talk at the DPF meeting at Rochester, N.Y., (September, 1971).
20. C.H. Albright, Phys. Rev. D2, 1330 (1970).
21. R. Jackiw and S. Weinberg, to be published in Phys. Rev.
22. M.S. Dixit, H.L. Anderson, C.K. Hargrove, R.J. McKee, D. Kessler, H. Mes, and A.C. Thompson, Phys. Rev. Letters 27, 878 (1971).
23. D.C. Curdy et. al., Phys. Letters 31B, 478 (1970).
24. S. Weinberg, to be published.

25. S. Weinberg, to be published.

26. E.S. Konopinski and H.M. Mahmoud, Phys. Rev. 92, 1045 (1953).

DISCUSSION

BROYLES: Why must the vector boson propagator have a term $\propto \frac{q_\mu q_\nu}{q^2+m^2}$?

WEINBERG: You could indeed try to adopt a propagator $\frac{g_{\mu\nu}}{q^2+m_w^2}$. However the $g_{\mu\nu}$ is not the sum over helicities of wave functions ε_μ of a spin 1 particle:

$$g_{\mu\nu} \neq \sum_\lambda \varepsilon_\mu^\lambda \varepsilon_\nu^\lambda$$

which it would have to be in order that the pole of the propagator is generated by an actual spin 1 particle.

DURR: Are there any predictions on double weak processes on the basis of your theory?

WEINBERG: I can't say anything about the K mass because I don't know anything about how to incorporate strange particles. In any case this is not a solution to all the problems of the world and it doesn't help you deal with the strong interactions. Even if I had a theory in which I could find and write down the diagrams and show that they converge for things like the K_1, K_2 mass difference, I would still really not know how to calculate it. I would have the problem of calculating the strong interaction effects there, which I don't know how to do. So, it's really premature to ask about the K_1, K_2 mass difference. You can ask about other things like double β decay. To lowest order of course the theory does not differ from the usual one and in higher orders all it does is say that the higher order corrections due to the weak interactions are small, small meaning not like 10^{-7}, but like 1/137.

SCHRADER: Can you say anything about a neutrino mass?

WEINBERG: Well, this theory as I have written it, because of the pattern of the symmetry breaking doesn't give a neutrino a mass, and indeed it is not natural for it to give the neutrino a mass. In the Konopinski-Mahmoud formalism the neutrino could get a mass but in fact again it seems rather natural that it doesn't. I have not yet completely explored the mechanism for the symmetry breaking there. It seems at first sight that it would not get a mass but I really don't know. I can't say anything more definitive. Certainly nothing here calls for a neutrino mass. It is quite natural that it doesn't get a mass. One of the things that the Konopinski-Mahmoud formulism says is that you shouldn't look for a muon-type neutrino with a much larger mass than the electron-type neutrino. They are in fact the same neutrino, so the 60 eV limit on the electron neutrino also applies to muon-type neutrino; but that 60 eV is an experimental number.

CURRENT HIGH ENERGY NEUTRINO EXPERIMENTS

A. K. Mann
University of Pennsylvania*
Philadelphia, Pennsylvania

and

National Accelerator Laboratory
Batavia, Illinois

I. Introduction

In the universe of elementary particles the neutrino is unique in that -- as far as we know -- it interacts with other particles only through the Fermi or weak interaction. Cross sections for neutrino collisions with matter are of the order of 10^{-38} cm^2 or less at moderate neutrino energies while the cross section for, say, electron scattering by matter is of order 10^{-30} cm^2 at similar energies.

Primarily for this reason the neutrino has remained a particularly deep mystery among the other mysteries in particle physics. The first scattering experiment using neutrinos (from a reactor) as projectiles was done in 1953 by Reines and Cowan[1] - about 23 years after the neutrino was originally suggested by Pauli[2] - to demonstrate that the neutrino did in fact exist and did interact with matter as weakly as had been hypothesized.[3] That pioneering experiment was followed in 1962 by the work of a Columbia University group[4] who observed the scattering of high energy (1 to 5 GeV) neutrinos produced by the AGS at the Brookhaven National Laboratory; this led to the discovery of two types of neutrinos, one associated with the electron, ν_e, and one associated with the muon, ν_μ.

In turn the Columbia experiment was followed by a series of neutrino experiments done at CERN[5] which confirmed the existence of two neutrinos and which provided data on the total neutrino cross section up to about 10 GeV as well as many details of quasi-elastic and inelastic neutrino

*Supported in part by the U.S. A.E.C.

scattering. More recently, very large bubble chambers using hydrogen at the Argonne National Laboratory[6] and a heavy liquid at CERN[7] have been exposed to the 1 to 10 GeV neutrinos available at those accelerators.

The maximum energy and intensity of the primary proton beams at Brookhaven, CERN and Argonne have severely limited exploitation of the possibilities opened by neutrino scattering experiments with accelerator produced neutrinos. Nevertheless, these earliest experiments, taken in conjunction with other information on the weak interaction, suggest that the weak processes we have thus far observed are the low energy limit of a more extensive and varied class of phenomena which will be encountered when experiments are done at higher energies. For example, all present observations of weak interactions - including neutrino scattering - are consistent with a local or point interaction. It is well-known, however, that a point interaction between leptons as in neutrino-electron scattering ($\nu_e + e \rightarrow \nu_e + e$) leads, according to current theory, to a cross section that will exceed the unitary limit $\pi\lambda\!\!\!^{-2}$ at a center of mass energy of about $G^{-1} \simeq 300$ GeV, where G is the weak interaction coupling constant. Thus we anticipate that at some center of mass energy, possibly just beyond our present horizon, we may find the rich, primary domain of the weak interaction; a domain that might contain one or more vector boson propagators of the weak interaction and perhaps new leptons and in which - at the very small distances involved - higher order weak processes may become important.

It is to explore these possibilities, among others, that new experiments are being mounted to study neutrino interactions at higher energy and higher intensity and with more massive detectors than previously. It is not intended here to give a complete review of the total experimental effort, but its flavor may perhaps be imparted by describing briefly the progress of two somewhat different experiments that are indicative of the main thrust of current high energy neutrino research.

II. Search for Short-Lived Sources of Neutrinos

The first of these experiments is directed at the

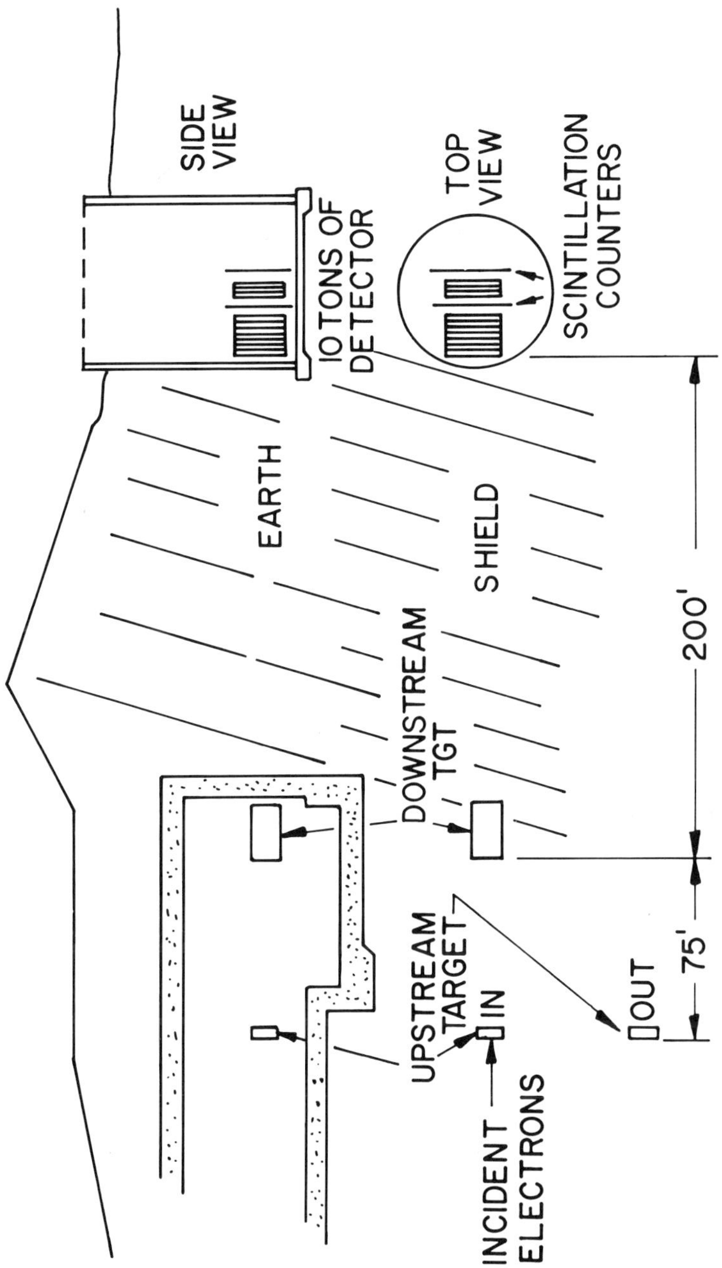

Fig. 1. Experimental arrangement to search for short-lived sources of neutrinos at SLAC.

general question: to what extent is our present understanding of the family of neutrinos complete? In particular, are there short-lived sources of neutrinos or neutrino-like particles of which we are now unaware? The experiment, which was carried out[8] at SLAC seeks to answer these questions in the following way.

The entire electron beam - about 2×10^{14} electrons/sec at an energy of about 18 GeV - impinges on either one of two targets as shown in Fig. 1. Almost all of the secondary particles produced by the incident electrons are absorbed by the 200 ft. of earth shielding between the targets and the detector which is situated in a hole 35 ft. below ground level. A fraction of the secondary pions and kaons decay in flight and give rise to approximately equal numbers of neutrinos and antineutrinos, some of which penetrate the earth shield and interact in the neutrino detector consisting of about 10 tons of thick aluminum plate spark chambers. The detector is triggered on by a coincidence between corresponding counters in the two scintillation counter arrays surrounding the furthest downstream spark chamber that occurs in conjunction with a current pulse from the electron accelerator.

The number of neutrinos (and antineutrinos) produced from the decay of the relatively long-lived pions and kaons ($\tau_\pi = 2.6 \times 10^{-8}$sec, $\tau_K = 1.2 \times 10^{-8}$sec) depends on the ratio of their mean decay length to the mean absorption length of the material they traverse after they are produced by the incoming electron beam. That ratio can be varied by the use of the two targets as shown in fig. 1. The short upstream target that can be inserted in or retracted from the electron beam line has a long free space behind it and should therefore yield many more neutrinos from pion and kaon decay than the long downstream target which incorporates appreciable absorbing material.

Suppose there exists an unknown particle able to decay into leptons with a short life, say, less than 10^{-10}sec. The yield of neutrinos from such particles - if they were produced by the very intense electron beam at SLAC - would be essentially constant and independent (except for solid angle effects) of the location or nature of the target presented to the electrons, because such short-lived particles would decay in either target before they could be absorbed. Hence the presence of a short-lived source of neutrinos could manifest itself through a lower ratio of UPSTREAM COUNTS (counts in the neutrino detector when the upstream target is in place in the electron beam) to DOWNSTREAM COUNTS (counts in the neutrino detector when the upstream target is retracted from the electron beam) than would be expected on the basis of a calculation that assumes pions and kaons to be the sole sources of decay neutrinos.

There are, of course, other ways in which neutrinos or neutrino-like particles might exhibit themselves in this experiment. They might appear to interact differently, giving rise, for example, to a different multiplicity distribution than is usual for the known neutrinos.

More specifically, leptons heavier than the muon, if they exist, would be a possible short-lived source of neutrinos. There have been a number of conjectures about such pointlike particles in the literature[9] in which are discussed the properties of a charged lepton L having its own neutrino ν_L associated with it. It is likely that the L has a mass $M_L \geq M_K$. For $M_K \leq M_L \leq 1$ GeV, the L will have a lifetime 10^{-11}sec $\leq \tau_L \leq 10^{-10}$sec and decay predominantly by the mode $L^{\pm} \rightarrow \pi^{\pm} + \nu_L$. There is no known particle that can decay into an L if $M_L > M_K$. L pairs with, say, $M_L = 0.5$ GeV would be photoproduced at SLAC at a rate of about 10^7 per sec assuming 2×10^{14} incident electrons per sec. The neutrinos ν_L and $\bar{\nu}_L$ from the rapid decay of these leptons interacting in the detector would produce an L and, possibly, a hadron; the L would in turn decay rapidly to π and $\nu_L(\bar{\nu}_L)$. Hence the visible particles in the final state of a reaction produced by a ν_L would be only energetic hadrons, which might be ascertained by their subsequent interactions in the detector. Even a small number ($\sim$10) of events of this type, i.e., without visible leptons in the final state, might serve as a positive signal from which the mass M_L could be roughly estimated.

Still another possibility lies in the pair production

of weakly interacting neutral particles through either a very small magnetic moment or a modest charge distribution.[10] There is no available evidence on the electromagnetic properties of such particles if their mass is greater than about 1 keV; there is, for instance, insufficient energy in stellar interiors to make a pair of 1 keV particles electromagnetically. Furthermore, weakly interacting neutral particles with mass less than about 2 or 3 MeV would not be apparent in the decays of pions or kaons even if they were coupled with the same strength as other leptons; they might, for example, appear in the decay $\pi^+ \to e^+ + e^0$ but this would not occur more frequently than $\pi^+ \to e^+ + \nu_e$ which is in turn less by a factor of 10^{-4} than the dominant mode $\pi^+ \to \mu^+ + \nu_\mu$. Of course, larger masses are possible with reduced coupling strength.

Such neutral particles would traverse the muon shield at SLAC with negligible interaction and be attenuated only by decays in flight if they are unstable. Roughly, a photoproduction cross section greater than $10^{-42}cm^2$ in conjunction with a mean lifetime τ in the region $3 \times 10^{-8}sec \leq \tau \leq 3 \times 10^{-6}sec$ would yield at least one decay of a neutral particle per day in the detector. A related possibility has also been discussed recently in the light of the present small counting rate observed in the search for solar neutrinos[11] which appears to be incompatible with the expected energy loss from the sun by neutrino emission.

Preliminary results of the experiment at SLAC are now available. They may be given as follows:

1. Multiplicity (prong) distribution. There is no significant difference in the prong distribution (1, 2 and $\geq$ 3 visible prong events) observed at SLAC and the prong distribution obtained in the heavy liquid bubble chamber[12] at CERN from pion and kaon decay neutrinos, especially since complete separation of 1 - and 2 - prong events is difficult in the thick plate spark chambers used at SLAC.

2. Total neutrino count rate. The observed neutrino interaction rate in the detector due to the downstream target is in good agreement with the rate predicted by assuming all neutrinos are produced by pion and kaon decay.

The predicted rate is uncertain primarily because of the difficulty in estimating the pion momentum spectrum produced by the incident electrons in the thick water target necessary to absorb the power in the electron beam at SLAC and, less importantly, because of the uncertain detection efficiency of the spark chamber - counter detector system.

There is also agreement within experimental error between the predicted and observed neutrino count rates from the up-stream target.

3. Neutrino count rate as a function of target position. Again assuming pions and kaons as the sole sources of neutrinos, the ratio of UPSTREAM CTS to DOWNSTREAM CTS may be calculated for this experiment as

$$\left[\frac{\text{UPSTREAM CTS}}{\text{DOWNSTREAM CTS}}\right]_{\text{CALC}} \simeq \frac{\text{frac. } \pi \text{ decays in 75 ft. free space}}{\text{frac. } \pi \text{ decays in dwnst'm tgt}}$$

$$\times \frac{\Omega_{UP}}{\Omega_{DOWN}} \times \text{frac. } \pi \text{ leaving upst'm tgt} \times \text{target scaling factor ratio}^{13} \simeq 6 \quad .$$

The observed ratio for total events and also for 1-, 2- and $\geq$ 3-prong events separately is in essential agreement with the calculated ratio.

4. Hadronic events. The data thus far provide no significant evidence for the production of events with single, lone hadrons and without a charged lepton, although several such events might have been expected from the photoproduction and subsequent decay of heavy leptons with $M_L \sim 500$ MeV.

The absence of a positive effect in this experiment will not, it is hoped, close the door to future experiments of similar nature. Order of magnitude improvements in detector solid angle and detector mass are possible; the capability of the detector can be increased by the use of a large area iron core magnet; and additional experience with thick targets suitable for absorbing the entire electron beam would eliminate much of the difficulty in normalizing the ratio of upstream to downstream count rates. These would permit a much more sensitive search for new neutrinos and neutrino-like particles to be made.

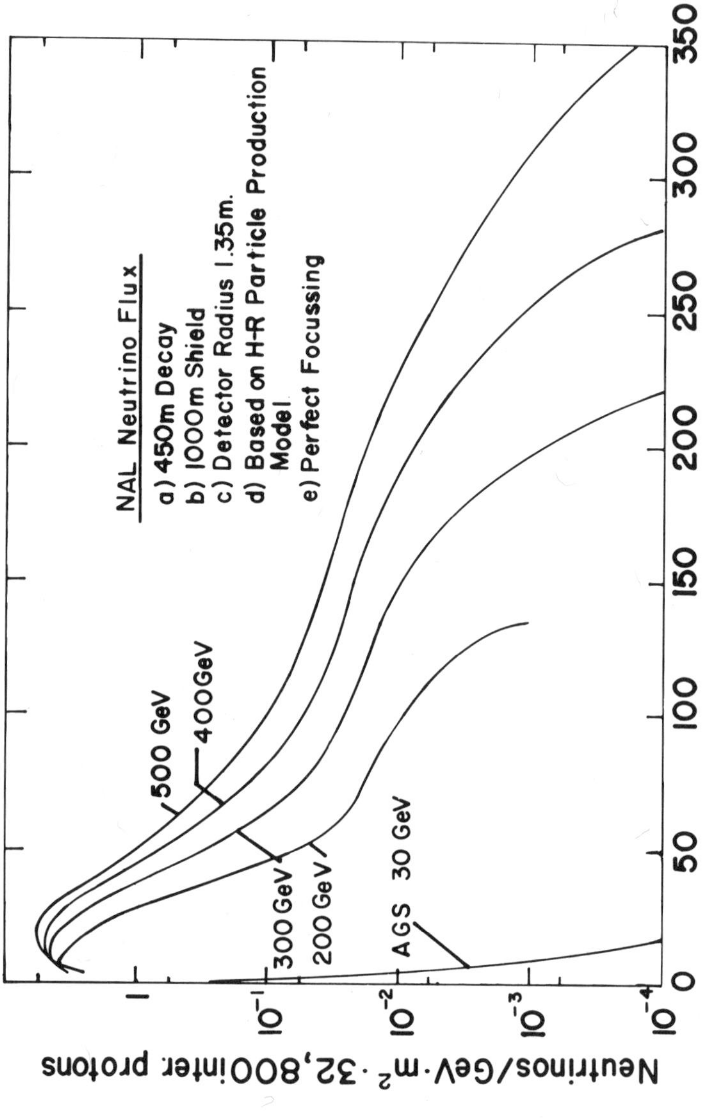

Fig. 2. Approximate spectrum of neutrinos expected at the National Accelerator Laboratory (see reference 14) compared with that from a lower energy accelerator.

III. Very High Energy Neutrino Scattering

It will soon be possible to study neutrino interactions in a hitherto unexplored neutrino energy region that will be opened by the National Accelerator Laboratory (NAL) at Batavia, Illinois. The unique advantage that the highest laboratory energy accelerator will provide for neutrino scattering experiments is indicated in fig. 2 which shows the intense, energetic neutrino flux that will be available at NAL. The exact spectrum depends on the details of the focusing system for the secondary hadrons but the spectrum shown[14] is a realistic one. One finds from fig. 2 that for 10^{13} protons per sec of energy 200 GeV on target there are $\sim 10^9$ neutrinos/(m^2 - sec) with energies between 20 and 100 GeV resulting from the decays of the secondary hadrons. A correlated advantage of the higher neutrino energy is the higher energy of the muons produced in the interactions of those neutrinos, which in turn leads to a cleaner separation of those muons from high energy hadrons by means of nuclear interactions. For example, the range of 1.5 GeV muon $\simeq$ 10^3gm/cm^2 which is approximately 10 collision lengths of iron. Thus the two major limitations of neutrino scattering experiments at lower energy - marginal neutrino intensity and inability to distinguish decisively between muons and hadrons produced by neutrinos - are to an appreciable extent obviated at higher energy.

An apparatus for producing and detecting neutrino interactions[15] is shown schematically in the fig. 3. It consists of a modular arrangement of lead plate - liquid scintillation counters and spark chambers which form a fine-grained ionization calorimeter followed by an iron core magnet - spark chamber assembly. The ionization calorimeter serves two functions: (i) it is a massive ($\sim$ 400 ton), high Z target for certain coherent or quasicoherent neutrino reactions, and (ii) it detects and measures the energy E_h of the hadronic-electromagnetic showers produced in inelastic neutrino reactions. The iron core magnet and its associated spark chambers determine the sign of the charge and the momentum of the high energy muons originating from neutrino interactions in the ionization calorimeter. The

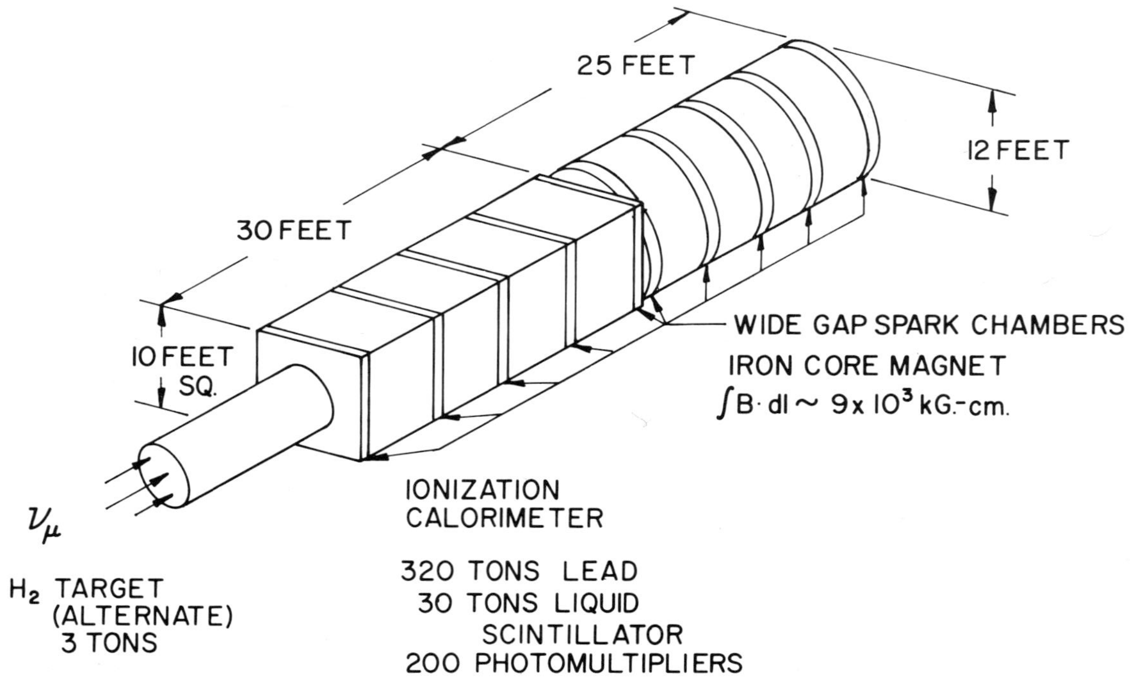

Fig. 3. Experimental arrangement for a neutrino scattering experiment at NAL.

energy of the neutrino initiating a given event is obtained from the sum $(E_h + E_\mu)$. A multi-ton liquid hydrogen target can be placed upstream of the ionization calorimeter to provide an alternate or additional region for neutrino interactions as indicated in fig. 3.

The neutrino beam and the target-detector arrangement just described open the possibility of studying experimentally pure leptonic processes other than muon decay which has thus far been the sole purely leptonic interaction available to experiment. On the basis of the present current-current interaction theory of weak interactions we expect neutrino production of three leptons (trilepton production), as shown in the diagrams of fig. 4, to take place. Observe that apart from the interaction of the virtual photon with the charge Z (which is well understood) the remainder of each of the diagrams in fig. 4 represents the general local weak interaction between lepton pairs. Study of trilepton production therefore yields the same information as the study of neutrino-electron scattering discussed earlier. There is no direct experimental evidence for either of these reactions and observation of one of them would demonstrate the existence of a self or diagonal current[16] which is an integral part of the current-current interaction theory. Indeed, it seems clear that if experiment does not confirm the theoretical predictions[17] concerning these reactions there will be required some profound modification of our present understanding of the weak interaction.

It is also worth noting that the absence or weaker strength of the diagonal current leading to $e^+ + e^- \rightarrow \nu_e + \bar{\nu}_e$ would have important repercussions with respect to present ideas on the aging of stars.[18]

The difficulties involved in an experiment to measure trilepton production and possible ways of surmounting them are easily summarized. The principal difficulty arises from the enormously more probable rate of occurrence of inelastic neutrino processes in the same target, which in this instance is the lead in the ionization calorimeter because coherence makes the lead more effective as a target. If the total cross section continues to rise linearly with laboratory

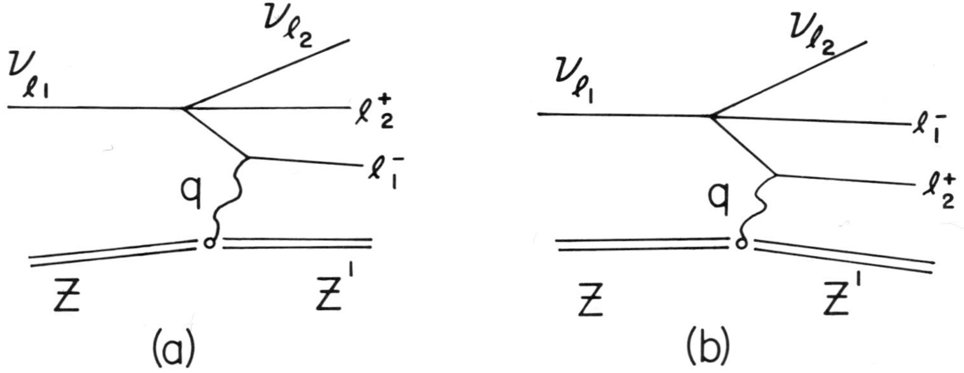

Fig. 4. Diagrams of the processes $\nu_\ell + Z \rightarrow \ell^+ + \ell^- + \nu_\ell + Z$.

neutrino energy above 10 GeV, the yield of inelastic events will exceed that from trilepton production by a factor of about 10^5. The design of the ionization calorimeter is based on the need to overcome that factor of $\sim 10^5$ by distinguishing coherent or quasicoherent processes such as trilepton production from inelastic events exhibiting energetic hadrons. The distinction relies in part on the presence of two leptons, say, two muons, alone in the final state of trilepton production, and on the high density of material in the calorimeter which absorbs energetic pions in the inelastic events before they decay in flight. Roughly, 10^{-3} of all pions with $E_\pi \geq 4$ GeV will decay in flight in the calorimeter. An additional suppression factor of $\sim 10^{-2}$ is expected from the electronic detection of an energy release from inelastic events in the ionization calorimeter beyond that due to two minimum ionizing particles; this is one of the reasons for the fine-grained construction of the calorimeter. Other, smaller suppression factors arise from comparison of the momentum and angular distributions of the two muons from trilepton production with those from background events. In short, it appears possible to obtain a moderately clean and abundant signal (several events per day) of trilepton production from which at least a rough comparison of the strengths of the diagonal and non-diagonal currents can be made.

A charged intermediate vector boson W with mass less than about 10 GeV, if it exists, would be produced copiously in the experimental arrangement described here, primarily through the reaction $\nu_\mu + p \rightarrow \mu^- + W^+ + p$. The Feynman diagrams for W production are shown in fig. 5; they lead to a cross section[19] for $M_W \simeq 5$ GeV roughly 10^3 larger than that for trilepton production and a cross section for $M_W \simeq 10$ GeV about equal to that for trilepton production. The W is expected to decay rapidly ($\tau_W \lesssim 10^{-16}$ sec) to two leptons, say, $\mu^+ + \nu_\mu$, at least part of the time[20] so that in some instances the final products of the reaction are the same as in trilepton production. Hence considerations similar to those in trilepton production apply to the signal to background ratio in W production. In the mass region

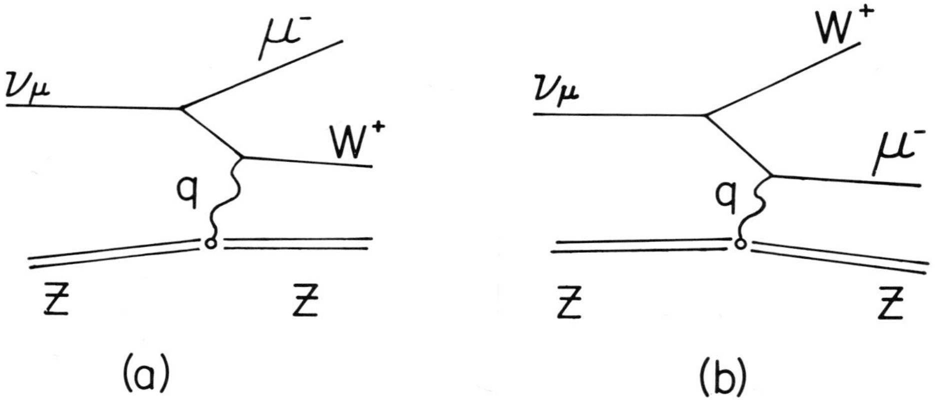

Fig. 5. Diagrams of intermediate vector boson (W) production by neutrinos.

$M_W \sim 10$ GeV where the yields of events from W and trilepton production would be roughly equal the two processes are distinguishable statistically because the very large transverse component of momentum of the μ^+ from W^+ decay is not reproduced in the transverse momentum of the μ^+ from trilepton production.

The mass range of the W that is accessible to a search by neutrino scattering at NAL may perhaps be extended through measurement of the total cross section for neutrino scattering as a function of neutrino energy. It has been shown[21] that, if strong interaction scaling continues to hold at high energy, a finite mass W will cause the total cross section to deviate from the linear rise with neutrino energy that is currently conjectured; this is illustrated in fig. 6 to show the magnitudes involved. Measurements of the total cross section relative to the energy independent quasielastic cross section ($\nu_\mu + n \to \mu^- + p$) in the forward direction (q^2 small) may be made with sufficient accuracy (either in the lead ionization calorimeter described above or in a less massive but more sensitive ionization calorimeter) to permit a search for a W up to a mass of about 15-20 GeV when NAL accelerates protons to 400 GeV.

The study of the abundant semileptonic processes induced by neutrinos, e.g., ν_μ + nucleus $\to \mu^-$ + anything, i.e., inelastic neutrino scattering, presents additional opportunities to test the weak interaction at high momentum and energy transfers. Perhaps the most interesting aspect of inelastic scattering, however, will be its use as a probe of nucleon structure. Neutrino inelastic scattering (summed over all final states) does not have the overall q^{-4} suppression that is present in electromagnetic scattering. Thus a new regime of momentum transfer (>100 GeV^2) will become available in high energy neutrino scattering. Thus a new regime of momentum transfer (>100 GeV^2) will become available in high energy neutrino scattering if no propagator of weak interaction exists to depress the cross section. Alternatively, a propagator will manifest itself through the replacement[21]

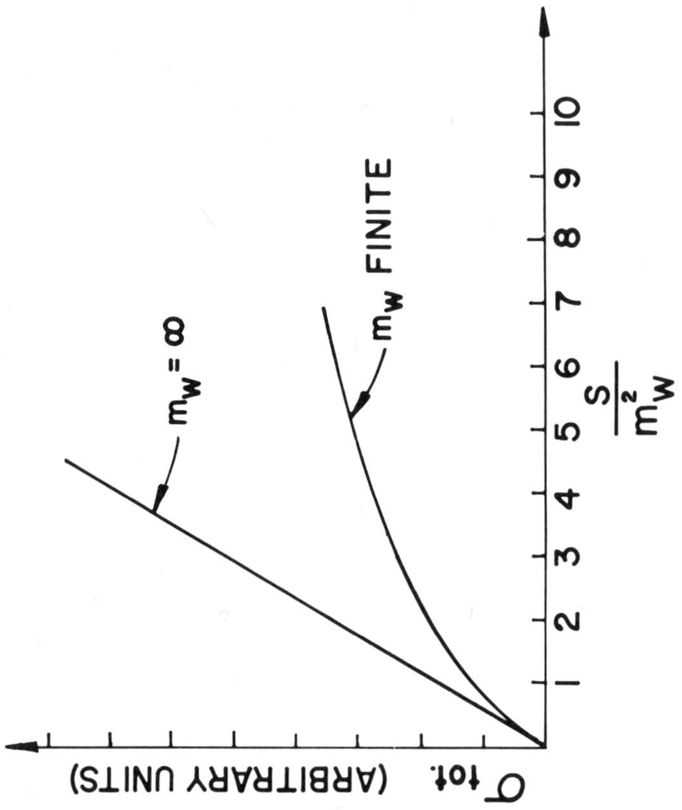

Fig. 6. Calculated curve from reference 24 showing how a finite mass W may affect the total neutrino cross section.

$$G^2 \rightarrow \frac{G^2}{(1 + q^2/M_W^2)^2}$$

in the basic formulas and thus $d^2\sigma/d\nu dq^2$ will be substantially modified at very large q^2 by a W mass of order 30 GeV. Such experiments and their implications are properly the subject of more extensive discussion than is possible here.

IV. Conclusions

The construction of accelerators of high energy and high intensity and the adroit use of lower energy accelerators give promise that the structure of the weak interaction will be deeply probed in the near future by neutrino production and scattering experiments.

Sensitive searches for unknown short-lived sources of neutrinos or neutrino-like particles are possible using intense proton or electron beams in "beam-dump" experiments. A definitive search for charged intermediate vector bosons produced by high energy neutrinos should be possible up to a boson mass in the vicinity of 25 GeV. In the absence of such bosons pure leptonic processes other than muon decay will be studied through trilepton production by neutrinos. Finally, semileptonic processes induced by high energy neutrinos will test our knowledge of weak interactions at high momentum and energy transfers and perhaps serve as a means of studying nucleon structure.

REFERENCES

1. F. Reines and C.L. Cowan, Phys. Rev. 92, 830 (1953).
2. W. Pauli, Jr., Rappt. Septieme Conseil Phys. Solvay, Bruxelles; 1933 (Gautier-Villars, Paris, 1934).
3. E. Fermi, Z. Physik, 88, 161 (1934); H.A. Bethe and R.E. Peierls, Nature 133, 532 (1934).
4. G. Danby et al, P.R.L. 9, 36 (1962).
5. G. Bernardini et al, Nuovo Cimento 38, 608 (1965); D.H. Perkins, Proc. of the Topical Conference on Weak Interactions, CERN 69-7, p. 1 (1969).
6. A. Tamosaitis, Proc. of Int'l. Conf. on Bubble Chamber Technology, Argonne National Laboratory, 1970 p. 765.
7. D. Perkins, Proc. of Int'l. Conf. on Bubble Chamber Technology, Argonne National Laboratory, 1970 p. 123; P. Mussett, ibid, p. 136.
8. Participants are: D. Fryberger, A. Rothenberg, M. Schwartz and T. Zipf, SLAC; D. Kreinick and A.K. Mann, University of Pennsylvania; D. Dorfan, University of California at Santa Cruz; J.M. Gaillard, CERN.
9. Ya B. Zel'dovich, Sov. Phys. Uspekhi, 5, 931 (1963); E.M. Lipmanov, Zurn. Eksp. Teor. Fiz., 16, 634 (1963); 19, 1291 (1964); E.W. Beier, Ph.D. Thesis, University of Illinois (1966), unpublished; K.W. Rothe and A.M. Wolsky, Nucl. Phys. 10B, 241 (1969); S.S. Gershtein and V.N. Folomeshkin, Sov. Journal Nuclear Physics 8, 447 (1969).
10. See, for example: J. Bernstein, M. Ruderman and G. Feinberg, Phys. Rev. 132, 1227 (1963); J. Bernstein and T.D. Lee, PRL 11, 512 (1963); Ph. Meyer and D. Schiff, Physics Letters 8, 217 (1964); W.K. Cheng and S.A. Bludman, Phys. Rev. 136 B1787 (1964).
11. R. Davis, Jr., Bulletin APS, 16, 847 (1971); J. Bahcall, Bulletin APS, 16, 1407 (1971).
12. M.M. Block et al, Physics Letters 12, 281 (1964).
13. The upstream and downstream targets are different physical targets and the pion yield from each must be related to that from a given (standardized) thickness of beryllium target.

14. Y.W. Kang and F.A. Nezrick, NAL Summer Study, Report FN-124, 1968.

15. Participants in this neutrino scattering experiment at NAL are: J.E. Pilcher, C. Rubbia and L. Sulak, Harvard University; A. Benvenuti, D. Cline, R.L. Imlay and D.D. Reeder, University of Wisconsin; A.K. Mann, University of Pennsylvania.

16. M. Gell-Mann, M.L. Goldberger, N.M. Kroll and F.E. Low, Phys. Rev. 179, 1518 (1969).

17. W. Czyz, G.S. Sheppy and J.D. Walecka, Nuovo Cimento 34, 404 (1964); J. Loevseth and M. Radomski, Phys. Rev. D3, 2686 (1971), K. Fujikawa, Ph.D. Thesis, Princeton University (unpublished) 1970; Phys. Rev., to be published.

18. H.Y. Chiu, Ann. Rev. Nucl. Sci. 16, 591 (1966).

19. R.W. Brown and J. Smith, Phys. Rev. D3, 207 (1971); R.W. Brown, R.H. Hobbs and J. Smith, Phys. Rev. (to be published).

20. Although see: D. Cline, A.K. Mann and C. Rubbia, PRL 25, 1309 (1970).

21. J.D. Bjorken and E.A. Paschos, Phys. Rev. 1, 3151 (1970).

DISPERSION INEQUALITIES AND THEIR APPLICATION TO THE PION'S ELECTROMAGNETIC RADIUS AND THE $K_{\ell 3}$ PARAMETERS*

S. Okubo
University of Rochester
Rochester, New York

I. INTRODUCTION AND SUMMARY OF MAIN RESULTS

The dispersion relation is an indispensable tool for analyzing various problems in high energy physics. In this paper, we shall consider applications of a new type of dispersion relation to various problems.

To be definite, let us assume that f(t) is a real analytic function of t in the entire complex t plane with a cut on the real axis at $t_o \leq t \leq \infty$. Notice that the reality condition implies

$$f^*(t^*) = f(t) \tag{1}$$

Therefore, f(t) is real for real values of t less than t_o (i.e. $t \leqslant t_o$). Without loss of generality, we can assume $t_o > 0$.

Now there are many situations in which we want to evaluate or find bounds on f(t) and its derivatives $f^{(n)}(t)$ when the following information is given:

Case (I)

Imf(t) is known on the cut $t_o \leqslant t < \infty$.

Case (II)

Arg f(t) is known on the cut $t_o \leqslant t < \infty$.

Case (III)

An upper bound for $|f(t)|$ is known on the cut. More specifically we have

$$|f(t)| \leqslant w(t) \tag{2}$$

*Work supported in part by the U.S. Atomic Energy Commission.

on the cut $t_o \leqslant t < \infty$ where $w(t)$ is a known non-negative function defined on the cut. A special case of this problem occurs when $|f(t)|$ itself is known on the cut. Then, we can set $w(t) = |f(t)|$ from the beginning.

Case (IV)

The only information given is that we have an inequality

$$\frac{1}{\pi} \int_{t_o}^{\infty} dt \; w(t) |f(t)|^2 \leqslant I^2 \tag{3}$$

where $w(t)$ is a known non-negative function and I is a given constant.

Case (I) corresponds to the familiar standard dispersion relation[1], and we need not mention any details except for the fact that dispersion relations generally require a few unknown subtraction constants.

Case (II) is known as the phase representation problem, and it occurs for example in Omnes-type equation. However, we shall not go into details since its general solution is fairly well-known.[2]

In this paper, we are mainly interested in discussing cases (III) and (IV). For such cases, we cannot hope in general to obtain an exact relation. However, we can find exact dispersive inequalities (as opposed to equalities) for $f(t)$ and $f^{(n)}(t)$ $(t < t_o)$. First, we shall discuss Case (III) since it is simpler. The problem has been solved by Geshkenbein[3] for the case $w(t) = |f(t)|$ and recently by Raszillier[4] for the general case. As we shall show in the next section, we can prove[5] inequalities such as

$$\log |f(o)| \leqslant \varepsilon \; , \tag{4}$$

$$\left| f'(o) - \left(\eta - \frac{\varepsilon}{2t_o}\right) f(o) \right| \leqslant \frac{1}{4t_o} \left(e^{\varepsilon} - e^{-\varepsilon} |f(o)|^2\right) \tag{5}$$

where ε and η are defined by

$$\begin{aligned} \varepsilon &= \frac{1}{\pi} (t_o)^{\frac{1}{2}} \int_{t_o}^{\infty} dt \; t^{-1} (t-t_o)^{-\frac{1}{2}} \; \log w(t) \; , \\ \eta &= \frac{1}{\pi} (t_o)^{\frac{1}{2}} \int_{t_o}^{\infty} dt \; t^{-2} (t-t_o)^{-\frac{1}{2}} \; \log w(t) \; . \end{aligned} \tag{6}$$

Equation (4) has also been derived by Truong and Vinh-Mau[6] for the case $w(t) = |f(t)|$, where it is simply the classical Jensen inequality.

It should be remarked that, in spite of apparent mismatch of scale dimentions for quantities involved in both sides of Eqs. (4) and (5), these equations are nevertheless correct. The reason for this peculiarity is due to the fact that both ε and η defined by Eq. (6) have so-called abnormal dimensions with respect to scale transformation, if $w(t)$ has a definite non-zero scale dimension. Actually, it is interesting to observe that the combination $\varepsilon - 2\eta\, t_o$ has zero dimension, irrespective of the dimensionality of $w(t)$.

As an application, let us consider the electromagnetic form factor of a charged pion $F_\pi(t)$ which is defined by

$$(4\, p_o\, p_o'\, V^2)^{\frac{1}{2}} < \pi^+(p')|\ j_\mu(o)|\pi^+(p)> \ = \ (p_\mu + p_\mu')F_\pi(t),$$

$$t = -\ (p-p')^2$$

where $j_\mu(x)$ is the electromagnetic current. Notice that conservation of the electric charge implies

$$F_\pi(o) = 1\ .$$

Obviously, $F_\pi(t)$ is a real analytic function of t with a cut on the real axis at $t_o \leqslant t < \infty$ ($t_o = 4m_\pi^2$). To obtain an upper bound $w(t)$, let us consider the electron-positron pair annihilation process into a pion pair

$$e^-\ e^+ \to \pi^-\ \pi^+$$

and designate the cross-section by $\sigma(t)$ where t is the square of the total energy in the barycentric system. Now, if we restrict ourselves to one-photon and two-photon exchange processes (neglecting three or more photon exchanges), then we find

$$\sigma(t) \geqslant \frac{\alpha}{6}\, t^{-\frac{5}{2}}\ (t-t_o)^{\frac{3}{2}}\ |F_\pi(t)|^2\ ,(t \geqslant t_o)\ ,\quad \alpha = \frac{e^2}{4\pi}\ .$$

To see this, notice that the one-photon and two-photon exchange processes produce two-pion final states with different charge conjugation parities $C = \mp 1$. Therefore, the one-photon and two-photon diagrams do not contribute an interference term to the total two-pion production cross-section. Hence, we may choose

$$w(t) = [\frac{6}{\alpha} t \sigma(t)]^{\frac{1}{2}} \; t^{\frac{3}{4}} (t - t_o)^{-\frac{3}{4}} \; .$$

If we wish, we may replace $\sigma(t)$ by the total cross-section $\sigma_T(t)$. Since $\sigma(t)$ can be easily measured experimentally and since $F_\pi(o) = 1$, we can obtain bounds on the electromagnetic radius (r_π) of the pion by combining the formula $F_\pi'(o) = \frac{1}{6}(r_\pi)^2$ with Eqs. (5) and (6). I turns out that the present experimental values of $\sigma(t)$, together with a reasonable extrapolation to large t, gives[5]

$$0.3 \leqslant r_\pi^2/r_\rho^2 \leqslant 1.0$$

where r_ρ is the value of r_π calculated by a simple ρ-dominance model. We also remark that the upper bound on r_π is relatively insensitive to the exact behavior of $\sigma(t)$ at large t.

Another application concerns the lifetime τ of $\pi^o \to 2\gamma$ and the shape parameter of the decay $\pi^o \to \gamma e\bar{e}$. If we set

$$(4\, p_o\, p_o'\, V^2)^{\frac{1}{2}} \langle\gamma(p')|j_\mu(o)|\pi^o(p)\rangle = \frac{i}{m_\pi} f(t)\; \varepsilon_{\mu\nu\alpha\beta}\, p_\nu p_\alpha' \varepsilon_\beta(p') \; ,$$

$$t = -\, (p-p')^2$$

then $f(t)$ is a real analytic function of t with a cut on $t_o \leqslant t < \infty$ $(t_o = 4m_\pi^2)$. Now, the π^o -lifetime τ is given by

$$\frac{1}{\tau} = \frac{\alpha}{16} m_\pi |f(o)|^2 .$$

Also, neglecting three or more photon exchange processes, we have[7]

$$\sigma(t) = \frac{1}{6m_\pi^2} \pi\alpha^2 \;(t-m_\pi^2)^3 t^{-3} \;|f(t)|^2 \; , \quad (t \geqslant m_\pi^2)$$

where $\sigma(t)$ is now the total cross-section for the pair annihilation process $e^+e^- \to \pi^o\gamma$. Hence, if we can measure this cross-section, then our inequalities Eqs. (4) and (5) give bounds on the π^o-lifetime τ and on the shape parameter $f'(o)$ of $\pi^o \to \gamma e\bar{e}$ decay. A similar inequality has been studied by Palmer.[7] Some other applications for $\gamma p \to \gamma p$ as well as $ep \to ep$ reactions will be given in the next section.

Finally, let us consider the class (IV) problem. In this case less information is specified than in the class (III) case. Nevertheless, we can still derive a useful inequality

$$|f(o)|^2 + |(2+\varepsilon-2t_o\eta)\, f(o)-4t_o f'(o)|^2 \leqslant \frac{1}{4t_o} I \quad \exp(-\varepsilon), \tag{7}$$

where ε and η are defined by Eq. (6) again. Notice that Eq. (7) immediately leads to the special case

$$|f(o)|^2 \leqslant \frac{1}{4t_o} I^2 \exp(-\varepsilon) \ . \tag{8}$$

This result was originally proved by Meiman[8] and our inequality Eq. (7) is its generalization.[9] Actually, in section III we shall prove a much more general inequality, involving f(o) and its derivatives together with $f(\lambda)$ where λ is any real point satisfying $\lambda < t_o$.

An interesting application concerns the parameters of the scalar form factor D(t) in the $K_{\ell 3}$ decay problem.

Setting

$$(4\ p_o\ p_o'\ V^2)^{\frac{1}{2}} < \pi^o(p')|\ V_\mu^{(4-i5)}(o)\ |\ K^+(p)>$$
$$= -\frac{1}{\sqrt{2}}\{\ (p + p')_\mu\ f_+(t) + (p - p')_\mu\ f_-(t)\} \quad ,$$
$$t = -\ (p - p')^2 \ ,$$

we fine

$$(4\ p_o\ p_o'\ V\ )^{\frac{1}{2}} <\pi^o(p')|\ \partial_\mu\ V_\mu^{(4-i5)}(o)\ |\ K^+(p)> = \frac{1}{\sqrt{2}}\ i\ D(t)\ ,$$
$$D(t) = (m_K^2 - m_\pi^2)\ f_+(t) + t\ f_-(t) \quad .$$

Now, D(t) is a real anlytic function of t with a cut on $t_o \leqslant t < \infty$ $[t_o = (m_K + m_\pi)^2]$. In principle, the value of $|D(t)|$ on the cut can be extracted from the experimental data on the cross-section for the reaction $e^+\nu \to K^+\pi^o$. Since such an experiment cannot be done in the forseeable future, we cannot apply method (III). However, we can circumvent this difficulty as follows. Let $A_\mu^{(\alpha)}(x)$ and $V_\mu^{(\alpha)}(x)$ $(\alpha = 1,...,8)$ be the octets of axial-vector and vector currents of the weak interaction physics, and set

$$A_{\alpha\beta}(t) = i \int d^4x \; e^{iq(x-y)} \langle 0|(\partial_\mu A_\mu^{(\alpha)}(x), \partial_\nu A_\nu^{(\beta)}(y))_+ |0\rangle,$$

$$V_{\alpha\beta}(t) = i \int d^4x \; e^{iq(x-y)} \langle 0|(\partial_\mu V_\mu^{(\alpha)}(x), \partial_\nu V_\nu^{(\beta)}(y))_+ |0\rangle$$

with $t = -q^2$. Then, the standard Kamefuchi-Umezawa-Lehmann-Källen representation is written typically as

$$V_{44}(t) = \int_{t_o}^{\infty} dt' \frac{\rho(t')}{t'-t} \quad , \quad [\, t_o = (m_K + m_\pi)^2] \quad ,$$

$$\rho(t') = \frac{1}{2} (2\pi)^3 \sum_\eta |\langle 0|\partial_\mu V_\mu^{(4-i5)}(0) \,|\eta\rangle|^2 \,\delta^{(4)}(q-p_\eta) \quad ,$$

$$t' = - q \quad .$$

The positivity of the spectral weight $\rho(t)$ implies the inequality[10]

$$|D(t)|^2 < \frac{64}{3} \pi^2 \, t \,(t-t_o)^{-\frac{1}{2}} (t-t_1)^{-\frac{1}{2}} \rho(t)$$

for $t \geqslant t_o$ with $t_1 = (m_K - m_\pi)^2$; here we have taken only πK intermediate states and we used the crossing relation to rewrite $\langle 0|\partial_\mu V_\mu^{(4-i5)}(0)|\pi(p')\, K(p)\rangle$ in terms of $D(t)$. Therefore, if $\rho(t)$ or its upper bound is somehow known, then we can still use the technique of Case (III), (i.e. inequalities Eqs. (4) and (5)) with $w(t) = (8\pi/\sqrt{3})\; t^{+1/2}(t-t_o)^{-1/4}(t-t_1)^{-1/4}\rho^{\frac{1}{2}}(t)$. However, the direct estimate of $\rho(t)$ or its upper bound is in general not easy unless we make some drastic approximations such as the kappa dominance. Nevertheless, we can reduce our problem to case (IV) as follows. Integrating the above inequality suitably, we find[10]

$$\frac{1}{\pi} \int_{t_o}^{\infty} dt \; w(t) \; | \; D(t)|^2 \leqslant V_{44}(0) = \int_{t_o}^{\infty} dt \; \frac{\rho(t)}{t}$$

where $w(t)$ is now given by

$$w(t) = \frac{3}{64\pi} \; t^{-2} \; (t-t_o)^{\frac{1}{2}} \; (t-t_1)^{\frac{1}{2}} \quad .$$

Hence, if an upper bound on $V_{44}(0)$ is found, we can use the technique of Case (IV). To find such a bound, we assume the chiral theory[11] of Gell-Mann, Oakes and Renner and of Glashow and Weinberg. In this case, we can prove the inequality,[12]

$$|V_{44}(0)| \leqslant |(A_{44}(0))^{\frac{1}{2}} - (A_{33}(0))^{\frac{1}{2}}| .$$

Since the dominant contribution to A_{33} and A_{44} is expected to come from one pion and one kaon intermediate state, respectively, this gives

$$|V_{44}(0)| \leqslant \frac{1}{\sqrt{2}} (m_K f_K - m_\pi f_\pi) .$$

This inequality can also be proved without making the pole dominance approximations for A_{33} and A_{44} if we appeal to asymptotic SW(3) symmetry.[13] At any rate, we can choose $I^2 = \frac{1}{\sqrt{2}} (m_K f_K - m_\pi f_\pi)$. Also, the soft pion theorem[14] requires that we have

$$D(\delta) / D(0) \simeq \frac{f_K}{f_\pi} \frac{1}{f_+(0)} \simeq 1.28 ,$$

$$\delta = m_K^2 - m_\pi^2 .$$

Now, the question we want to ask is as follows. The analytic function D(t) satisfies the condition of type (IV); moreover, we know its value $D(\delta)$ at $t = \delta$. Now, the problem is to find a bound on D'(o). This problem has been completely solved;[13] we find

$$0.008 \leqslant \Lambda_0 \leqslant 0.019 .$$

Here we have set $\Lambda_0 = m_\pi^2 D'(0)/D(0) = \lambda_+ + \xi m_\pi^2 (m_K^2 - m_\pi^2)^{-1}$, and we have assumed $|f_+(0)| \leqslant 1$ because of the Ademollo-Gatto theorem.[15] So far, the best experimental value for Λ_0 is given by[16]

$$\Lambda_0 = - 0.11 \pm 0.03$$

which contradicts our theoretical bound. Similarly, we can find a bound[13] for the second-derivative $m_\pi^4 D''(0)/D(0)$, which is again outside the present experimental value[16] of 0.0085 ± 0.0065.

We can apply the same technique to the $\pi^0\pi^0$ scattering problem. If a_0 and a_2 are the S and D $\pi^0\pi^0$ scattering lengths, then we can prove the exact bound

$$a_0 + 10\, a_2 \geqslant -1.11 \quad .$$

Moreover, we remark that an upper and a lower bound on strong interaction corrections to the Lamb-shift[17] and to the anomalous magnetic moment[18] of the muon have been derived by similar methods.

II. Inequalities for Case III

This problem has been solved by Geshkenbein[3] for the case $w(t) = |f(t)|$ and generalized by Raszillier.[4] To begin, let us consider a function

$$g(t) = \exp\{\frac{1}{\pi}\,(t_0-t)^{\frac{1}{2}} \int_{t_0}^{\infty} dt'\,(t'-t)^{-1}\,(t'-t_0)^{-\frac{1}{2}} \log w(t')\} \qquad (9)$$

where the integral is assumed to exist.

Now $g(t)$ is a real analytic function of t in the cut-plane with the cut at $t_0 \leqslant t < \infty$ and satisfies

$$|g(t)| = w(t) \quad , \qquad (t \geqslant t_0) \qquad (10)$$

on the cut. Moreover, $g(t)$ has no zeroes at all. Hence, the function defined by

$$B(t) = \frac{f(t)}{g(t)} \qquad (11)$$

is a real analytic function of t in the cut plane, and on the cut it satisfies

$$|B(t)| \leqslant 1 \quad . \qquad (12)$$

Also, at infinity ($t = \infty$), the Lindelöf-Phragmén theorem[19] requires $|B(t)| \leqslant 1$, provided that both $w(t)$ and $f(t)$ behave non-exponentially at infinity. Now, by the maximum modulus theorem, we must have the inequality (12) in the entire complex plane, i.e.,

$$|f(t)| \leqslant |g(t)| \quad . \qquad (13)$$

In particular, if we set $t = 0$, this reproduces Eq. (4).

Next, Eq. (5) is easily derived as follows. Any analytic function $B(t)$ in the cut plane, which satisfies the inequality

Eq. (12), must automatically satisfy the following inequality for $B'(t)$:

$$|4(t_o - t)\ B'(t)| \leqslant 1 - |B(t)|^2 \tag{14}$$

for real values of t less than t_o, (i.e. $t < t_o$). This implies that we have

$$|4(t_o-t)(g(t)f'(t) - g'(t)f(t))| \leqslant |g(t)|^2 - |f(t)|^2 \tag{15}$$

for $t < t_o$. Setting $t = 0$, this gives Eq. (5) of the previous section.

Now, let us prove Eq. (14). We follow the method originally used by Ciulli.[20] It is convenient to map the complex cut t-plane into the interior of the unit circle $|z| = 1$ by the following conformal mapping:

$$(t-t_o)^{\frac{1}{2}} = i\ (t_o-\beta)^{\frac{1}{2}}\ \frac{1+z}{1-z} \tag{16}$$

where β is an arbitrary real number satisfying the condition $\beta < t_o$. Then, the upper and lower cuts in the t-plane are mapped into the lower and upper semi-circle $|z| = 1$, respectively. Also, three points $t = \infty$, β, and t_o are mapped into $z = 1$, 0, and -1, respectively. We remark that $g(t)$ of Eq.(9) is now rewritten as

$$g(t) = \exp\ \{ \frac{1}{2\pi} \int_0^{2\pi} d\theta\ \frac{e^{i\theta} + z}{e^{i\theta} - z}\ \log W(\theta)\ \} \tag{17a}$$

where

$$W(\theta) \equiv w(t) \quad , \quad t = t_o + (t_o - \beta)\cot^2\frac{\theta}{2}\ . \tag{17b}$$

Also, defining

$$B_1(z) \equiv B(t) \tag{18}$$

we find that $B_1(z)$ is analytic inside the unit circle $|z| < 1$, and satisfies $|B_1(z)| \leqslant 1$ there. Now, if $B_1(z)$ is not identically a constant with unit modulus, then the function $B_2(z)$ defined by

$$B_2(z) = \frac{1}{z}\ \frac{B_1(z) - B_1(o)}{1-B_1{}^*(o)B_1(z)}$$

is again analytic inside the unit circle $|z| = 1$ and satisfies

$|B_2(z)| \leqslant 1$ on the circle. Therefore, by the maximum modulus theorem we have $|B_2(z)| \leqslant 1$ inside the circle. Now, setting $z = 0$, we find that $B_1(z)$ must satisfy

$$|B_1'(0)| < 1-|B_1(0)|^2 . \tag{19}$$

If $B_1(z)$ is a constant with unit modulus, Eq. (19) is also satisfied; therefore, it is valid for any function $B_1(z)$ satisfying our condition. Using Eqs. (16) and (18), we can immediately derive Eq. (14) with $t = \beta$. Since "β" is arbitrary, we can replace it by t to obtain Eq. (14).

The special case $w(t) \equiv |f(t)|$ is interesting. In that case, we find $|B(t)| = 1$ on the cut, so that $|B_1(z)| = 1$ on the unit circle $|z| = 1$. Then, g(t), given by Eq. (17a), is known as the outer function while $B_1(z)$ is said to be inner.[21] The structure of the inner function $B_1(z)$ is well-known. Neglecting the so-called singular measure function[21] S(z) on the physical grounds, we can represent $B_1(z)$ as

$$B_1(z) = z^m \prod_{n=1}^{\infty} \frac{\lambda^*{}_n(\lambda_n-z)}{|\lambda_n|(1-\lambda_n^* z)} \tag{20}$$

where λ_n $(n = 1,2,...)$ are the zero points of $F(z) \equiv f(t)$ indide the open circle $|z| < 1$. The right hand side of Eq. (20) is known as a Blaschke product. In particular, if f(t) has no zero points in the cut-plane, then we expect $B(t) \equiv 1$; therefore, Eqs. (13) and (15) will become equalities instead of inequalities. Actually, a slightly more careful analysis shows that the equality in Eq. (13) is possible if and only if f(t) has no zero point, while the equality in Eq. (15) is still possible if f(t) has one (but not more than one) zero.

Returning to the general case, we shall now consider applications. Since we mentioned the electromagnetic radius of the pion in the Introduction, we will not repeat that calculation here. As another application, let us consider virtual Compton scattering on a proton target. Setting

$$i \frac{p_o V}{m} \int d^4x \quad e^{iq(x-y)} < p|(j_\alpha(x), j_\beta(y))_+ \quad |p> =$$

$$= t_1(\nu,q^2)\ [\delta_{\alpha\beta}\, q^2 - q_\alpha q_\beta] \tag{21}$$

$$+ t_2(\nu,q^2)\ [\nu^2 \delta_{\alpha\beta} + \frac{1}{m^2} q^2 p_\alpha p_\beta + \frac{1}{m}\nu (p_\alpha q_\beta + p_\beta q_\alpha)]$$

with $\nu = -\frac{1}{m}(pq)$, we know[22] that $t_2(\nu,q^2)$ satisfies unsubtracted dispersion relation ($m \equiv m_N$, $\mu \equiv m_\pi$);

$$t_2(\nu,q^2) = \frac{4m^2}{(q^2)^2 - 4m^2\nu^2}\beta(q^2) + \frac{2}{\pi}\int_{\nu_o}^{\infty} d\nu' \frac{\nu'}{\nu'^2 - \nu^2} \mathrm{Im}\, t_2(\nu',q^2). \tag{22}$$

Here, $\beta(q^2)$ and ν_o are given by

$$\beta(q^2) = \frac{1}{m}[q^2 G_M^2(q^2) + 4m^2 G_E^2(q^2)] \cdot (q^2 + 4m^2)^{-1},$$

$$\nu_o = \frac{1}{2m}(q^2 + 2m\mu + \mu^2) \tag{23}$$

and $G_M(q^2)$ and $G_E(q^2)$ are the magnetic and electric form factors of the proton.

Since $\mathrm{Im}\, t_2(\nu',q^2) \geqslant 0$ for $q^2 \geqslant 0$ and $\nu' \geqslant \nu_o$, we know that $t_2(\nu,q^2)$ can have exactly two zeros in the complex ν plane if we have $t_2(\nu_o,q^2) > 0$, while it has no zero point at all if $t_2(\nu_o,q^2) < 0$. For simplicity we set

$$f(t) = (t - \frac{1}{4m^2}q^4)\, t_2(\nu,q^2), \quad t \equiv \nu^2 . \tag{24}$$

Then, $f(t)$ is a real analytic function of t with a cut at $t_o \leqslant t < \infty$ ($t_o = \nu_o^2$). Also, we have $f(t) = -\beta(q^2)$ at $t = \frac{q^4}{4m^2}$. Therefore, setting $t = \frac{q^4}{4m^2}$ in Eq. (13), we find an inequality

$$\beta(q^2) \leqslant \exp\{\frac{2}{\pi}(\omega_o^2 - 1)^{\frac{1}{2}} \int_{\omega_o}^{\infty} d\omega \frac{\omega}{\omega^2 - 1}(\omega^2 - \omega_o^2)^{-\frac{1}{2}}$$

$$\times \log|(\frac{q^4}{4m^2})(\omega^2 - 1)\, t_2(\nu,q^2)|\} \tag{25}$$

where we changed the integration variable from ν' to a new variable

$$\omega = \frac{2m\nu}{q^2} = -\frac{(pq)}{q^2} \quad , \quad (q^2 > 0) \tag{26}$$

and ω_o is given by

$$\omega_o = 1 + \frac{1}{q^2}(\mu^2 + 2m\mu) \; . \tag{27}$$

We remark that the equality in Eq. (25) is valid if we have $t_2(\nu_o, q^2) < 0$. Now, we let $q^2 \to \infty$. Anticipating the familiar scaling property of Im $t_2(\nu, q^2)$, we set

$$t_2(\nu, q^2) = \frac{2m\pi e^2}{(q^2)^2} \frac{1}{\omega} G_2(\omega, q^2) \; . \tag{28}$$

Then, the standard Bjorken scaling law is expressed as

$$\lim_{q^2\to\infty} \text{Im } G_2(\omega, q^2) = F_2(\omega) \; . \tag{29}$$

In this paper we shall assume a stronger scaling law, assuming that $|G_2(\omega, q^2)|$ scales also. If we define $K(\omega, q^2)$ by

$$|G_2(\omega, q^2)| = \gamma\, \omega(\omega^2 - \omega_o^2)^\rho \; K(\omega, q^2) \tag{30}$$

then the assumed scaling law will be

$$\lim_{q^2\to\infty} |G_2(\omega, q^2)| = \gamma\, \omega(\omega^2 - 1)^\rho \; K(\omega) \; . \tag{31}$$

The multiplicative factor $\gamma(\omega^2 - \omega_o^2)^\rho$ is chosen so that we have

$$K(1) = 1 \; . \tag{32}$$

Having made these preparations, we now insert Eq. (30) into Eq. (25) to obtain

$$\beta(q^2) \leqslant \frac{1}{m} \; 2\pi e^2 \gamma \; (\omega_o^2 - 1)^{\rho+1} \; \zeta \; ,$$

$$\zeta = \exp\{\frac{2}{\pi}(\omega_o^2 - 1)^{\frac{1}{2}} \int_{\omega_o}^{\infty} d\omega \; \frac{\omega}{\omega^2 - 1} \; (\omega^2 - \omega_o^2)^{-\frac{1}{2}} \log K(\omega, q^2)\} \tag{33}$$

after some calculations. Now, we assume that we can interchange the limit $q^2 \to \infty$ and the integral in Eq. (33). Then, noting that $\omega_o \to 1$ (see Eq. (27)) and $K(\omega,q^2) \to K(\omega)$, we find

$$\lim_{q^2\to\infty} \zeta = \lim_{q^2\to\infty} \exp\{\frac{2}{\pi} (\omega_o^2-1)^{\frac{1}{2}} \int_1^\infty d\omega\ \omega(\omega^2-1)^{-\frac{3}{2}} \log K(\omega)\}$$

Since $K(1) = 1$ (see Eq. (32)), we expect that the integral will converge at $\omega = 1$; therefore, we will have $\zeta \to 1$ since the the multiplicative factor $(\omega_o^2 - 1)^{\frac{1}{2}}$ goes to zero. Thus, we find finally

$$\lim_{q^2\to\infty} m(q^2)^{\rho+1} \beta(q^2) < C \quad , \tag{34a}$$

$$C = 2\pi\ e^2\gamma\ [2(\mu^2 + 2m\mu)]^{\rho+1} \quad . \tag{34b}$$

If we note

$$m\ \beta(q^2) \approx [F_1^{e.m.}(q^2)]^2$$

for $q^2 \to \infty$, then Eq. (34a) is immediately recognizable as a weaker form (i.e. inequality instead of the equality) of the conjectured Drell-Yan relations.[23] If we have $t_2(\nu_o,q^2) < 0$ for large q^2, then the equality in Eq. (34a) is valid and gives the Drell-Yan type relation. Experimentally, the dipole formula for electromagnetic form factors of the nucleon requires that $\rho = 3$.

Another inequality holds for the $\gamma p \to \gamma p$ reaction. Setting

$$f_1(\nu) = \frac{1}{4\pi}\ \nu^2\ t_2(\nu,o) \quad , \tag{35}$$

$f_1(\nu)$ represents the spin non-flip part of the Compton scattering amplitude. It is normalized[24] such that

$$\text{Im}\ f_1(\nu) = \frac{1}{4\pi}\ \nu\ \sigma_T(\nu)$$

where $\sigma_T(\nu)$ is the total Compton scattering cross-section. Also, the Thomson limit implies

$$f_1(o) = -\frac{\alpha}{m_N} \quad , \quad (\alpha = \frac{1}{4\pi}\ e^2) \quad . \tag{36}$$

Noting that $f(t) \equiv f_1(\nu)$ (with $t = \nu^2$) is a real analytic function of t with a cut at $t_o \leqslant t < \infty$ [$t_o = (m_N + m_\pi)^2$], the inequality Eq. (4) immediately gives

$$\frac{1}{m_N} \alpha \leqslant \exp\{\frac{2}{\pi} \nu_o \int_{\nu_o}^{\infty} d\nu\, \nu^{-1}(\nu^2-\nu_o^2)^{-\frac{1}{2}} \log |f_1(\nu)| \} \quad , \tag{37}$$

$$\nu_o = m_N + m_\pi \quad .$$

Note that Eq. (37) will become an equality if we have $f_1(\nu_o) < 0$.

Next, let us consider the spin-flip amplitude $f_2(\nu)$, which is a crossing off function of ν satisfying[24]

$$f_2'(0) = - \frac{\alpha}{2m_N^2} \mu^2 \tag{38}$$

where μ is the anomalous magnetic moment of the proton. Applying the same technique to the crossing even function $\frac{1}{\nu} f_2(\nu)$ and setting $\nu = 0$ we find

$$\frac{\alpha}{2m_N^2} \mu^2 < \exp\{\frac{2}{\pi} \nu_o \int_{\nu_o}^{\infty} d\nu\, \nu^{-1}(\nu^2-\nu_o^2)^{-\frac{1}{2}} \log |\frac{f_2(\nu)}{\nu}| \quad . \tag{39}$$

Multiplying Eqs. (37) and (39) and noting[24] that the forward elastic Compton scattering cross-section is given by

$$[\frac{d}{dt} \sigma(\nu,t)]_{t=0} = |f_1(\nu)|^2 + |f_2(\nu)|^2 \geqslant 2|f_1(\nu)|\cdot|f_2(\nu)| \quad ,$$

we find

$$\frac{\alpha^2}{m_N^3} \mu^2 \leqslant \exp\{\frac{2\nu_o}{\pi} \int_{\nu_o}^{\infty} d\nu\, \nu^{-1}(\nu^2-\nu_o^2)^{-\frac{1}{2}}\log|\frac{1}{2\nu} (\frac{d}{dt}\sigma(\nu,t))_{t=0}|\} \quad . \tag{40}$$

We can derive a similar inequality for the forward pion-nucleon scattering amplitudes $A^{(\pm)}$ and $B^{(\pm)}$, but we shall not go into the details of these calculations.

III. Discussion of Type (IV) Problems

In this section we shall discuss Case (IV) problems. Meiman[8] originally derived the inequality Eq. (8) by means of the Szegö theorem; recently this result was generalized[9], in the form of Eq. (7). Here, we shall follow the method of reference 13 and will derive an even more general inequality.

It is convenient to use the conformal mapping Eq. (16) in order to define F(z) as

$$F(z) \equiv f(t) \quad . \tag{41}$$

Then, F(z) is an analytic function of z for $|z|<1$, and the condition Eq. (3) can be rewritten as

$$\frac{1}{2\pi} \int_0^{2\pi} d\theta \, \tilde{W}(\theta) \, | F(e^{i\theta})|^2 < I^2 \tag{42}$$

where

$$\tilde{W}(\theta) = (t-\beta) \, (t-t_o)^{\frac{1}{2}} \, (t_o-\beta)^{-\frac{1}{2}} \, w(t) \quad , \tag{43}$$

$$t = t_o + (t_o-\beta) \cot^2 \frac{\theta}{2}$$

Moreover, let us set

$$\phi(z) = \exp \left\{ \frac{1}{4\pi} \int_0^{2\pi} d\theta \, \frac{e^{i\theta}+z}{e^{i\theta}-z} \, \log \tilde{W}(\theta) \right\} \quad . \tag{44}$$

Actually, the function $\phi(z)$ is related to the function g(t), defined by Eq. (9) or Eq. (17a), by

$$4(t_o-\beta) \, g(t) = (1+z)^{-1} \, (1-z)^3 \, [\phi(z)]^2 \quad . \tag{45}$$

At any rate, noting that we have[21]

$$| \phi(e^{i\theta})|^2 = \tilde{W}(\theta) \tag{46}$$

almost everywhere, we can rewrite Eq. (42) as

$$\frac{1}{2\pi} \int_0^{2\pi} d\theta \, | G(e^{i\theta})|^2 \leqslant I^2 \tag{47}$$

if we set

$$G(z) \equiv \phi(z) \, F(z) \quad . \tag{48}$$

Now, for any two functions g(z) and h(z) which are analytic inside the unit circle $|z| < 1$, let us define an inner product (g,h) by

$$(g,h) = \frac{1}{2\pi} \int_0^{2\pi} d\theta \, g^*(e^{i\theta}) \, h(e^{i\theta}) \quad . \tag{49}$$

Then, all analytic functions $g(z)$, which are analytic inside $|z| < 1$ with the finite norm $||g|| = (g,g)^{\frac{1}{2}}$, form a Hilbert space[21] denoted by H^2. Now, the inequality Eq. (47) can be rewritten as

$$(G,G) \leqslant I^2 , \tag{50}$$

i.e. $G(z)$ belongs to H^2 with its norm less than I. At this point the question is how to find bounds on $G(o)$, $G'(o)$, etc. in terms of the norm of G. This can be done as follows.

Let $h_n(z)$ $(n = 0,1,2...)$ be an orthonormal set in H^2. Then, the standard Bessel inequality assures us that

$$\sum_{n=0}^{N} | (h_n, G)|^2 \leqslant (G, G) \leqslant I^2 \tag{51}$$

where N is an arbitrary non-negative integer. Now, let λ be an interior point (i.e. $|\lambda|<1$), and define $\psi(z)$ as

$$\psi(z) = (1-|\lambda|^2)^{\frac{1}{2}} \; z^{N+1} \; (1-\lambda^* z)^{-1} \quad . \tag{52}$$

Then, it is easy to check that $\psi(z)$ belongs to H^2. Also, the N + 2 functions $h_n(z)$ $(0 \leqslant n \leqslant N + 1)$ defined by

$$\begin{aligned} h_n(z) &= z^n \qquad (0 \leqslant n \leqslant N) , \\ h_{N+1}(z) &= \psi(z) \qquad (n = N + 1) \end{aligned} \tag{53}$$

form an orthonormal set in H^2; i.e.

$$(h_n, h_m) = \delta_{nm} , \; (n,m = 0,1,2...,N+1) . \tag{54}$$

Moreover, the elementary Cauchy formula gives us

$$\begin{aligned} (z^n, G) &= \frac{1}{n!} G^{(n)}(o) , \\ (\psi, G) &= (1-|\lambda|^2)^{\frac{1}{2}} \lambda^{-(N+1)} \{G(\lambda) - \sum_{n=0}^{N} \frac{1}{n!} \lambda^n G^{(n)}(o)\} . \end{aligned} \tag{55}$$

Therefore, the Bessel inequality Eq. (51) now yields[13]

$$\sum_{n=0}^{N} |\frac{1}{n!} G^{(n)}(o)|^2 + (1-|\lambda|^2)|\lambda|^{-2(N+1)} |G(\lambda) - \sum_{n=0}^{N} \frac{\lambda^n}{n!} G^{(n)}(o)|^2 \leqslant I^2 . \tag{56}$$

In particular, this gives

$$|G(0)| \leq I \quad , \tag{57}$$

$$|G(0)|^2 + |G'(0)|^2 \leq I^2 \quad .$$

These are the best bounds you can obtain since it is easy[13] to construct G(z) satisfying the equality in Eq. (56). Rewriting these equations in terms of f(t) and choosing $\beta \equiv 0$ from the beginning in Eq. (16), we can reproduce Eqs. (8) and (7) of the Introduction. For the $K_{\ell 3}$ problem, we let N = 1 or 2 and choose λ to correspond to the soft pion point $t = \delta = m_K^2 - m_\pi^2$. However, since the problem has been discussed in some detail in the Introduction, we shall not repeat the result here.

As another application, let us consider the π^0-π^0 scattering amplitude F(s,t,u). Since the crossing relation demands that F(s,t,u) is completely symmetric and that we have

$$s + t + u = 4 \tag{58}$$

(we choose units such that $m_\pi = 1$ for simplicity), we can simply write

$$F(s,t) \equiv F(s,t,u) \quad . \tag{59}$$

Then, as is well-known[25], F(s,t) satisfies a twice-subtracted dispersion relation. Choosing t = 2, we find

$$F(s,2) - F(4,2) = \frac{2}{\pi} (s-4)(s+2)$$

$$\times \int_4^\infty ds' \; \frac{(s'-1) \; \mathrm{Im} \; F(s',2)}{(s'-s)(s'+s-2)(s'-4)(s'+2)} \quad . \tag{60}$$

Here, we shall follow the method given by Martin.[26] Noting the partial wave expansion

$$F(s,t) = 2\left(\frac{s}{s-4}\right)^{\frac{1}{2}} \sum_{\substack{\ell=0 \\ \ell=\text{even}}}^{\infty} (2\ell+1) \; f_\ell(s) \quad P_\ell\left(1+\frac{2t}{s-4}\right) \tag{61}$$

for $s \geq 4$, and $0 \leq t \leq 4$ and the unitarity condition

$$|f_\ell(s)|^2 \leq \mathrm{Im} \; f_\ell(s) \quad , \quad (s \geq 4) \tag{62}$$

we find first that F(s,2) is an increasing function of s in the region $2 \leqslant s \leqslant 4$ and second[26] that by means of the Bunyanski-Cauchy-Schwarz inequality we find

$$|F(s,0)|^2 \leqslant k(s,t) \text{ Im } F(s,t) \tag{63}$$

for $s \geqslant 4$ and $4 \geqslant t \geqslant 0$; here k(s,t) is given by

$$k(s,t) = 2\left(\frac{s}{s-4}\right)^{\frac{1}{2}} \sum_{\substack{\ell=0 \\ \ell=\text{even}}}^{\infty} \frac{2\ell+1}{P_\ell\left(1+\frac{2t}{s-4}\right)} .$$

Using the inequality Eq. (63) and noting F(4,2) > F(2,2) = F(2,o), we find that Eq. (60) (with s = 2) gives us

$$F(4,2)-F(2,0) \geqslant \frac{16}{\pi} \int_4^\infty ds \frac{(s-1)}{(s-2)s(s+2)(s-4)} \frac{|F(s,0)|^2}{k(s,2)} . \tag{65}$$

Moreover, we note that F(s,o) is analytic function of s with a cut at $4 \leqslant s < \infty$ and at $-\infty \leqslant s \leqslant 0$. We can map this twice cut plane inside the unit circle $|z| < 1$; then we can utilize our inequality to obtain (with $I^2 = F(4,2) - F(2,0)$),

$$\begin{aligned} &F(4,2)-F(2,0) \geqslant |F(2,0)|^2 \exp\{ \frac{1}{2\pi} \int_{-\pi}^{\pi} d\alpha \log w(\alpha)\} , \\ &w(\alpha) = \frac{1}{2} \frac{1+2v}{(1+v)(2+v) \sum\limits_{\ell=0,\ell=\text{even}}^{\infty} (2_\ell+1)/ P_\ell\left(\frac{v+1}{v-1}\right)} , \\ &v = [1 + \tan^2 \frac{\alpha}{2}]^{\frac{1}{2}} . \end{aligned} \tag{66}$$

Common and Wit[26a] make the unnecessary assumption that F(4,2) is negative, i.e. F(4,2) < 0. Then, of course, Eq. (66) gives their result;

$$\exp\{ - \frac{1}{2\pi} \int_{-\pi}^{\pi} d\alpha \log w(\alpha)\} \geqslant | F(2,0)| .$$

However, this assumption is not only unnecessary but also detrimental in obtaining a better lower bound for F(4,2) itself; therefore, we will not assume it. At any rate, they numerically evaluate the integral and find

$$\exp\{ - \frac{1}{2\pi} \int_{-\pi}^{\pi} d\alpha \log w(\alpha)\} \simeq 8.90 .$$

Therefore, we find that Eq. (66) gives

$$F(4,2) \geqslant 0.112\ |F(2,0)|^2 + F(2,0)\ .$$

Since this involves a quadratic form in F(2,0), we can find the minimum of the right hand side with respect to F(2,0) and thereby obtain an exact bound

$$F(4,2) \geqslant -\ 2.22\ . \tag{67}$$

This bound is better than those bounds which were obtained by Common and Wit via more elaborate calculations.

Moreover, we remark that, if we define the scattering length a_ℓ by

$$a_\ell = \lim_{s\to 4}\ f_\ell(s)\ \left(\frac{4}{s-4}\right)^{\ell+\frac{1}{2}} \tag{68}$$

then, we find

$$F(4,t) = 2 \sum_{\substack{\ell=0,\\ \ell=\mathrm{even}}}^{\infty} a_\ell\ \frac{(2\ell+1)!}{(\ell!)^2}\ \left(\frac{t}{4}\right)^\ell\ ,\ (|t|<4). \tag{69}$$

On the other hand, we know[27] ($\ell\geqslant 2$, ℓ=even)

$$0 \leqslant a_{\ell+2} \leqslant \frac{1}{16}\ \frac{(\ell+1)(\ell+2)}{(\ell+\frac{3}{2})(\ell+\frac{5}{2})}\ a_\ell\ . \tag{70}$$

Combining Eqs. (69) and (70), we find

$$F(4,2) \leqslant 2a_o + 20\ a_2 \tag{71}$$

which, in view of the inequality (67) gives

$$a_o + 10\ a_2 \geqslant -\ 1.11\ . \tag{72}$$

If $10a_2$ is negligible[26a] in comparison to a_o, then we find an approximate inequality $a_o \geqslant -1.11$, which is better than those given in references (26a) and (28). We can perhaps improve this bound if we use more refined methods of Martin (see ref. 26).

Finally, we remark that we can apply our inequalities to study[29] the scale dimensions of the scale-breaking interactions in the $(3,3^*) \oplus (3^*,3)$ model of scale and chiral symmetric breaking. However, we will not go into the details of those calculations.

Acknowledgement

The author would like to express his gratitude to Dr. D. N. Levin for reading this manuscript.

REFERENCES

1. e.g. M. L. Goldberger and K. M. Watson, "Collision Theory" John-Wiley and Sons, N. Y. (1964).
2. e.g. G. Frye and R. L. Warnock, Phys. Rev. 130, 478(1963); M. Sugawara and A. Tubis, ibid 130, 2127 (1963).
3. B. V. Geshkenbein, Yad. Fiz. 9, 1232 (1969)[English Translation, Soviet Journ. of Nucl. Phys. 9, 720 (1969)].
4. I. Raszillier, Lett. Nuovo Cimento 2, 349 (1971) and preprint, Institute of Physics, Bucharest (1971).
5. D. N. Levin, V. S. Mathur and S. Okubo, Phys. Rev. to be published. In this paper, we rediscovered the same results of reference (3) without being aware of its prior existence.
6. T. N. Truong and R. Vinh-Mau; Phys. Rev. 117, 2494 1969).
7. D. R. Palmer, Nuovo Cimento, to be published.
8. N. N. Meiman, Zh. Eksperim. i. Theor. Fiz. 44, 1228(1963) [English Translation: Soviet Phys. J.E.T.P. 17, 830 (1963)].
9. S. Okubo, Phys. Rev. D4, 725 (1971) and D3, 2807 (1971).
10. L. F. Li and H. Pagels, Phys. Rev. D3, 2191 (1971) and D4, 255 (1971).
11. M. Gell-Mann, R. J. Oakes and B. Renner, P ys. Rev. 175, 2195 (1968); S. L. Glashow and S. Weinberg, Phys. Rev. Lett. 20, 224 (1968).
12. See reference (9). Actually, another possible solution $(A_{44})^{\frac{1}{2}} + (A_{33})^{\frac{1}{2}} \lesssim (V_{44})^{\frac{1}{2}}$ has been rejected in view of the fact that it contradicts the exact SU(3) limit.
13. S. Okubo and I. F. Shih, Phys. Rev. D4, 2020 (1971); I. F. Shih and S. Okubo, Phys. Rev. D4, to appear in Dec. 1 (1971) issue.
14. C. Callan and S. B. Treiman, Phys. Rev. Lett. 16, 153 (1966); M. Suzuki, ibid 16, 212 (1966); V. S. Mathur, S. Okubo and L. K. Pandit, ibid, 16, 311 (1961). In view of a SU(3) consideration, we choose the the soft pion point at $t = m_K^2 - m_\pi^2$. See R. Dashen and M. Weinstein, Phys. Rev. Lett. 22, 1337 (1969).

15. M. Ademollo and R. Gatto, Phys. Rev. Lett. 13, 264 (1965); H. R. Quinn and J. D. Bjorken, Phys. Rev. 171, 1660 (1968).

16. L. M. Chounet, J. M. Gaillard and M. K. Gaillard, CERN preprint (1971).

17. S. D. Drell, A. C. Finn, and A. C. Hearn, Phys. Rev. 136, 1439 (1964).

18. D. R. Palmer, Phys. Rev. D4, 1558 (1971).

19. Here, we use a version of the Phragmén and Lindelöf theorm due to Nevalinna, Eindeutige Analytische Funktionen, (Anfl. Springer Verlag. Berlin 1953, p. 44); E. Hille, Analytic Function Theory, Vol. II, (Ginn and Co. Boston, p. 412, 1962).

20. S. Ciulli, quoted in reference (4). Unfortunately, the present author did not have an opportunity to see this paper.

21. K. Hoffmann, Banach Spaces of Analytic Functions, (Prentice Hall, Englewood Cliffs, N. J. 1962); H. Helson, Lectures on Invariant Sub-spaces, (Academic Press, New York 1964).

22. H. Harari, Phys. Rev. Lett. 17, 1303 (1966). However, the normalization of $t_2(\nu,q^2)$ is different by a factor π from the one used in this paper.

23. S. D. Drell and T. M. Yan, Phys. Rev. Lett. 24, 181 (1970).

24. M. Damashek and F. J. Gilman, Phys. Rev. D1, 1319 (1970).

25. Y. S. Jin and A. Martin, Phys. Rev. 135, B1395 (1964).

26. A. Martin, "High-Energy Physics and Elementary Particles," International Atomic Energy Agency, Vienna, 1965, p. 155; L. Lukaszuk and A. Martin, Nuovo Cimento, 52A, 122 (1967).

26a. A. K. Common and R. Wit, Nuovo Cimento 3A, 179 (1971).

27. A. Martin, Nuovo Cimento 47, 265 (1969); A. K. Common, ibid 63, 863 (1969); F. J. Yndurain, ibid 64, 225 (1969).

28. B. Bonnier and R. Vinh Mau, Phys. Rev. 165, 1923 (1967).

29. D. N. Levin, S. Okubo and D. R. Palmer, Phys. Rev. D4, 1847 (1971); D. N. Levin and D. R. Palmer, Phys. Rev. D, to appear in the Dec. 15, 1971 issue.

PROSPECTS FOR THE DETECTION OF HIGHER ORDER WEAK PROCESSES AND THE STUDY OF WEAK INTERACTIONS AT HIGH ENERGY

David Cline
University of Wisconsin
Madison, Wisconsin

1. INTRODUCTION

The first observation of weak interactions is now over 75 years old.[1] An impressive array of understanding of a vast number of phenomena has been achieved for low energy processes, and yet some of the simplest questions that can be asked about the basic nature of the weak interaction can not presently be answered. In many ways we know less about this interaction than we do about the strong interaction. Apparently Heisenberg was the first to recognize the significance of the dimensionality of the coupling constant of the lowest order current-current interaction,[2] the lowest order interaction being

$$H_{eff} = \frac{G}{\sqrt{2}} j_\lambda j_\lambda^+ \quad , \tag{1}$$

where j_λ, j_λ^+ are appropriate currents and G is the coupling constant. G has the dimensions of $(\text{length})^2$ or $(1/m)^2$ with a numerical value

$$G = (1.01 \times 10^{-5})/(m_p)^2 \quad .$$

In order to form a dimensionless parameter for the weak interaction it is frequently suggested to use s and to form the parameter[3]

$$\lambda = Gs, \tag{2}$$

s being the only parameter of the scattering process that sets a length (or m^{-2}) scale (s is the center of mass energy squared).

There is at present no experimental information that sets the length scale of the weak interactions. However there are two theoretical suggestions as to what the length scale might be.

1. The 'length' at the unitarity limit. If the weak

*Supported in part by the United States Atomic Energy Commission under Contract AT(11-1)-881, COO-881-325.

interaction was pointlike all two body cross sections would rise like

$$\frac{G^2 s}{\pi}$$

and being pointlike only the S wave interaction is allowed. However, the unitarity limit for the cross section for S wave scattering goes as π/s; thus at a large value of s the weak interaction cross section must be modified to avoid a unitarity violation (at the energy $\sqrt{s_u} = \frac{1}{G}$). The length associated with this value of s (which was called the 'fundamental length' by Heisenberg) is

$$\ell_f = \frac{1}{\Lambda_u} = \frac{1}{\sqrt{s_u}} = \sqrt{G} \sim 10^{-17} \text{ cm} \quad . \tag{3}$$

Note that, by definition, the dimensionless coupling constant

$$\lambda_u = G s_u \simeq 1, \tag{4}$$

thus indicating that the weak interactions actually become 'strong' at these very high energies. It appears that the intrinsic strength and the range of force of the weak interactions are therefore intimately tied together. The interaction is strong in the sense that the S wave cross section is as large as any S wave cross section can be. (In strong interactions the low partial waves are strongly absorbed and thus the S wave cross section probably does not stay at the unitarity limit; thus at the unitarity limit the weak interaction cross section would likely exceed the strong interaction S wave cross section; however, the actual cross section would only be $\sigma \sim \frac{G}{\pi} \sim 10^{-33}\text{cm}^2$ compared to $\sim 3 \times 10^{-26}\text{cm}^2$ for hadron scattering cross sections, because of the large number of angular momentum states excited in the hadron scattering.)

2. A second way to set the 'length' scale for weak interactions is to imagine that the exchange of a massive boson is responsible for the weak force between two particles.[4] The mass (M_W) of this hypothetical boson then sets the scale

$$\ell_W \sim \frac{1}{M_W} \tag{5}$$

and the coupling constant for the W coupling to say two leptons is semiweak and given by

$$g^2 \simeq GM_W^2 \simeq \lambda_W \ .$$

Thus the larger the mass M_W, the stronger the semiweak interaction becomes. This illustrates again that the fundamental nature of the weak interaction is presently indeterminate, there being a tradeoff between the strength and the range of the interaction. Experimentally it is, therefore, necessary to determine either the fundamental dimensionless coupling constant or directly measure the range of the interaction. Clearly measurement of a distance of 10^{-17} cm is a very ambitious undertaking since momentum transfers of $\sim(300)^2$ GeV/c^2 would be required. Nevertheless as discussed later we might contemplate observation of momentum transfers of $(30)^2$ within the decade, in forthcoming neutrino experiments allowing a probe of distance down to $\sim 10^{-16}$cm.

There have been other suggestions as to a fundamental length of weak interactions in terms of the exchange of scalar bosons and a variety of other postulated particles.[5] These particles were invented to provide a renormalizable theory of weak interactions.[5]

Recently a dispersion theoretic approach has been applied to the question of the high energy behavior of weak interactions starting with the posthumous paper by Pomeranchuk.[6,7] Other calculations have followed this lead.[8] There are no firm conclusions to be drawn from such analyses but some very interesting speculation about the processes that may dominate the weak interactions at high energy are made. Also, as shown by Pomeranchuk,[6] if the weak interaction becomes long ranged at high energy with a cross section that approaches that of strong interactions, such a behavior cannot set in before an energy of the unitarity energy $\sqrt{s_u}$. Dolgov, Okun and Zakharov have attempted a dispersion theoretic estimate of the lower limit of the contribution from higher order weak diagrams for lepton-lepton collisions.[8]

Other theoretic attempts at handling the higher order weak interactions have focused on a summation of the contributions from <u>all</u> higher order diagrams.[9,10] The first such attempt known to us was made by Feinberg and Pais and more recently by Arbuzov.[9]

An interesting proposal for modifying the weak interaction was made by Gell-Mann, Goldberger, Kroll and Low.[11] Their proposal would lead to a modification of the universality of first order weak interactions such that the diagonal and nondiagonal lepton-lepton processes would proceed with different rates.

Many other suggestions have been made for calculating the higher order diagrams or for formulating a renormalizable theory of weak interactions. (See Refs. 12, 13, 14, 15, 16 for an incomplete list).

A promising way to separate (or estimate) the range and 'intrinsic' strength of the weak interaction is through the observation of a certain class of higher order weak interaction processes. While the validity of such calculations is certainly not proved, as order of magnitude estimates these calculations make some sense, especially when applied to pure leptonic systems.[15,16,17,18,19] If higher order weak processes are suppressed in all systems relative to first order processes then the observation of higher order weak processes will likely be carried out with low energy weak interaction processes such as a rare decay mode of K-mesons because of the possible large abundance of such decay particles.

At the same time study of high energy weak interactions bring us closer to the unitarity limit where we expect surprises. These studies will likely be carried out with high energy neutrino beams or colliding lepton beams. In fig. 1 we attempt to summarize the present and projected range of energies available for weak interaction studies as well as the present range of transition rates that have been studied for K decays; in particular, in this figure we attempt to show the regions in these variables where new surprises in the weak interaction might be expected. The moral to be gained from this graph is that already experiments have covered a large range of energy and transition rates and we are close to the regions where surprises might be expected.

A short summary of the experimental measurements needed to 'unravel' the range and 'intrinsic' strength of the weak interaction is in order. The 'intrinsic' range and 'intrinsic' strength are assumed to be tied together in such a way that

$$G \sim g^2 \cdot 1/(m_\ell)^2 \quad , \tag{6}$$

where g is the intrinsic coupling strength and m_ℓ is a mass

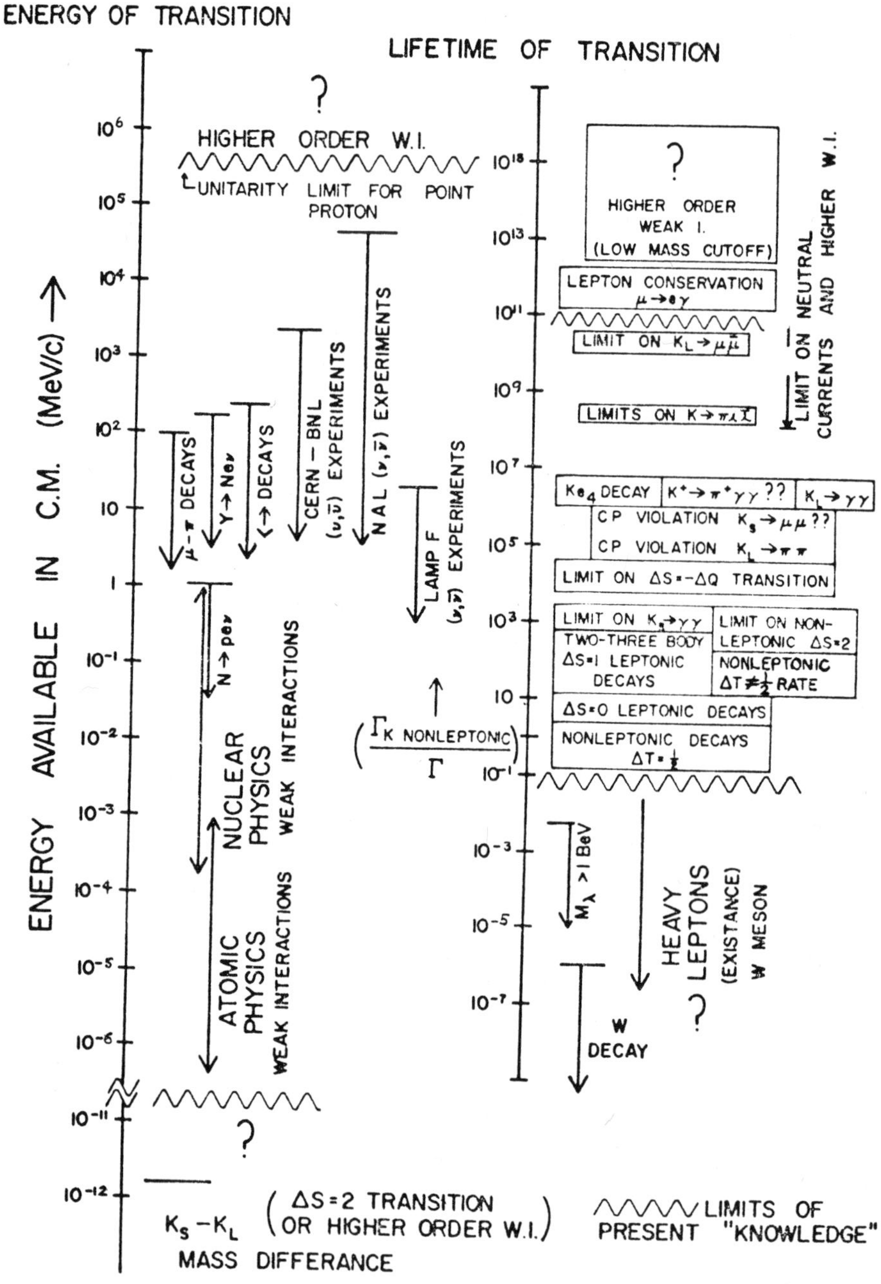

Figure 1. Weak interactions at high and low energies.

that characterizes the range of forces.

There are basically three ways to detect or measure the value of m_ℓ

1. Study <u>high momentum</u> transfer processes observing the m_ℓ in the form factor

$$\frac{d\sigma}{dq^2} \propto \frac{1}{\left[1 + \left(\frac{m_\ell}{q}\right)^2\right]^2} \qquad (7)$$

2. Study <u>very high energy</u> scattering; in the vicinity $\sqrt{s} \sim m_\ell$ where higher partial waves will enter the weak interactions and a 'break down of locality' will occur.

3. Observe processes that can only proceed by 2nd or higher order weak interactions and assume (on the basis of the perturbation theory algorithm) that the rate for such processes related to that for first order processes is, order of magnitude,

$$\frac{\Gamma(\text{2nd order})}{\Gamma(\text{1st order})} \sim G^2 m_\ell^4 \ . \qquad (8)$$

In a more careful perturbation calculation the ratio of second to first order rates becomes[19]

$$\xi^2 = \frac{G^2 \Lambda^4}{32\pi^4} \quad , \qquad (9)$$

where Λ is a cut off mass that is used to remove the divergence of the integrals associated with second order contributions. For nonleptonic or semileptonic processes these calculations assume that the range or size of the strong interactions does not provide a cutoff to the integral.[15,20] Such an assumption can be justified on the grounds of current algebra or the quark model or any model where the weak current couples to pointlike objects inside the hadron (like the parton model).[21,22] However, this assumption does seem to violate simple minded intuition that the hadrons can not generally support high momentum transfers. Recent observations of inclusive processes where hadrons appear to be capable of supporting high momentum transfers,[24] can be explained by parton or quark pointlike structures.[22,23] However, it is not clear that pointlike structure is <u>necessary</u> to explain this phenomena (nor in fact that it is really sufficient) and more mundane explanations of the deep inelastic scattering have been proposed.[25] Therefore, it is not presently clear that the higher order processes are not cut off by the strong interaction in semileptonic or nonleptonic processes. For this reason it is very important that leptonic processes be studied.

Experimentally techniques 1 and 2 require high energy particles and the possibilities for such studies are only now becoming available with the advent of high energy machines such as NAL and the CERN 300 GeV machine. In practice such studies will likely be carried out using high energy neutrino beams.

The direct observation of higher order weak processes will likely depend on the intervention of a selection rule in first order weak interactions that are violated by the higher order processes. However, in some cases it may be necessary to separate higher order weak processes from first order contributions by observing the nonlocality generated by the higher order process.[26,27] Generally, therefore, the detection of higher processes will only be as sensitive as the validity of the selection rule. So far the best obeyed selection rules appear to be the absence of neutral currents in semileptonic processes and the $|\Delta S| < 2$ rule for nonleptonic processes.[28] In the next section we review the present status of the selection rules obeyed by the weak interaction.

It is interesting to note the different dependence on m_ℓ in techniques 1 - 3. For 1 and 2 the larger m_ℓ the more difficult it becomes to 'measure' m_ℓ (or to detect a deviation from $m_\ell \to \infty$). However, for the higher order corrections, especially for lepton-lepton collisions, the larger m_ℓ the easier it is to 'measure' m_ℓ. Of course perturbation intuition may fail here but if it does not then these techniques are complementary and should all be pursued. For example, it is difficult to foresee in the near future experiments that attain momentum transfers of $(300)^2$ GeV/c^2 and therefore $m_\ell \sim 300$ GeV would be hard to observe by techniques 1 or 2. However, for $m_\ell \sim 300$ GeV the higher order corrections become maximal and might be detected eventually in e^+e^- collisions as discussed below.

In table 1 we have attempted to summarize the present

TABLE 1

Present Information on the Weak Interaction Cutoff

Λ	PROCESS	COMMENTS	AUTHORS
~ 2600 GeV	$\nu+\bar{\nu} \rightarrow W+\overline{W}$	Intermediate Boson theory	Gell Mann et al.[11]
~ 320 GeV	$\nu+e \rightarrow \nu+\mu$	Simple unitary limit	
< 100 GeV	$K^+ \rightarrow \pi^+ \ell\ell$	Cut off of divergent integral	Ioffe and Shabalin[15]
~ 30 GeV	$\nu+\ell \rightarrow \nu+\ell$	Crossing symmetry included in calculation	Applequist and Bjorken [7]
≤ 14 GeV	$K_L \rightarrow \mu\mu$	Cut off of divergent intergrals using Bjorken technique	Ioffe and others[17] (LRL Experiment)[29]
$\Lambda \sim 8$ GeV	$K_S \rightarrow \pi\pi$	Soft π and K techniques	Glashow, Schnitzer and Weinberg[31]
$\Lambda \sim (4\text{-}8)$ GeV	$K_L - K_S$ Mass difference	Bjorken technique and cut off of divergent intergrals	Ioffe et al.[15,17] Mohapatra et al.[16]
$\Lambda \simeq$ Small	Rare Electro-magnetic decays of K mesons	Electromagnetic processes with virtual photon diverge quadratically	Geshkenbein and Ioffe[30]
$\Lambda \simeq$ Small	Nonleptonic decays	$f \sim G\Lambda^2 \sim 10^{-5}\text{-}10^{-6}$	
$\Lambda > 2$ GeV	W production	Assume $\Lambda \sim M_W$	CERN bubble chamber and counter experiments

guesses for the limit on Λ from various viewpoints; the low values of Λ all come from semileptonic processes or nonleptonic processes. This table might be viewed in the following way; there are hints that the weak interaction cutoff is low and therefore something interesting is expected to occur in weak interaction processes for $\sqrt{s} \lesssim 10$ GeV. Also if the weak force is transmitted by an intermediate vector boson the mass is expected to be relatively low compared to the unitarity limit. However, these speculations are based on calculations that in all cases involve hadrons in the weak process. It may still be that the low values of Λ in table 1 are (i) determined by the strong interaction range or (ii) that perturbation theory is not relevant. To answer the first question will require the study of leptonic processes at large s. Probably the answer to question (ii) will require study of weak interaction processes very near $s \sim 1/G$.

The plan of this paper is essentially spelled out in the index. We first review the status of various weak interaction selection rules and discuss briefly the prospects for detecting intermediate vector bosons in the near future. The rest of the paper is broken up into sections that are classified by the kinds of particles that participate in the weak process. Each section deals with the processes suitable for detecting higher order weak processes or the high energy behavior of the weak interaction for that particular system.

1a. Status of Various Selection Rules

The selection rules in weak interactions are not presently required by any basic theory; the rules being almost completely empirical. For this reason it is not known how exact such rules should be, and in fact some selection rules are known to be broken at the 5% level in the amplitude. However, some selection rules are suspected to be exact in first order weak interactions, but perhaps broken in higher orders. If this is true then the observation of a violation of the rule would be a signature for higher order processes; but, it need not be since the rule might simply be broken by the first order weak interaction. Since the observation of the violation of CP invariance, we know that sometimes very small violations in weak amplitudes (or super weak) can occur, and perhaps small violations of other selection rules might equally be observed. However, in the case of the absence of neutral semileptonic currents ($\Delta Q = 0$, $\Delta S \neq 0$ processes), the upper limit on the violation has now been shown to be three orders of magnitude lower than the CP violation rate;[29] perhaps indicating that the absence of neutral currents is a better selection rule than CP invariance.

In table 2 the current upper limits on the amount of violation for weak amplitudes for the selection rules is presented for:

$\Delta Q \neq 0$	leptonic processes
$\Delta Q \neq 0$	semileptonic processes
$\Delta S = \Delta Q$	semileptonic processes
$\Delta S < 2$	semileptonic processes
$\Delta S < 2$	nonleptonic processes

A notable point in this table is the absence of any useful limit on the $\Delta Q \neq 0$ selection rules for purely leptonic systems. Remarkably, the only well tested selection rule is the $\Delta Q \neq 0$, semileptonic rule, and only for the $\Delta S \neq 0$ subclass.

The $\Delta T = 1/2$ selection rule is now known to be broken by about 5% in the amplitude for several processes suggesting that the rule is only approximate in <u>all</u> cases. We, therefore, neglect this rule in table 2. Similarly, second class current in semileptonic amplitudes may come in at the same level.

One moral that might be drawn from table 3 is that when searching for higher order weak processes, violations of the ($\Delta Q \neq 0$, semileptonic) rule would be more likely to pay off because the other selection rules have yet to be tested to a sensitive level. For example, if the higher order processes come in at the relative amplitude level of 10^{-6}, this is 4-5 orders of magnitude in the amplitude lower than these selection rules have been tested, but only one or two orders below the ($\Delta Q \neq 0$, $\Delta S \neq 0$ semileptonic) rule. Even if the second order process comes in (1-2) orders of magnitudes below a primitive neutral current, it might still be possible to separate the higher order process as discussed below.

1b. <u>Dectection of Intermediate Vector Bosons</u>

The discovery of one or more bosons that couple semiweakly to leptons and hadrons and thus are candidates for the

TABLE 2

Selection Rule	Approximate limit on the Ratio of $\left(\frac{\text{Violating Amplitude}}{\text{Nonviolating Amplitude}}\right)$	Processes that Violate the Rule
$\Delta Q \neq 0$, leptonic	~ 1	$\nu_\mu + e^- \to \nu_\mu + e^-$
		$e^+e^- \to \nu_\mu\bar{\nu}_\mu$
		$e^+e^- \to \mu^+\mu^-$
$\Delta Q \neq 0$, semileptonic	$\sim 4 \times 10^{-5}$	$K^\circ_L \to \mu^+\mu^-$
$\Delta S \neq 0$	$\sim 10^{-3}$	$K^+ \to \pi^+e^+e^-$
	$\sim 3 \times 10^{-3}$	$K^+ \to \pi^+\nu\bar{\nu}$
$\Delta S = 0$	$\sim 5 \times 10^{-1}$	$\nu_\mu + n \to \pi^- p\ \nu_\mu$
$\Delta S = \Delta Q$, semileptonic	$\sim 10^{-1} - 4 \times 10^{-2}$	$K^\circ \to \pi^+e^-\nu$
	$\sim 10^{-1}$	$K^+ \to \pi^+\pi^+e^-\nu$
	$\sim 10^{-1}$	$\Sigma^+ \to ne^+\nu$
$\Delta S < 2$, semileptonic	~ 1	$\Xi^- \to ne^-\nu$
	~ 1	$\Omega^- \to ne^-\nu$
$\Delta S <2$, nonleptonic	3×10^{-2}	$\Xi^- \to \pi^- n$
		$\Xi^\circ \to \pi^- p$
		$\Omega^- \to \pi^- n$
		$\Omega^- \to \pi^- \Lambda$

'mediators' of weak interactions would go a long ways towards answering the basic questions about weak interactions posed in the introduction. Thus the search for such hypothetical but crucial states is of great importance and experimenters are well aware of this as can be proved by looking at the current proposals for experiments at the NAL.[32]

With the advent of high intensity neutrino beams at NAL or CERN it should be possible to produce, in a massive detector adequate numbers of W vector bosons to discover such a particle if the mass is below $\sim$ 12-15 GeV.[33] It also appears that the boson can be detected independent of the relative branching fraction into leptonic and hadronic final states and, therefore, a conclusive search can be made in this mass range.[34]

Higher mass bosons might be detected in hadronic or photonic interactions at NAL or CERN up to the mass of 30-40 GeV, provided the cross sections for the production are comparable to the estimates of Lederman and Pope and provided the boson decays via the leptonic decay mode.[35] We emphasize that in the range of 15-40 GeV it will likely be impossible to conclusively exclude the existence of the intermediate vector boson because of the uncertainty of production cross sections and decay rates. Thus, up to $\sim$ 15 GeV an exhaustive search can be made and if conditions are favorable a W of mass 15-40 GeV could be detected.

The observation of a scalar charged meson is virtually impossible due to the expected small production cross section and the suppression of the leptonic decay mode.[36] If neutral vector bosons exist (perhaps producing so far undetected neutral leptonic current processes) and have any mass above the kaon mass, they likely would not have been detected up to the present. A neutral W° could be produced in e^+e^- collisions, but sensitive experimental searches have yet to be carried out in these processes.[37] It has been proposed to search for the existence of W° bosons using the process $e^+e^- \to \mu^+\mu^-$.[38] This search should be sensitive to the existence of any W° boson with mass below 8 GeV using colliding beam facilities such as SPEAR.[38]

2. Lepton-Lepton Collisions

Without the obscuring effects of the strong interactions,

lepton-lepton scattering provides a 'clean' study of weak interactions. Experimentally, the detection of weak lepton-lepton processes is just coming into the range of experimental feasibility. There are basically three kinds of processes that may yield practical and interesting results:

$$\nu_\ell + z \to \ell + \bar{\ell} + \nu_\ell + z \tag{10}$$

$$\nu_\ell + \ell \to \nu_\ell' + \ell' \tag{11}$$

$$e^+e^- \xrightarrow{\text{weak}} \mu^+\mu^- \quad . \tag{12}$$

Study of the first two processes is becoming feasible because of the advent of high energy-high intensity neutrino beams at NAL and CERN. The s available to such processes, however, is likely to be limited to the range

$$s \sim 2m_e E_\nu \lesssim 5 \times 10^{-1} \text{ GeV}^2 \ .$$

For processes like 10 the requirements of coherence limits the mass of the three leptons to equally small values. Process 12 is the only one where values of s can be obtained where surprises and perhaps departures from the standard lowest order weak interaction theory may occur. In this case, s values in the vicinity of

$$s \sim 10 - 64 \text{ GeV}^2$$

might be attained with storage ring machines that are presently being constructed.

Unfortunately, since weak interactions are in general overwhelmed by electromagnetic interactions in process 12, a special dispensation is required to observe weak interactions. It has been recently speculated that such a dispensation may occur under special circumstances at colliding beam facilities such as SPEAR.[38]

2a. Deviations from the Universal V-A Theory in Lowest Order--the Diagonal Coupling

Gell-Mann, Goldberger, Kroll and Low[11] have suggested a theory of weak interactions in which the leading divergences occur only in the diagonal interactions (i.e. $(\nu_e e)(\nu_e e)$ terms), which are thus speculated to be quite unconnected with the off diagonal interactions (i.e. $(\nu_e e)(\nu_\mu \mu)$ terms). Thus, higher order weak corrections may be manifested in a resulting difference between the diagonal and off diagonal coupling constants, which in turn would be observable in $s \to 0$ processes. In order to test this idea it will be necessary to compare processes like

$$\nu_\mu + \mu^- \to \nu_\mu + \mu^- \qquad (13)$$

$$\nu_e + e^- \to \nu_e + e^- \qquad (14)$$

with processes like

$$\nu_\mu + e^- \to \mu^- + \nu \quad . \qquad (15)$$

Fortunately, these processes will likely be measured in the near future and the issue can be resolved.

Observation of process (14) may be accomplished in neutrino experiments currently underway at CERN using the Gargamelle bubble chamber or in early experiments at NAL using the 15' bubble chamber filled with neon.[39]

Reaction (13) is the most problematic since free muon targets do not exist in nature. A convenient substitute for this process is the process[40]

$$\nu_\mu + z \to \mu^+\mu^-\nu_\mu z \quad . \qquad (16)$$

This process can likely be detected also at NAL and the Harvard-Penn-Wisconsin Collaboration experiment (E1A) has been designed with this process in mind. I will not go into detail concerning the projected experimental difficulties in studying this process since Professor Mann has described this in his talk. If this process can be separated from background at NAL, it should be possible to make a 10% measurement of the cross section. Incidentally, the calculations of the rate for process (16) are presently only good to $\sim$ 10%.[40]

We must emphasize, however, that the bulk of the events detected at NAL, even though the neutrinos are high energy, will likely have a low $\mu^+\nu_\mu$ invariant mass and thus the study of process (13) via (16) is at small s.[33,40] Neverless, it should soon be possible to experimentally compare the diagonal and off-diagonal coupling constants at <u>low s</u> and thus decide on the GGKL conjecture.

2b. <u>Pseudo Neutral Leptonic Currents</u>

(i) Spacelike

At present there is no evidence to support the absence of first order neutral currents coupled only to leptons (see table 2). Recently it has been conjectured by Weinberg and others that such currents could exist in a renormalizable theory of weak and electromagnetic interactions.[13] The most convenient processes to use to search for neutral leptonic currents in first order are

$$\nu_\mu + e^- \rightarrow \nu_\mu + e^- \tag{17}$$

$$\nu_\mu + z \rightarrow e^+e^-\nu_\mu z \quad . \tag{18}$$

Again process (17) is on the verge of detectability in present or near future experiments. For example, process (17) can perhaps be detected in the present CERN studies with Gargamelle if the cross section is no less than $\sim$ 5 times smaller than the present limit on this process.[41] The present limit on the cross section for (17) relative to the cross section expected for process (14) (on the basis of the universal V-A theory) is[42]

$$\frac{\sigma(\nu_\mu + e^- \rightarrow \nu_\mu + e^-)}{\sigma(\nu_e + e^- \rightarrow \nu_e + e^-)} < 0.4 \quad . \tag{19}$$

The lower limit of this ratio predicted by the theory of Weinberg is[13]

$$\frac{\sigma(\nu_\mu + e^- \rightarrow \nu_\mu + e^-)}{\sigma(\nu_e + e^- \rightarrow \nu_e + e^-)} \geq 0.125 \quad . \tag{20}$$

The search for process (17) in the neon bubble chamber at NAL is likely to be even more definitive. The study of process (18) is problematic because of the large background of Dalitz pairs in neutrino collisions.

If process (17) is not detected at the level of first order weak in bubble chambers it becomes interesting to see at what level the higher order corrections may come in and if the resulting cross section can be measured by massive target-counter techniques. An estimate of the cross section for process (17) proceeding through second order weak processes and assuming that the weak interaction cutoff is at the unitarity limit ($\Lambda \sim \sqrt{s_u}$) gives[19]

$$\sigma(\nu_\mu + e^- \rightarrow \nu_\mu + e^-) \simeq 1.5 \times 10^{-44} (E_\nu) \text{cm}^2/\text{GeV} \quad ,$$

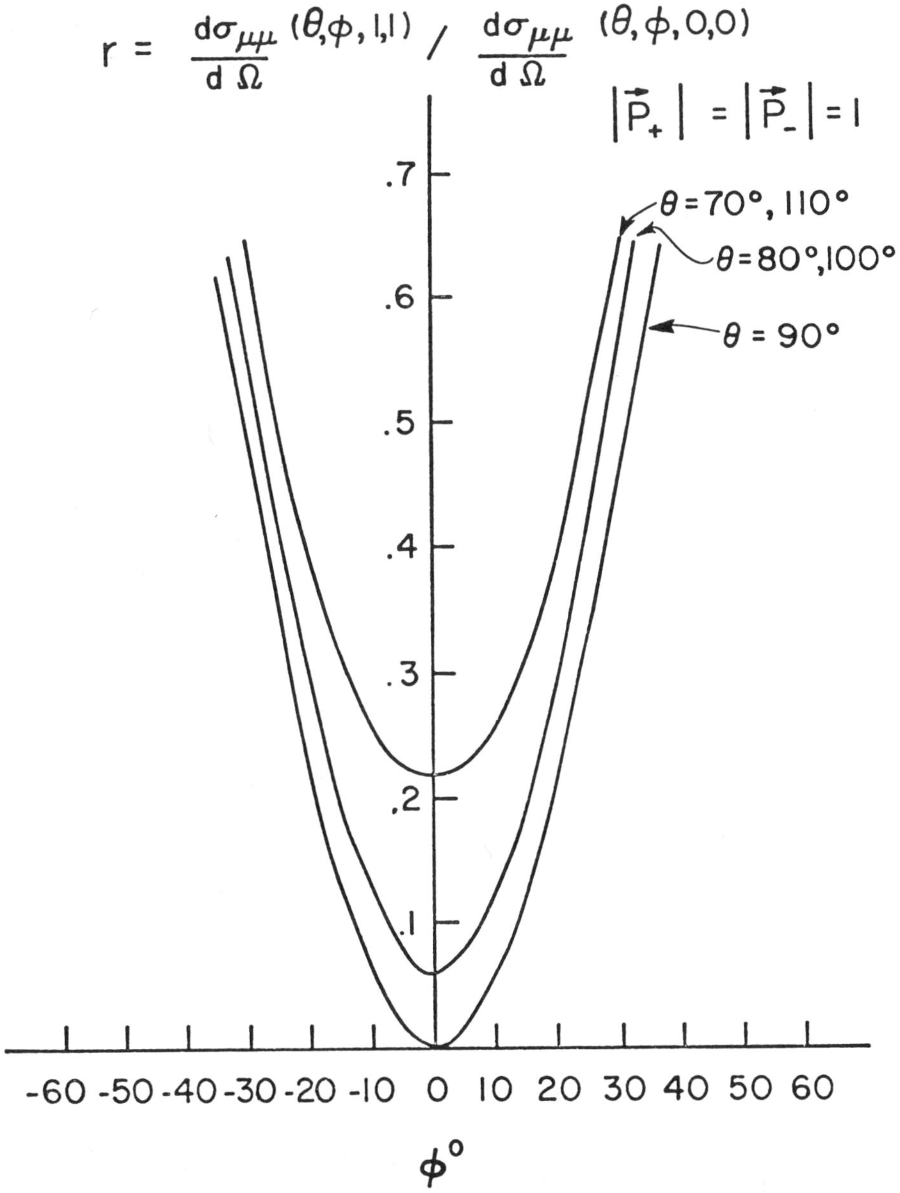

r = dσ_μμ/dΩ (θ,φ,1,1) / dσ_μμ/dΩ (θ,φ,0,0)
|P₊| = |P₋| = 1
θ=70°, 110°
θ=80°,100°
θ = 90°
.7
.6
.5
.4
.3
.2
.1
-60 -50 -40 -30 -20 -10 0 10 20 30 40 50 60
φ°

Figure 2

where E_ν is the ν_μ energy in GeV. Using full design intensity of the NAL machine and a 500 ton Pb detector approximately 2 events of type (17) would be produced per day. Thus, in principle, a purely leptonic higher order weak process could be detected at NAL, provided the unitarity limit provides the weak interaction cutoff. We do not mean, to imply, however, that it is presently known how to separate these two events/day from the large background, but only that the process seems in principle detectable under favorable circumstances. Note, however, that even at this level the ratio of cross sections is

$$\frac{\sigma(\nu_\mu + e^- \to \nu_\mu + e^-)}{\sigma(\nu_e + e^- \to \nu_e + e^-)} \sim 10^{-3}$$

and thus the resulting limit on first order weak neutral currents would only be at best $\sim 3 \times 10^{-2}$ in the amplitude. Thus, it appears difficult to put limits on the absence of first order neutral leptonic currents to the level that $\Delta S \neq 0$ semileptonic neutral currents have reached.

(ii) Timelike

Process (12) can proceed via weak interactions in several speculative ways: (1) direct channel production of a W° on the mass shell; (2) a first order weak neutral current coupling of the form (ee) ($\mu\mu$); (3) an induced neutral current coming from higher order weak interactions.

Experimentally, the detection of any of these weak processes requires a suppression of the dominant electromagnetic amplitudes and a unique signature for the weak process. It appears that a sizable suppression of the first order electrodynamic contribution can be obtained if the initial leptons in process (12) are highly polarized in opposite transverse directions. A 'hole' appears in the angular distribution of the outgoing muons at favored values of θ and ϕ ($\cos\theta = \hat{p}_\mu \cdot \hat{p}_{e^-}$, $\cos\phi\sin\theta = \hat{p}_\mu \cdot \hat{a}$, where $\hat{a}$ is a unit vector along the e^- polarization vector).[38,43] This 'hole' is illustrated in fig. 2 as the ratio of the differential cross section for reaction (12) for completely polarized initial leptons to the cross section for unpolarized initial leptons, and in fig. 3 in a projection drawing of the differential cross section for the two cases. At the bottom of the 'hole' should be a sensitive place to

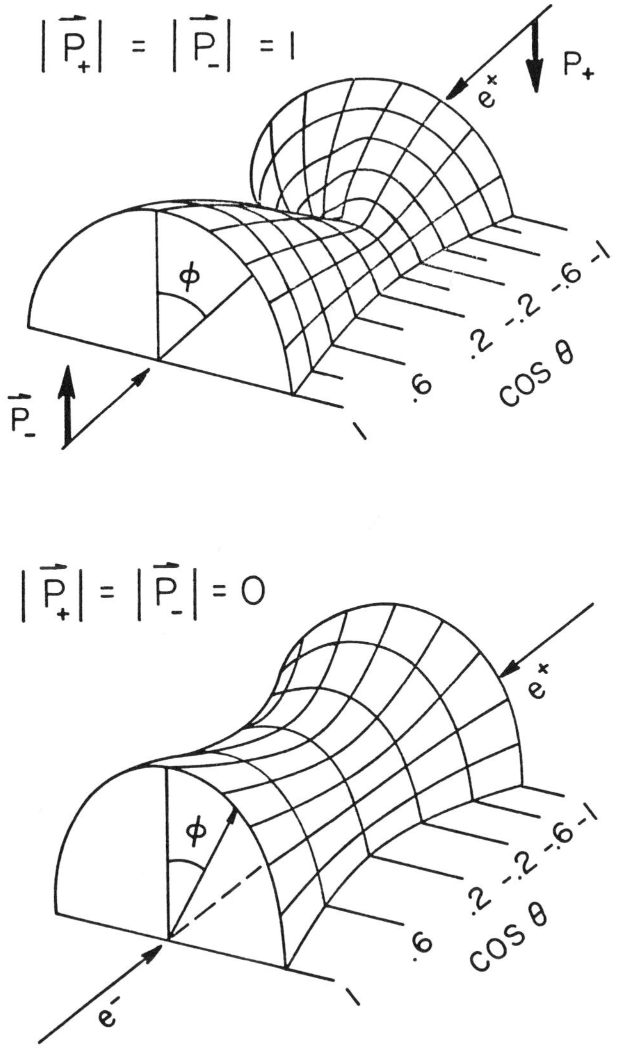

Figure 3

search for any anomalies in process (12) including a weak interaction process.[38] In particular the μ longitudinal polarization will likely be sensitive to interference between first order EM and perhaps weak amplitudes. The polarization will be enhanced in the 'hole'.[38,44] It is too early to conclusively conclude that amplitudes can be uniquely extracted in this way, but there seems to be an intriguing possibility here that should be pursued. It seems very likely that the existence of a W° boson with mass below ∿ 8 GeV could be directly observed in this way.[38] Careful theoretical calculations of this polarization and the background from higher order EM processes would be very useful in planning experiments.

3. Semi-leptonic Processes

a. Second Order Weak K Decays

The studies of K meson decays over the past two decades have provided a rich field for the study of nature and the weak interaction. Nearly every symmetry principle of particle physics has been successfully tested or found to be violated using K meson decays. The primary reason for this richness of the K meson system is due to the large mass of the Kaon relative to the leptons and π mesons. It is fortunate indeed that K mesons exist. Higher order corrections could, in principle, show up in any K decay including the nonleptonic decays. If the intrinsic coupling constant were large then the higher order corrections might be of comparable magnitude to the first order processes. For this reason exhaustive searches for rare decay that is observed with an anomalous rate relative to the best theoretical guesses for the rate based on first order theory, is a candidate for evidence concerning higher order weak processes. In fig. 4 is shown the branching fraction levels to which exhaustive searches for rare decays have been made. In this figure are examples of processes with the lowest branching ratios that have been presently studied. As a rough rule of thumb exhaustive searches for rare K^+ decay modes have been extended down to a branching ratio of $\sim 10^{-5}$ to 10^{-6}.[45] For K_L° decays the corresponding branching ratio is $\sim(10^{-3}$ to $10^{-4})$ and for K_L° mesons the branching ratio is only $\sim(10^{-2}$ to $10^{-3})$. For K^- mesons the branching ratio is $\sim 10^{-2}$; however, CP invariance requires the K^+ and K^- decay ratios to be the

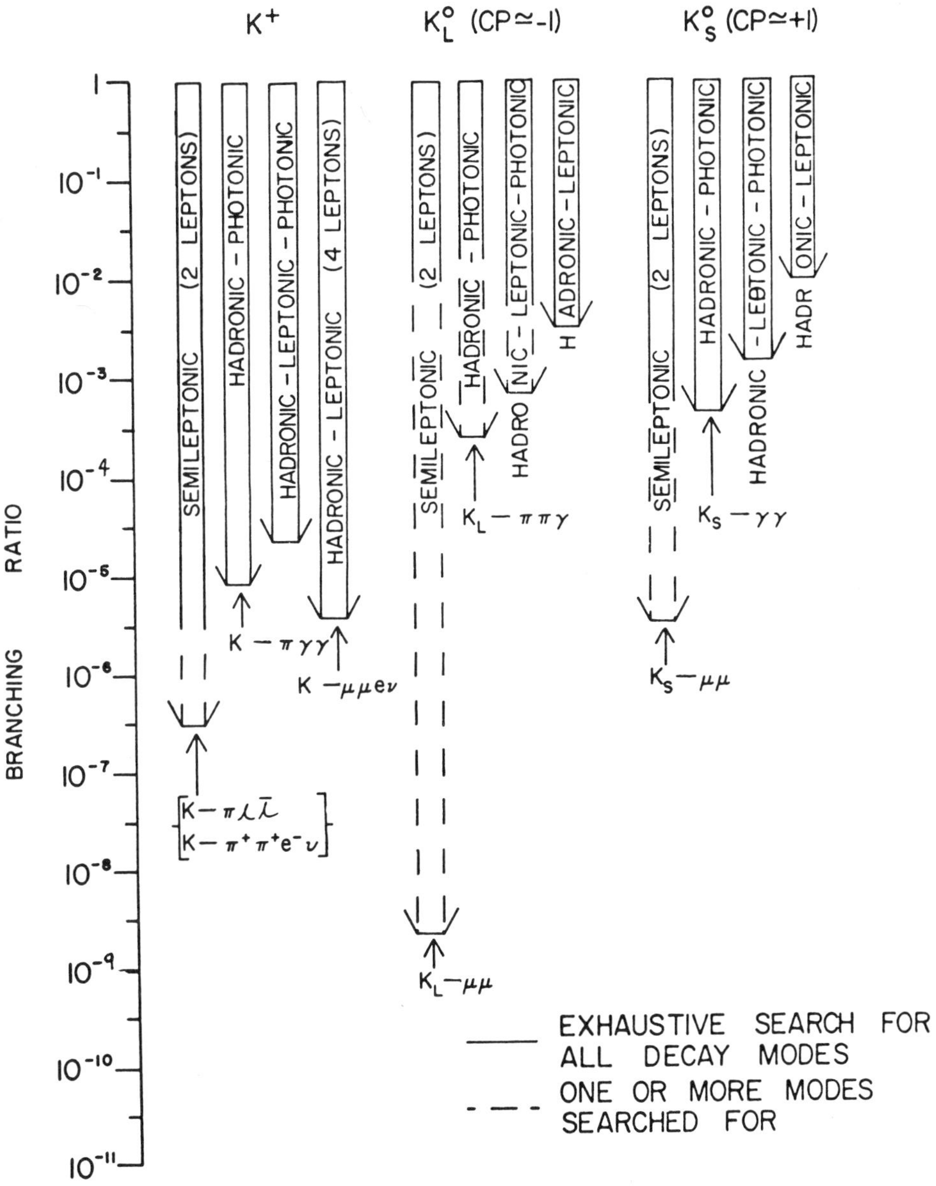

Figure 4. K meson rare decay mode searches.

same and the results from K^+ decays can then be inferred for K^- decay. In some cases it is possible to relate K_L° and K^+ decays of K_S° and K^+ decays and therefore the results for K^+ decays can be applied to the K_L°, K_S° decays.

Recently searches for special individual rare decay modes have been extended down to the branching ratio of $\sim (10^{-8}$ to $10^{-9})$.[29] Although only a few experiments of this kind have been attempted we may hope that the branching ratios region of 10^{-6} to 10^{-10} will be searched considerably more in the future. The advent of high intensity $K^\pm$ and K° beams at the AGS and the Bevatron will be the key factor in these studies.

The study of rare decay modes of K mesons therefore naturally divides into two parts. Studies of the branching ratio region of 10^{-2} to 10^{-6} where nearly exhaustive searches for all rare decay modes have been made and the branching ratio of 10^{-6} to 10^{-10} where studies are just beginning.

It appears that no important surprises are found in the K decay processes observed down to the level of $\sim 10^{-6}$. It seems likely that the higher order processes are not important in this region.

At lower levels the search for HOW processes has been associated with the $\Delta Q \neq 0$ selection rule and this seems to be the logical place to push for definitive evidence of HOW processes. The most important decay processes in this respect are

$$K_L^\circ \to \mu^+\mu^- \tag{21}$$

$$K_S^\circ \to \mu^+\mu^- \tag{22}$$

$$K^+ \to \pi^+e^+e^- \tag{23}$$

$$\to \pi^+\mu^+\mu^- \tag{24}$$

$$\to \pi^+\nu\bar{\nu} \tag{25}$$

$$K_L^\circ \to \pi^\circ e^+e^- \tag{26}$$

In the first four cases the decay can also proceed through a first order weak and first or second order electromagnetic transition. Unless interference is invoked between the HOW and the electromagnetic processes, these processes can only be used to search for HOW amplitudes. In both processes 21 and 23 the present experiments have approximately reached the level where the E.M. processes should be seen. These processes will probably not be useful to pursue the search to lower levels unless something is amiss in our present understanding of the electromagnetic corrections.

Processes 25 and 26 are likely to provide the most sensitive way to unambiguously search for HOW processes and push lower the limit $\Delta Q = 0$, $\Delta S \neq 0$ currents. The first order weak-electromagnetic amplitude for process 25 is expected to be highly suppressed due to the zero charge of the neutrino. However since the neutrino is likely to have distribution of charge the amplitude does not vanish. A crude guess is that the rate for this process should be at least down by $q^4 \langle r^2 \rangle^2$, where r is the electromagnetic radius of the neutrino. The best guess for $\langle r^2 \rangle$ is $\sim 10^{-32}\text{cm}^2$ and for $q^2 \sim m_\pi^2$ we obtain a suppression factor of 10^{-12} in the rate.[46] Thus process 25 should be safe as a signature for HOW or neutral currents down to a branching ratio of $\sim 10^{-18}$.

The electromagnetic contribution to process 26 is likely to be strongly suppressed because CP invariance forbids the single photon intermediate state contribution to this process.[26] The lowest order E.M. process will then be due to diagrams with two photon intermediate states. We can crudely estimate the lower limit due to such contributions using a recent experimental limit on $K^+ \to \pi^+\gamma\gamma$.[45]

$$\frac{\Gamma(K_L^\circ \to \pi^\circ e^+e^-)}{\Gamma(K_L^\circ \to \text{all})} \lesssim \alpha \, \frac{\Gamma(K^+ \to \pi^+\gamma\gamma)}{\Gamma(K^+ \to \text{all})} \sim 10^{-5} \cdot 2 \times 10^{-5} \sim 10^{-10}.$$

Using current theoretical estimates for the rate of $K^+ \to \pi^+\gamma\gamma$ we find a branching factor of $\sim 10^{-12}$ or less.[45] The contribution coming from CP violation in the first order weak process is expected to be much smaller.

Experimentally, process 25 has been searched for in two experiments each covering a different region of the available phase space.[45,52] The best limit for the process that is independent of the behavior of the matrix element is $\sim 4 \times 10^{-5}$ at the 90% confidence level.[45] If a phase space or V-A matrix element is assumed the limit is reduced by an order of magnitude.[52] It seems feasible to search for this process, in the near future down to the level of $\sim 10^{-10}$.

Process 26 has yet to be searched for in any definite way. Considering all factors this process is likely the best candidate for a realistic search for HOW process if the branching ratios are below 10^{-9}.

It is possible to estimate the rate for processes 25 and 26 due to HOW in perturbation theory as discussed in the introduction. Primakoff has estimated that[19]

$$\frac{\Gamma(K \to \pi\nu\bar{\nu}, \pi^\circ \ell\ell)}{\Gamma(K \to \pi\ell\nu)} \sim 8\xi^2 \cos^2\theta_c \quad ,$$

where θ_c is the Cabbibo angle. If these processes are not detected before 10^{-12} in this ratio, Λ the resulting cutoff would be reduced to $\sim$ 1 GeV.

3b. Interference Between Second Order Weak Amplitudes and Others

A possibly more sensitive technique to search for HOW is to observe a large sample of events of the kind

$$K^+ \to \pi^+ e^+ e^- \tag{23}$$

that likely proceeds dominantly through first order weak-first order E.M. processes. An asymmetry in the momentum spectrum of the e^+ and e^- could come about because of the HOW amplitude interfering with the lowest order process. Estimates of this effect have been presented in reference 27. Until process 23 is experimentally observed, it is impossible to estimate the experimental feasibility of this approach.[47]

3c. Production of Leptons in Hadron Collisions (NN $\to$ (ℓ,ν) + hadrons)

If (ℓ,ν) lepton pairs were observed in hadron collisions direct evidence for weak transitions in these processes would be obtained. Lederman has suggested that at a high energy pp colliding beam facility it might be possible to observe such processes.[48] He has used an analogy with the process pp $\to$ (ℓ,ℓ) + hadrons and attempted to extrapolate available data at low energies to these very high C.M. energies. Provided this all works, we might expect that high mass (ℓ,ν) pairs would be produced. In fact it might be possible to obtain events where

$$m^2_{\ell\nu} \sim s_u \ .$$

Since the lepton system is at the same s as the unitarity limit we might expect appreciable (perhaps observable) HOW amplitudes.

4. Non-Leptonic Processes

a. Violation of Selection Rules

As can be seen from table 3, the only important selection rule for nonleptonic processes seems to be the $\Delta S < 2$ rule. The only obvious way to search for HOW non-leptonic amplitudes is to search for $\Delta S \geq 2$ transitions. The only experimentally detected non-leptonic processes with $\Delta S > 0$ are kaon and hyperon decays. The only $\Delta S \geq 2$ kaonic process is the interaction responsible for the K°_S - K°_L mass difference. It is presently thought that the mass difference is due to HOW which break the $\Delta S < 2$ rule. Unfortunately, the mass difference is only one very small number and it has not yet been calculated reliably. The search for other HOW amplitudes is likely to be best accomplished by looking for the decays of $|S| > 1$ hyperons into S = 0 final states. For example:

$$\Xi^- \to n\pi^- \quad (\Delta S = 2) \tag{27}$$

$$\Xi^\circ \to p\pi^- \quad (\Delta S = 2) \tag{28}$$

$$\Omega \to \pi^- n \quad (\Delta S = 3) \tag{29}$$

$$\to \pi^- \Lambda \quad (\Delta S = 2) \tag{30}$$

With the advent of high energy proton beams it becomes feasible to produce copious high energy hyperon beams. Process (28) is the easiest to detect because of the two charged particles in the final state and the characteristic Q value of the process relative to $\Lambda \to \pi^- p$ decay. There is an approved experiment at NAL which will likely be sensitive to this process.[49] It has been estimated that a branching ratio limit of $\sim 10^{-8}$ can be reached within a modest running time if the NAL machine runs at design intensity.[50] March estimates that a limit of $\sim 10^{-10}$ might eventually be achieved.[50]

Theoretical estimates of the possible HOW contribution to these processes seem to be nonexistent and would be appreciated.

4b. CP Violation as 2nd Order Weak

In the Wolfenstein superweak theory of CP Violation, the violation occurs in the mass matrix with $\Delta S = 2$. It seems to

TABLE 3

Rate for Selected Deep Inelastic Scattering Events with $q^2 > 200\ (GeV/2)^2$

(Based on the Parton Model)*

E_ν	Quad Focus H-R	No Focus H-R	Quad Focus CKP	No Focus CKP	Quad Focus H-R H_2 Target ($2\frac{1}{2}$ Tons)
135-145	12	5	2	1	.2
145-155	67	28	10	4	.9
155-165	125	53	16	7	1.6
165-175	172	77	17	8	2.2
175-185	238	103	19	8	3.1
185-195	280	118	18	8	3.7
195-205	308	132	16	7	4.0
205-215	300	128	13	6	3.9
215-225	280	120	12	5	3.6
225-235	280	120	11	5	3.6
235-245	235	104	10	4	3.1
245-255	200	81	8	3	2.6
Total Events/Day					32.5 H_2 Target Rate
(192 Ton Detector)	2497	1070	152	66	
(20 Ton Detector)	260	107	16	6.6	

*Folding in the correct detection efficiency may drop all of these rates by factors of at least 2.

us quite possible (but we know of no theoretical suggestions along this line) that the CP violation is a direct manisfestation of HOW processes.

5. High Energy Neutrino Scattering

Clearly the most likely place to observe departures from the expectations of conventional, lowest order weak theory is at large s, in neutrino scattering. It is fortunate indeed that under certain circumstances the hadronic systems in such collisions will likely behave as though they were massive, pointlike scattering centers. Thus we expect that very high momentum transfers can be achieved in early experiments at NAL and the CERN SPS.

As before we expect HOW process to lead to violations of certain selection rules in neutrino processes. In addition it may be possible to directly observe the nonlocality that HOW process may produce.

5a. Electromagnetic Charge Radius of the Neutrino

The small distance behavior of weak interactions will be sensitively probed by observing the charge radius of the neutrino. The best guess for this radius leads to a cross section ratio of[46]

$$\frac{\sigma(\nu_\mu + N \rightarrow \nu_\mu + N)}{\sigma(\nu_\mu + N \rightarrow \mu^- + N)} \sim 10^{-5} \ .$$

We would also expect by analogy that the contribution to deep inelastic ν_μ scattering would also behave the same way with

$$\frac{\sigma(\nu_\mu + N \rightarrow \nu_\mu + N)}{\sigma(\nu_\mu + N \rightarrow \mu^- + \text{all})} \sim 10^{-5} \ .$$

The process

$$\nu_\mu + N \rightarrow \nu_\mu + (\text{all}) \tag{31}$$

could also arise from $\Delta S = 0$, $\Delta Q = 0$ first order semileptonic currents and from HOW induced neutral currents. Thus, we expect that the search for such induced currents will not be confused by EM processes (i.e. the ν_μ charge radius) unless the resulting cross section is only $\sim 10^{-5}$ of the charged current cross sections.

The measurement of the charge radius is in itself an

interesting experiment. In order to separate the charge rad- from the neutral currents the Z^2 behavior of the electromagnetic process would need to be observed.

5b. Deep Inelastic 'Neutral' Currents

The SLAC experiments have given evidence that hadrons can 'act' point like if appropriate processes are studied (inclusive processes).[24] Using high energy neutrinos, and hitting these 'pseudo point like hadrons' allows very high momentum transfers in the lepton-lepton system. To the extent that the hadrons act point-like, the HOW divergent intergals may truly be cutoff by the weak interactions and not the hadronic size. It is thus possible that if the weak interactions cutoff is near $\sqrt{s_u}$ the HOW amplitudes may be relatively much larger than in the case of semi-leptonic decay processes. Thus, these processes may be almost 'lepton-lepton like'.

Experimentally it would be necessary to study the processes

$$\nu_\mu + N \rightarrow \nu_\mu + (\text{all}) \tag{31}$$

and separate this from the large background of events

$$\nu_\mu + N \rightarrow \mu^- + (\text{all}) \quad . \tag{32}$$

In particular it would be necessary to prove that there is no μ^- in the final state. It is likely that this can be easily done in a Ne bubble chamber or the detector for E1A at NAL if the ratio of cross sections for these reactions is 10^{-2} - 10^{-3}.[39] Going to smaller ratios would likely require a major change of the experimental setup for E1A or the use of the Ne bubble chamber with an External Muon Identifier to reject a larger fraction of events of type (32).

Primakoff has estimated the ratio of these cross sections to be[19]

$$\frac{\sigma(\nu_\mu + N \rightarrow \nu_\mu + \text{all})}{\sigma(\nu_\mu + N \rightarrow \mu^- + \text{all})} = 3\xi^2$$

for the integrated cross section. This ratio would likely be larger if only large q^2 $(= (p_\nu - p_\mu)^2)$ events were used. For $\Lambda \sim \sqrt{s_u}$ we obtain a theoretical ratio of $\sim 10^{-3}$. Thus, if the weak interaction cutoff is at $\sqrt{s_u}$ and if the hadronic system in reaction (31) does not provide a cutoff of the divergent integral and if the cutoff procedure is valid, then the HOW induced process (19) will likely be observed at NAL.

5c. Breakdown of Locality in Deep Inelastic Scattering

We now turn to a brief discussion of the possibility of direct locality tests in deep inelastic processes of the type[53]

$$\nu_\mu + N \to \mu^- + (\text{all}) \tag{32}$$

and thus the direct observation of the 'range' of weak interactions. We use the ordinary definitions of the variables for process 32

$$q^2 = 4E_\nu E_\mu \sin^2\theta_\mu/2$$

$$\nu = E_\nu - E_\mu$$

$$x = q^2/2\nu m_p ; \; y = \nu/E_\nu \; .$$

If scale invariance holds the differential cross section can be expressed entirely as a function of x and y. We assume that scale invariance holds and proceed to discuss locality tests (which test the locality at the lepton-lepton vertex if these assumptions are valid). We must distinguish two kinds of nonlocality in this regard.

(i) Type 1. In the (ν-μ) system an orbital angular momentum of > 0 is observed. Tests for this kind of nonlocality were pointed out long ago by Lee and Yang.[51] These tests take on a particular significance when high momentum transfer collisions are studied. The most general expression for the differential cross section for inelastic neutrino scattering, if locality holds, is of the form

$$\frac{d^2\sigma}{dxdy} = G(q^2,x) \; f(y;x,q^2)$$

with $\quad f = \sum\limits_{n=0} a_n y^n$ and $a_n = 0$ for $n > 2$.

(ii) Type 2. This is the type of nonlocality that comes from a meson propagating from the leptonic vertex to the hadronic vertex. The mesonic propagator is then expected to modify the differential cross section for deep inelastic scattering. If scale invariance holds it would then be possible to write the differential cross section as a product of

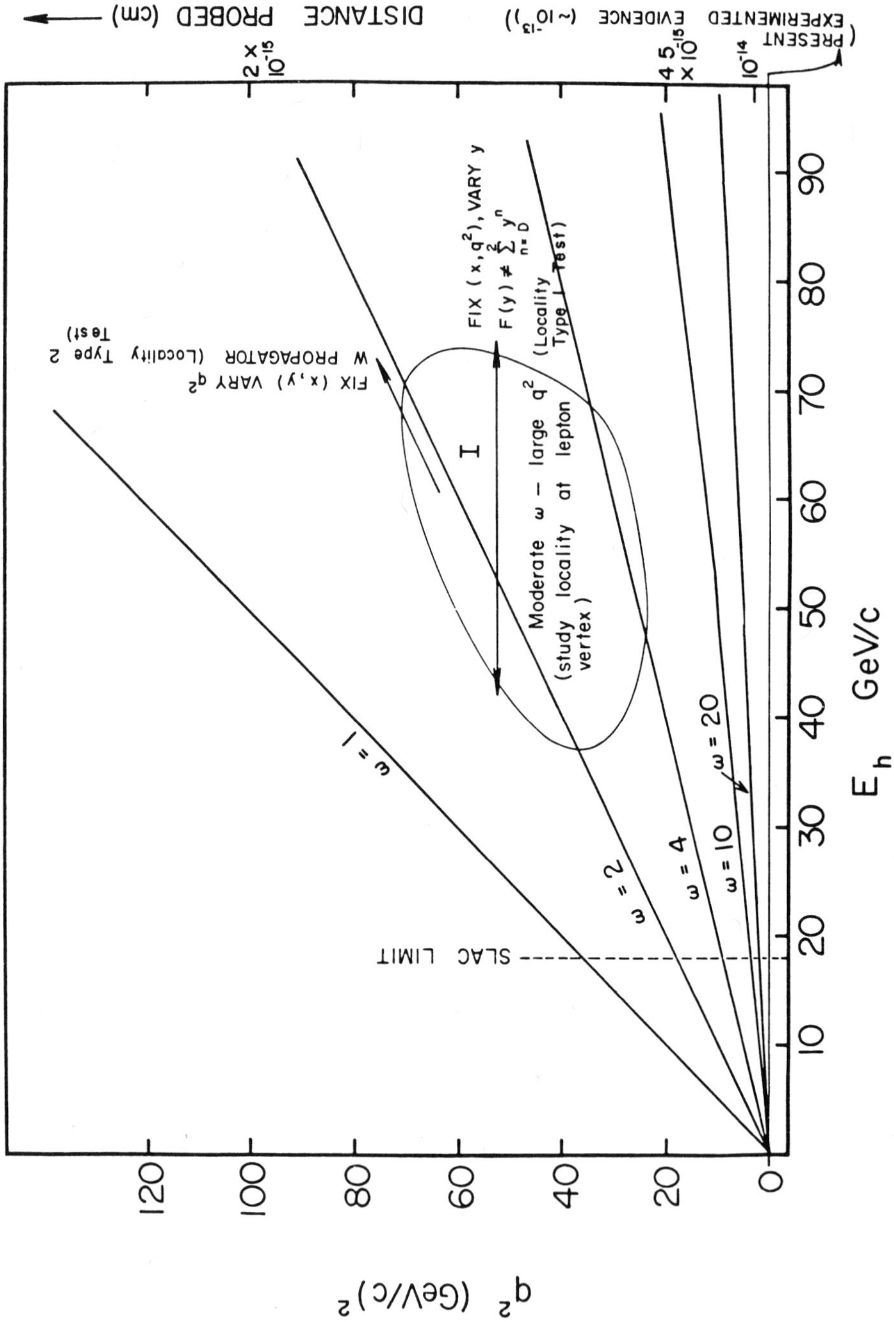

Figure 5. Interesting regions of deep inelastic neutron scattering.

three functions (taking the diffraction model)

$$\frac{d^2\sigma}{dxdy} = \frac{G^2}{\pi} \; M \; E \; [\nu\beta] \; [1 - y + y^2] \; [f(q^2)]$$

where, in particular we take the meson mass to be the W mass,

$$f(q^2) = \frac{1}{(1 + q^2/m_W^2)^2} \; .$$

This might allow us to search well above the mass range covered by the direct production of W's by neutrinos. If scale invariance is badly broken it would be difficult to use deep inelastic scattering to probe this form of nonlocality.

In fig. 5 is shown graphically the type of measurements that would be used to test for a breaking of the two types of locality. We have assumed that the NAL machine only runs at 200 GeV for this graph. In one case (q^2,x) would be fixed and the behavior of the resulting cross section with y would be studied. If y^3 or higher powers of y are needed to explain the data, evidence for nonlocality of type 1 would be obtained. In the second case (x,y) would be fixed and the resulting q^2 behavior of the cross section will be studied.

In fig. 5 is also shown the possible sensitivity of this probe of locality. Present tests of type 1 locality have reached the level of $\sim 10^{-13}$cm (in K-decay) whereas the experiment proposed here offers the possibility of studying distances of the order of 10^{-15}cm. An increase of two orders of magnitude in the locality check would clearly be of great interest.

We now briefly turn to the question of event rates for the deep inelastic process. We use as an example the predicted rates for E1A.[33] This detector which is schematically illustrated in fig. 6 will have a target mass of $\sim$ 400-500 tons. This is to be compared with the large H_2 bubble chamber at NAL with a target mass of $\sim$ 1 ton and the Ne filled chamber with a mass of $\sim$ 20 tons.

In table 3 we present the expected rates/day for events where $q^2 > 200$ GeV/c^2, under a variety of assumptions concerning the incident neutrino beam for 500 GeV/c protons in the machine. Even in the most pessimistic case an adequate number

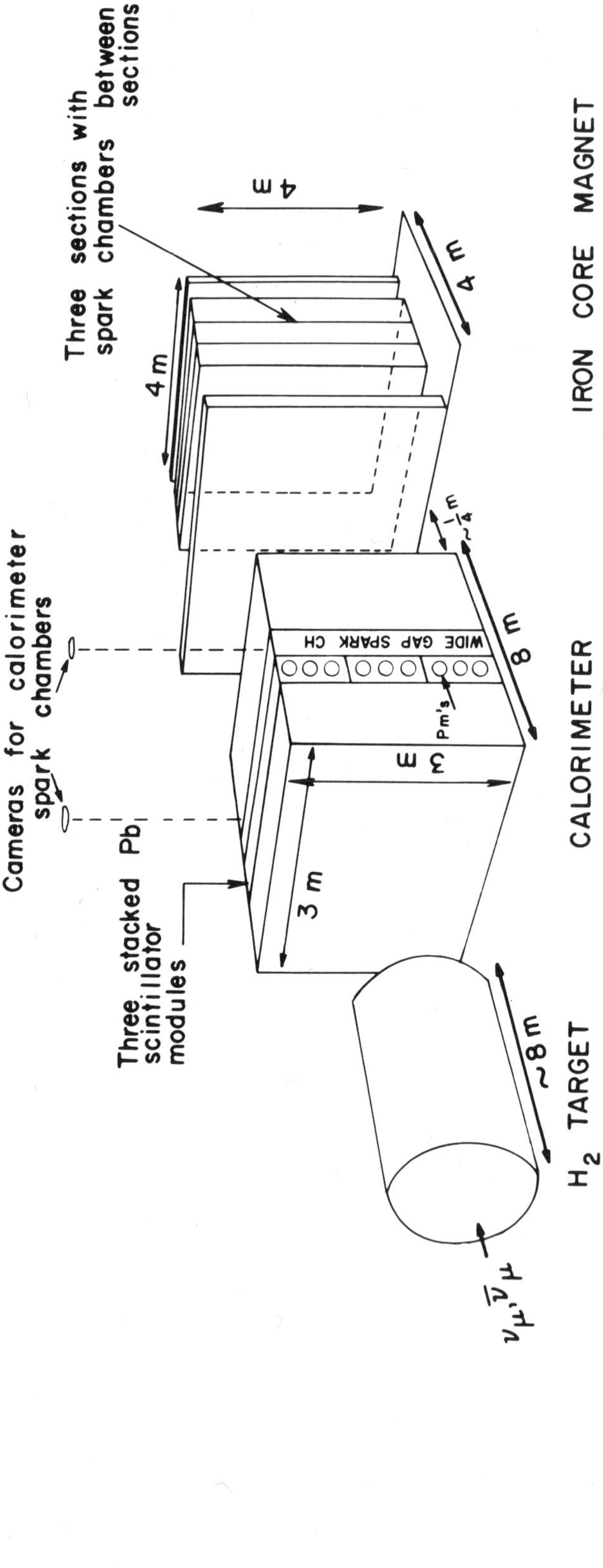

Figure 6. Harvard-Pennsylvania-Wisconsin neutrino detector (schematic).

of events can be obtained to carry out the locality test described above. Thus it seems likely that a definitive statement can be made concerning the range of weak interactions down to $\sim 10^{-15}$cm. With good luck and a 1000 GeV NAL proton beam perhaps 10^{-16}cm could be reached.

6. Summary and Conclusions

The short ranged behavior of the weak interaction is not presently known. Within the framework of conventional theory a pointlike interaction leads to divergent integrals which must be cutoff. It is probably necessary to consider different cutoffs depending on the type of process being investigated. For example, the cutoffs might be arranged as Λ_{NL}, Λ_{SL}, Λ_L denoting the nonleptonic, semileptonic and leptonic processes, respectively. We suggest that a further subdivision of the semileptonic taking into account the quasi-point-like behavior of the hadrons in deep inelastic processes. We denote this cutoff as Λ_{SLDI} for semileptonic-deep inelastic. Possibly this cutoff is more directly related to the Λ_L whereas the Λ_{SL} is more directly related to Λ_{NL}. However, arguments based on the Bjorken technique would likely not differentiate these cutoffs.

Within this framework we can summarize the conclusions of this paper

1. The search for $\Delta Q = 0$ semileptonic decay processes limits $\Lambda_{SL} \lesssim 15$ GeV. Reducing this limit further will require the search for $\Delta Q = 0$ processes that have strongly suppressed electromagnetic corrections. Two processes were suggested where the electromagnetic correction is likely sufficiently small to allow a limit on Λ_{SL} of ~ 1 GeV. The search for these processes requires new high intensity K beams.
2. The search for $\Delta Q = 0$ leptonic processes, in principle, allow an upper limit to be set on Λ_L of $\sim$(100-300) GeV. The experimental detection of such processes will be very difficult.
3. The search for $\Delta Q = 0$ semileptonic-deep inelastic processes will probably allow an upper limit of ~ 100 GeV to be set on Λ_{SLDI}. The experiment looks feasible at NAL either using the Ne bubble chamber or the

massive calorimeter-target detector.

4. A lower limit on Λ_{SLDI} can likely be set by observing the resulting nonlocality (type 2). We guess that $\Lambda_{SLDI} > 30$ GeV can be obtained at NAL with the large calorimeter-target detectors.
5. The existence of a W° with mass less than 8 GeV and a $W^{\pm}$ with mass less than (11-15) GeV can be determined using $e^+e^- \to \mu^+\mu^-$ and neutrino production, respectively. First order neutral leptonic currents at high Q^2 might also be detected in $e^+e^- \to \mu^+\mu^-$.
6. A breakdown of locality of type 2 in the weak interaction might be detected at high Q^2 using deep inelastic neutrino scattering.
7. A crude limit can be set on Λ_{NL} by searching for $\Delta S \geq 2$ decays.

Thus within this conventional picture it would be possible to bracket Λ_{SLDI} by $\Lambda_{SLDI} < 100$ GeV and $\Lambda_{SLDI} > 30$ GeV. This is about the best we can hope for. If $\Lambda_{SLDI} \sim \Lambda_{SL}$ then the present limits on Λ_{SL} would lead to interesting-observable nonlocal effects in the neutrino experiments.

The most exciting possibility is of course that totally new phenomena dominate weak interactions at large s and Q^2. In this regard neutrino microscopy also offers the exciting possibility of probing nature in the new region of small distances.

We wish to thank Profs. J. D. Bjorken, A. K. Mann, C. Rubbia, and S. Treiman for helpful discussions. This is not to imply that these people share the same optimistic viewpoint as expressed in this paper.

REFERENCES

1. H. Becquerel, Compt. rend. 122, 420 (1896).
2. W. Heisenberg, Physik 101, 533 (1936).
3. See, for example, L. B. Okun, "Weak Interactions of Elementary Particles", (Addison-Wesley Publishing Company, Inc. 1965), Chapter 18.
4. H. Yukawa, Proc. Phys-Math Soc., Japan 17, 48 (1935).
5. See, for example, W. Kummer and G. Segré, Nucl. Phys. 64, 585 (1965) and N. Christ, Phys. Rev. 176, 2086 (1968).
6. I. Ya Pomeranchuk, Soviet Journal of Nuclear Physics 11, 477 (1970).
7. T. Appelquist and J. D. Bjorken, "On Weak Interactions at High Energies", SLAC preprint.
8. A. D. Dolgov, L. B. Okun and V. I. Zakharov, "Weak Interactions of Colliding Lepton Beams with Energy (10^2 - 10^3) GeV", preprint from the Institute for Theoretical and Experimental Physics of the USSR State Committee on Utilization of Atomic Energy, Moscow 1971.
9. G. Feinberg and A. Pais, Phys. Rev. 131, 2724 (1963).
10. B. A. Arbuzov, "Weak Interactions" and references herein, Proceedings of the 1970 CERN School of Physics, CERN 71-7.
11. M. Gell-Mann, M. Goldberger, N. Kroll and F. E. Low, Phys. Rev. 179, 1518 (1969).
12. T. D. Lee and C. G. Wick, Nuclear Phys. B9, 209 (1969) B10, 1 (1969) and Phys. Rev. D2, 1033 (1970).
13. S. Weinberg, Phys. Rev. Letts. 19, 1264 (1967); see also G. T. Hooft, "Predictions for Neutrino-Electron Cross Sections in Weinberg's Model of Weak Interactions", University of Utrecht preprint (1971) and B. W. Lee, "Experimental Tests of Weinberg's Theory of Leptons,"Stony Brook preprint (1971).
14. A. Shabalin, Soviet Journal of Nuclear Physics 9, 615 (1969).
15. B. L. Ioffe and E. P. Shabalin, Soviet J. of Nucl. Phys. 6, 6031 (1968).
16. R. Mahapatra, J. Rao and R. E. Marshak, Phys. Rev. 171, 1502 (1968).

17. B. L. Ioffe, Soviet Physics JETP 11, 1158 (1960).
18. R. S. Willey and J. M. Tarter, "Higher Order Terms in the Current-Current Theory of Weak Interactions", University of Pittsburgh Preprint NYO-3829-47 (1970).
19. H. Primakoff, "Weak Interactions", Proceedings of the 1970 Brandeis Summer School, MIT Press, Cambridge (1970).
20. J. D. Bjorken, Phys. Rev. 148, 1467 (1966).
21. J. D. Bjorken, Phys. Rev. 179, 1547 (1969).
22. R. P. Feynman, Phys. Rev. Letts, 23, 1415 (1969).
23. J. D. Bjorken and E. A. Paschos, Phys. Rev. 185, 1975 (1969).
24. E. Bloom, et al., Phys. Rev. Letts. 23, 930 (1969).
25. T. D. Lee, "Scaling Properties in Inelastic Electron Scattering with a Fixed Final Multiplicity", Columbia University preprint.
26. A. Pais and S. Treiman, Phys. Rev. 176, 1974 (1968).
27. S. K. Singh and L. Wolfenstein, Nucl. Phys. B24, 77 (1970).
28. D. Cline, "The Experimental Search for Weak Neutral Currents", Proceedings of the 1967 Herceg Novi Summer School.
29. A. R. Clark, et al., Phys. Rev. Letts. 26, 1667 (1971).
30. B. V. Geshkenbein and B. L. Ioffe, Soviet J. of Nucl. Phys. 12, 552 (1971).
31. S. Glashow, H. J. Schnitzler and S. Weinberg, Phys. Rev. Letts. 19, 205 (1967).
32. There are presently at least 15 proposals for experiments at NAL to search for charged intermediate vector bosons.
33. See, for example, NAL proposal ElA.
34. D. Cline, A. K. Mann and C. Rubbia, Phys. Rev. Letts. 25, 1309 (1970).
35. L. Lederman and Pope, Columbia University preprint.
36. T. D. Lee, Phys. Rev. Letts. 25, 1144 (1970); see also M. S. Turner and B. C. Barish, California Tech. preprint 68-331.
37. R. Gatto, "Theoretical Aspects of Colliding Beam Experments", Proceedings of the International Symposium on Electron and photon Interactions at High Energy held at Hamburg, 1965, Vol. I, p. 106.

38. D. Cline, A. K. Mann and D. D. Reeder, "Search for Forbidden Angular Configurations in the Reaction $e^+e^- \to \mu^+\mu^-$', proposal submitted to SLAC for SPEAR, SPEAR proposal #7.
39. For example in the Wisconsin-CERN NAL experiment E28.
40. W. Czyz, G. C. Sheppey and J. D. Walecka, Nuovo Cimento 34, 404 (1964); K. Fujikawa, Thesis at Princeton, 1970 and Løveth and Radomski, Stanford preprint, 1971.
41. This is my best guess based on the expected ratio of ν_μ to ν_e and the cross section for $\nu_e + N \to e^- + N$ which will provide a serious background.
42. C. H. Albright, Phys. Rev. D2, 1271 (1970).
43. B. Ya Zel'dovich and M. V. Terent'ev, Soviet J. of Nucl. Phys. 7, 650 (1968).
44. J. D. Bjorken, private communication.
45. D. Ljung, "Experimental Search for Rare K^+ Decay Modes", thesis, University of Wisconsin (1972).
46. J. Bernstein and T. D. Lee, Phys. Rev. 136, B1787 (1964).
47. This process is currently being searched for down to the level of $\sim 2 \times 10^{-7}$ by the Hawaii-Wisconsin Collaboration. The best limit from this experiment so far is 7.5×10^{-7} at the 90% confidence level, D. Clarke, private communication.
48. L. Lederman, "Large Q^2 Experiments at Isabelle", BNL report CRISP 71-29, Isabelle project (1971).
49. E8 - Wisconsin - Michigan Collaboration on a Neutral Hyperon Beam.
50. R. H. March, "How to Find (or Unfind) $\Xi^\circ \to p\pi^-$', private communication.
51. T. D. Lee and C. N. Yang, Phys. Rev. 126, 2239 (1962).
52. J. H. Klems, R. W. Hildebrand and R. Stiening, Phys. Rev. Letters 24, 1086 (1970).
53. A. Pais, Annals of Physics 63, 361 (1971).

BREAKING NAMBU-GOLDSTONE CHIRAL SYMMETRIES*

Heinz Pagels
The Rockefeller University

Here we will describe some of the implications of breaking Nambu-Goldstone Chiral symmetries. Most of the work described here, in particular the nonanalytic character of expansions of matrix elements in the symmetry breaking parameters, and the theory of SU(3) violation was worked out in collaboration with Ling-Fong Li at Rockefeller University.[1,2,3]

Before describing our work it is useful to develop the framework and assumptions about thinking of SU(3)×SU(3) and SU(2)×SU(2) as a symmetry of the strong interactions. In particular it is important to recognize that there are two ways of realizing symmetries. What is referred to here is the distinction between a Hamiltonian symmetry and a vacuum symmetry (which need not be the same as the Hamiltonian symmetry).

I. SU(3)×SU(3) as a symmetry of the strong interactions

A. Exact Symmetry

Let us suppose there exists a Hamiltonian

$$H = H_0 + \lambda H' \quad , \tag{1.1}$$

where H_0 is assumed to be SU(3)×SU(3) invariant, and H' breaks this symmetry with strength λ. What we mean by this statement is that in a world with $\lambda=0$, H commutes with the 16 generators of SU(3)×SU(3)

$$[H, {}^aQ^{A,V}] = 0 \qquad a = 1,2,\ldots 8 \quad , \tag{1.2}$$

where 16 vector and axial vector charges obey the algebra proposed by Gell-Mann[4],

$$[{}^aQ^V, {}^bQ^V] = if^{abc}\, {}^cQ^V \quad ,$$

*Work supported in part by the U.S. Atomic Energy Commission under Contract Number AT(11-1)-3505.

$$[{}^aQ^V, {}^bQ^A] = if^{abc}\,{}^cQ^A \quad ,$$
$$[{}^aQ^A, {}^bQ^A] = if^{abc}\,{}^cQ^V \quad . \qquad (1.3)$$

If $\lambda \neq 0$ then we say that SU(3)×SU(3) is broken -- dynamical breaking -- although it is assumed that the algebra at equal times (1.3) is still preserved.

As a second assumption we will assume in the world with $\lambda=0$ that the physical vacuum state is SU(3) invariant by which we mean

$$ {}^aQ^V|0> = 0, \quad {}^aQ^A|0> \neq 0 \quad . \qquad (1.4)$$

One might wonder what the connection between a vacuum symmetry and a Hamiltonian symmetry is. Relevant to this question are two theorems which we will state but not prove here.

<u>Coleman's Theorem</u>[5] ("a symmetry of the vacuum is a symmetry of the world") asserts that if $Q = \int j_0(x)d^3x$ is a space integral of local current density (as is assumed for the Gell-Mann generators) and if $Q|0>= 0$ then $[Q,H]=0$. This implies that a vacuum symmetry be manifest for all physical states. For this case of interest here, that the vacuum is SU(3) invariant implies that physical states can be classified according to the irreducible representations of SU(3). It is well known that the actual hadron spectrum of single particle states falls into approximate supermultiplets of dimension $\underline{8}$, $\underline{10}$ etc. The converse of Coleman's theorem however, is not true. What is relevant here is the Goldstone theorem.

<u>Goldstone's Theorem</u>[6] asserts that if $[Q,H]=0$ then either $Q|0> = 0$ (converse of Coleman's Theorem) <u>or</u> there exist massless, spinless Nambu-Goldstone particles and $Q|0> \neq 0$. (In this case Q does not even properly exist because of the massless particles). Since we have assumed ${}^aQ^A|0> \neq 0$ the Goldstone theorem requires in this case that there exist massless, pseudoscalar single particle states which are identified with the octet multiplet π, K, η. So if the Hamiltonian and vacuum symmetry are not the same we have Goldstone states and this is what is called a "spontaneous" breaking of the (Hamiltonian) symmetry. Of course in the real world with $\lambda\neq 0$ the Goldstone particles acquire a mass $\mu^2 \sim \lambda$. The reason this is

such an attractive picture is (i) that the π,K,η are indeed the lowest mass hadron octet and (ii) we see multiplets of SU(3) symmetry, not of SU(3)×SU(3) symmetry, which would imply parity doubling. Because the π,K,η have the lowest masses we have some indication that SU(3)×SU(3) is not violently broken and λ is small.

B. Breaking SU(3)×SU(3) symmetry

Now we consider what happens if $\lambda \neq 0$ and the Hamiltonian (and vacuum) symmetry is broken. It is remarkable that although SU(3)×SU(3) symmetry of H_0 was proposed by Gell-Mann ten years ago we have little idea today about the dynamics and transformation properties of the breaking term $\lambda H'$. The reason for this is that in physical matrix elements which are measured experimentally the Hamiltonian and vacuum symmetry breaking become entangled and hard to separate. Let us consider two extremes: that H' is SU(2)×SU(2) invariant and that H' is SU(3) invariant,

(i) If H' is SU(2)×SU(2) invariant then the following numbers vanish:

(a) $m_\pi^2 / m_K^2 = 0.075$

(b) $1 - \dfrac{Mg_A}{gf_\pi} = 0.08 \pm 0.02$

(c) $\Sigma_N / M = 0.12 \sim 0.04$.

If H is SU(2)×SU(2) invariant then the pion mass vanishes by the Goldstone theorem. If we chose to compare this with the kaon mass, which is the next heaviest hadron, then we obtain (a). The number (b) is the correction to the Goldberger-Treiman relation where M is the nucleon mass, $g_A = 1.23$, $g^2/4\pi \simeq 14.37$, $f_\pi \cong \mu_\pi / \sqrt{2}$, the pion decay constant. Σ_N is the sigma term in π-N scattering which is essentially the amount the nucleon mass gets shifted if we turn on SU(2)×SU(2) violating forces. It has been estimated by Cheng and Dashen to be 110MeV, although it could be smaller by a factor of 3 or 4. On the basis of these numbers, which are all comparable, we would estimate that SU(2)×SU(2) is a good symmetry of H to 10%-5%. It would be nice to have more such experimental numbers.

(ii) If H' is SU(3) invariant then we have an abundance of measurements of SU(3) violations from the hadron spectrum. Examples are the splitting of the decuplet and the vector mesons,

$$\text{(a)} \quad \frac{M_{\Sigma*}-M_{N*}}{M_{N*}} = 0.1 \qquad \text{(b)} \quad \frac{M^2_{K*}-M^2_{\rho}}{M^2_{\rho}} = 0.3 \quad .$$

From these numbers one would conclude that SU(3) is violated by 30%-10%. It would seem that SU(2)×SU(2) is slightly better symmetry than SU(3) but they are obviously competitive, complicating analysis.

We have no S-matrix formulation fo these ideas about Nambu-Goldstone symmetries and their realization. It would be desirable to have such a non-Hamiltonian formulation, for it would surely deepen our understanding of this phenomenon.

II. Perturbation Theory About a Nambu-Goldstone Symmetry

If we suppose that SU(3)×SU(3) (or SU(2)×SU(2)) is an approximate Hamiltonian symmetry there remains the task of implementing this assumption in actually calculating matrix elements in order to make contact with the real world. What is usually assumed is that one can compute the S-matrix of the real world $S_{\alpha\beta}(\lambda)$ by doing a power series expansions in the parameter λ,

$$S_{\alpha\beta}(\lambda)= S_{\alpha\beta}(0) + S'_{\alpha\beta}(0)\lambda + \frac{1}{2!} S''_{\alpha\beta}(0)\lambda^2+ \ldots. \tag{2.1}$$

and assuming that terms of order λ^2 and higher can be ignored to a first approximation. However in case the symmetry limit is realized by massless Nambu-Goldstone mesons the S-matrix is not analytic in λ near $\lambda=0$ so the expansion (2.1) is not possible. Even though we assume the S-matrix in the symmetric world, $S_{\alpha\beta}(0)$, exists the approach to this limit can go like $\lambda\ln\lambda$ or $\sqrt{\lambda}$. The reason for this nonanalytic behavior is that as the symmetry limit is taken $\lambda\to 0$ the ground state meson masses vanish $\mu^2\to 0$ by the Goldstone theorem and the strong interactions will have a long range component. This produces infrared type behavior in the matrix elements which is known to be nonanalytic. There is no infrared divergence for chiral symmetries because the pseudoscalar ground state mesons have P-wave couplings thwarting a divergence. Still the limit is approached nonanalytically.

A speculative possibility, which probably does not apply to chiral symmetries, is that the nonanalyticity in $S_{\alpha\beta}(\lambda)$ is so severe or path dependent as $\lambda\to 0$ that we don't even get back to the S-matrix $S_{\alpha\beta}(0)$ computed with $\lambda=0$. Such a situation we would refer to as chiral hysteresis.

To see how this nonanlyticity in λ is present more explicitly let us extract a single ground state meson loop from a general S-matrix element

$$S_{\alpha\beta}(\lambda) = \int \frac{d^4q\delta^{ab}}{(q^2-\mu^2)} T^{ab}_{\alpha\beta}(\mu^2,q\ldots.) \quad , \tag{2.2}$$

where q_μ = meson loop momentum and $T^{ab}_{\alpha\beta}$ is the amplitude for the virtual processes $\pi^b(q)+\beta\to\pi^a(q)+\alpha$ with π^a the Goldstone meson of mass μ^2. Since to leading order $\lambda\propto\mu^2$ we may to this order differentiate (2.2)

$$\frac{\partial S_{\alpha\beta}(\lambda)}{\partial\lambda} \propto \int \frac{d^4q\delta^{ab}}{(q^2-\mu^2)^2} T^{ab}_{\alpha\beta}(\mu^2,q\ \ldots) \quad , \tag{2.3}$$

where for our purposes we have ignored the dependence of the virtual process on λ as is true to first order in the meson coupling. As $\lambda\to 0$ $\mu^2\to 0$, and if $T^{ab}_{\alpha\beta}(\mu,0\ \ldots) \neq 0$, then (2.3) implies $\partial S_{\alpha\beta}(\lambda)/\partial\lambda \sim \ln\lambda$ as $\lambda\to 0$, reflecting the infrared divergence. If $T^{ab}_{\alpha\beta}(0,0\ldots) = 0$, then we can differentiate again and the nonanalyticity will show up in higher order. To see in what order the nonanalyticity is manifest for a particular matrix element must, of course, be examined in detail for each matrix element.

A familiar example of the nonanalytic character of perturbation theory about SU(2)×SU(2) symmetry is given by the isovector nucleon electromagnetic radii. We would like to thank Professor M.A.B. Beg for this example. From a dispersion relation for $F^V_{1,2}(t)$ the isovector charge and anomalous moment form factors of the nucleon one obtains for the mean square radius $\frac{1}{6}\langle r^2\rangle = F'(0)$

$$F'^V_{1,2}(0) = \frac{1}{\pi}\int_{(2\mu)^2}^{\infty} \frac{dt}{t^2}\,\mathrm{Im}F^V_{1,2}(t) \quad . \tag{2.4}$$

If one computes the absorptive parts appearing in this dispersion integral from the 2π state with threshold at $t_0=4\mu^2$, with μ the pion mass, one finds from the unitarity condition and taking the SU(2)×SU(2) limit for which the pion mass vanishes that

$$\operatorname{Im}F_1^V(t) \underset{t\to 0}{\to} C_1 t \quad , \quad \operatorname{Im}F_2^V(t) \underset{t\to 0}{\to} C_2\sqrt{t} \quad . \tag{2.5}$$

One concludes from this behavior and the dispersion integral (2.4) that

$$\underset{\mu^2\to 0}{F_1^{V'}(0)} = \frac{C_1}{\pi}\ln\mu^2 \quad ; \quad \underset{\mu^2\to 0}{F_2^{V'}(0)} = \frac{C_2}{\pi}\frac{1}{\mu} \quad , \tag{2.6}$$

which explicitly exhibit nonanalytic behavior in the SU(2)× SU(2) limit. The behavior at the threshold t=0 from 4 or more pions is much gentler than (2.5) and will not produce infrared divergences.

III. Breaking Chiral Symmetry[2]

One might despair because with this nonanalytic behavior one can not apply the techniques of power series perturbation theory, with which we are all familiar. However, the nonanalyticity comes not as a curse but a blessing. This is because if a matrix element is nonanalytic to leading order in λ then to this leading order we can calculate it exactly. So the leading order in symmetry breaking is determined exactly by the symmetry itself. We refer to such results on matrix elements as chiral limit theorems.

The reason for this is that the nonanalytic behavior arises from the long range components of the strong interaction in the symmetry limit. But the long range component here refers to low energy Goldstone bosons in the symmetry limit. Such amplitudes can be calculated exactly using the techniques of current algebra establishing low energy theorems.

We have seen from the above example on the nucleonic radii that the nonanalyticity arises when dispersion integrals diverge at the production thresholds of the ground state bosons as $\lambda\propto\mu^2\to 0$. Unitarity, supplemented with current algebra low energy theorems for the matrix elements in the unitarity condition exactly control the threshold behavior. Hence we can exactly determine the coefficients of the leading nonanalytic term. The following examples illustrate these ideas.

A. Σ term in π-N scattering[9] (with W. Pardee)

Here we consider an example of a chiral limit for perturbation theory about chiral SU(2)×SU(2). We define the Σ term by

$$\bar{u}(p')\mu^2 f_\pi \Sigma(t) u(p) = \frac{1}{3}\sum_{a=1}^{3} \langle N(p')|[{}^aQ^A,[{}^aQ^A,\mathcal{H}'(0)]]|N(p)\rangle \quad , \quad (2.7)$$

$$t = (p'-p)^2 \quad ,$$

where μ is the pion mass, $f_\pi \simeq \mu/\sqrt{2}$ the decay constant and $\mathcal{H}'(x)$ the SU(2)×SU(2) violating Hamiltonian density. Dashen and Cheng[7] have shown that at $t=2\mu^2$ one can relate the sigma term to scattering data,

$$\mu^2 f_\pi \Sigma(2\mu^2) = (\text{on shell } \pi\text{-N scattering data}) + 0(\mu^4 \ln\mu^2) \quad ,$$

although this result is not central to our analysis. Perhaps of greater interest than $\Sigma(2\mu^2)$ is $\Sigma(0)$ since this is the quantity related to dimensions of field operators, etc. If we make the weak assumption $t\Sigma(t)\to 0$, $t\to\infty$ then this quantity is related to $\Sigma(2\mu^2)$ by a dispersion integral

$$\Sigma(2\mu^2) - \Sigma(0) = \frac{2\mu^2}{\pi} \int_{4\mu^2}^{\infty} \frac{dt \, \mathrm{Im}\Sigma(t)}{t(t-2\mu^2)} \quad , \quad (2.8)$$

with a threshold corresponding to the production of a pion pair. If we retain just the 2π state in the unitarity condition for $\mathrm{Im}\Sigma(t)$ then this absorptive part is just equal to

$$\sum_{a=1}^{3} \langle 0|[{}^AQ^a,[{}^AQ^a,\mathcal{H}'(0)]]|2\pi\rangle\langle 2\pi|\bar{N}N\rangle_{J=0}$$

times two body phase space. In the SU(2)×SU(2) limit in which the pion mass vanishes these matrix elements can be computed from current algebra exactly as $t\to 0$. One finds $\mathrm{Im}_{2\pi}\Sigma(t)\to C\sqrt{t}$, $t\to 0$ with C a determined constant. From the 4π state, $\mathrm{Im}_{4\pi}\Sigma(t)\to C' t^2\sqrt{t}$, so this state produces no divergence in the integral (2.8). Our conclusion is

$$\Sigma(2\mu^2) - \Sigma(0) \underset{\mu^2\to 0}{=} \frac{3}{8\pi}\left(\frac{g_A}{2f_\pi}\right)^2 \frac{\mu}{f_\pi} + 0\,(\mu^2 \ln\mu^2) \quad , \quad (2.9)$$

since the 2π state requires that the integral diverge like μ^{-1}. The first term represents a 14MeV correction to the value of $\mu^2 f_\pi \Sigma(2\mu^2)$.

B. Decay constants.

If we consider perturbation theory about chiral SU(3)× SU(3) so that the symmetry is realized by $m^2_{K,\pi,\eta} \to 0$ we obtain a chiral limit theorem on $f_K/f_\pi = 1+0(\lambda \ln\lambda)+0(\lambda)$. This result is

$$f_K/f_\pi - 1 = \frac{3(m_K^2 - m_\pi^2)}{64\pi^2 f_\pi^2} \ln\left(\frac{64\pi^2 f_\pi^2}{3(m_K+m_\pi)^2}\right) + 0(\lambda) \quad , \tag{2.10}$$

where we have set the mass scale of the logarithm from the solution to an eigenvalue problem. Of course one can change this mass scale and affect only the terms of order λ. With $f_K/f_\pi f_+(0)=1.28$, $f_+(0)\simeq 0.95$ we have $f_K/f_\pi - 1 \simeq 0.22$ while from (2.10) we obtain 0.19 for this quantity.

C. Dashen-Weinstein $K_{\ell 3}$ Decay Formula[10] (with R. Dashen and M. Weinstein)

Dashen and Weinstein established a result for the $K_{\ell 3}$ decay matrix element

$$D(t) = (m_K^2 - m_\pi^2)\, f_+(t) + t\, f_-(t) \quad , \tag{2.11}$$

where $f_\pm(t)$ are the usual form factors for $K_{\ell 3}$ decay. Their result was based on chiral SU(3)×SU(3) power series perturbation theory and hence will be modified by our considerations. We find as a modification

$$D'(m_K^2) = \frac{1}{2}\left(\frac{f_K}{f_\pi} - \frac{f_\pi}{f_K}\right) + N'(m_K^2) + 0(\lambda^2 \ln\lambda) \quad , \tag{2.12}$$

$$N'(m_K^2) = \frac{3I(m_K^2, m_\pi^2)}{(16\pi)^2 (f_K f_\pi)^2} (m_K^2 - m_\pi^2)(m_\pi^2 f_\pi + m_K^2 f_K)(f_\pi + f_K) \quad , \tag{2.13}$$

$$I(m_K^2, m_\pi^2) = \int_{t_0}^{\infty} \frac{dt}{t^2} \left[\frac{(t-t_0)(t-t_1)}{t^2}\right]^{\frac{1}{2}}, \quad t_{0,1} = (m_K \pm m_\pi)^2 \quad . \tag{2.14}$$

Here the first term in (2.12) is $0(\lambda \ln\lambda)$ and is model independent. One might guess that $N'(m_K^2)$ given by (2.13) is only $0(\lambda^2)$ from the factors of meson masses in the numerator. But in the SU(3)×SU(3) limit $m^2_{K,\pi}\to 0$ and we see $I\to m^{-2}$ so that $N'(m^2_K)$ is actually of $0(\lambda)$. The result for $N'(m^2_K)$ has been computed in a $\overline{3}3 + 3\overline{3}$ model of Hamiltonian symmetry breaking. However it is only 15% of the first term of $0(\lambda \ln\lambda)$; so as far as numerical results are concerned we need consider only the leading term.

From our result, $D'(m^2_K)\simeq\frac{1}{2}\ (f_K/f_\pi - f_\pi/f_K)\simeq 0.28$, and the Callen-Treiman-Mathur-Okubo-Pandit relation, which is $D(m^2_K)/D(0)=f_K/f_\pi f_+(0)\simeq 1.28$, we can make some conclusions regarding the behavior of D(t) (see figure 1). Experiments indicate that D(t) for low t>0 is falling.[11] With more data one can go near to the edge of the Dalitz plot at $t_1=(m_K-m_\pi)^2$. In order to achieve the value at $t=m^2_K$ for both the function and its slope we predict D(t) must exhibit considerable curvature. In particular, a linear fit to the data between $0<t<(m_K-m_\pi)^2$ should be completely inadequate. One should use a quadratic fit or, even better, a fit to D(t) that takes into account the cut beginning at $t_0=(m_K+m_\pi)^2$.

IV. Breaking SU(3) Symmetry

If a matrix element is nonanalytic to leading order we can prove that the threshold dominates the dispersion integral. This threshold dominance was the basis of the chiral limit theorems. If a matrix element is analytic to leading order we will introduce as a new assumption that the low energy Goldstone boson states dominate the dispersion integral in the symmetry limit. One can not prove this assumption of threshold dominance as in the nonanalytic case. Here we will explore the consequences of making this assumption.

This hypothesis of threshold dominance for matrix elements analytic to leading order has a primary consequence the identification of the Coleman and Glashow tadpole[12] as Goldstone boson pair states. However we can go beyond the tadpole model in proving octet enhancement rather than assuming as is done in the tadpole model. In other words, chirality + threshold dominance of Goldstone boson pair states = > octet enhancement.

A. Ground-State Meson Masses[3]

First we will show how the hypothesis of threshold dominance leads to an eigenvalue problem for the ground state

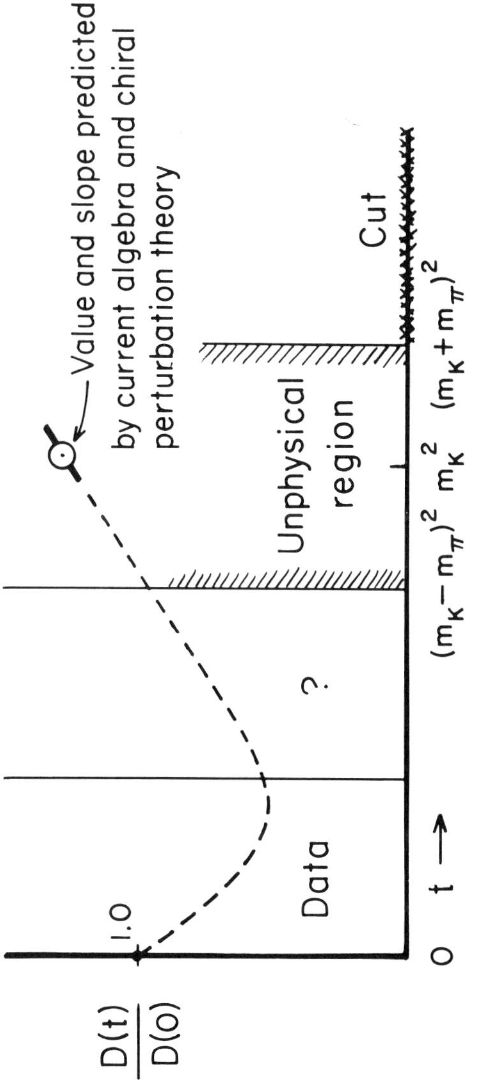

Figure 1. $K_{\ell 3}$ Decay Scalar Form Factor D(t).

meson mass splittings. The matrix element of

$$d^{abc}(t) = \langle M^a(p_2)|-i\partial_\mu V_\mu^b(0)|M^c(p_1)\rangle, t=(p_1-p_2)^2 \quad , \tag{4.1}$$

where $\langle M^a|$ is a single meson state and $V_\mu^b(x)$ is the vector current. The mass difference of the pseudoscalar bosons is then

$$d^{abc}(0) = if^{abc}(m_a^2-m_c^2) + 0(\lambda^3 \ln\lambda) \tag{4.2}$$

and we will work only to leading order in symmetry breaking. The main point of this approach is that we can compute exactly to leading order in chiral breaking the behavior of the absorptive part at threshold $t\to 0$ from unitarity (see fig. 2a),

$$\underset{t\to 0}{\mathrm{Im}d^{abc}}(t) = -\frac{i}{2}\underset{t\to 0}{d^{bef}}(t)\Phi_2^{ef}(t)M_{ac}^{ef}(t) \quad . \tag{4.3}$$

Here $d^{bef}(0)$ is just given by (4.2), $\Phi_2^{ef}(t)$ is two body phase space for the massless mesons and $M_{ac}^{ef}(t)$ is the (symmetrized) boson-boson scattering amplitude in the chiral limit.

If we now introduce the hypothesis of threshold dominance in the form of a cut-off dispersion integral

$$d^{abc}(0) \simeq \frac{1}{\pi}\int_0^{4M^2}\frac{dt}{t}\,\mathrm{Im}d^{abc}(t) \tag{4.4}$$

and use our exact result on the absorptive part (4.3), there results the eigenvalue problem

$$\begin{aligned}
m_K^2-m_\pi^2 &= \frac{5a}{8}(m_K^2-m_\pi^2) - \frac{3a}{8}(m_K^2-m_\pi^2) \quad ,\\
m_K^2-m_\eta^2 &= -\frac{3a}{8}(m_K^2-m_\eta^2) - \frac{3a}{8}(m_K^2-m_\pi^2) \quad ,\\
m_{\pi^+}^2-m_{\pi^0}^2 &= -\frac{a}{2}(m_{\pi^+}^2-m_{\pi^0}^2) \quad ,\\
m_{K^+}^2-m_{K^0}^2 &= \frac{a}{4}(m_{K^+}^2-m_{K^0}^2) + \frac{\sqrt{3}}{2}a\Delta_{\pi\eta} ,\\
\Delta_{\pi\eta} &= \frac{\sqrt{3}}{4}a(m_{K^+}^2-m_{K^0}^2) \quad .
\end{aligned} \tag{4.5}$$

The eigenvalue $a=M^2/f_\pi^2 4\pi^2 \; 0$, so its sign is determined from our theorem (4.3) and $\Delta_{\pi\eta}$ is the π^0-η transition mass. If we reject the trivial solution to (4.5) then

$$\det \begin{vmatrix} 8-5a & 3a \\ 3a & 8+3a \end{vmatrix} = 0$$

or $a=4/3$ or $a=-2$. The eigenvalue $a=-2$ which corresponds to 27 type splittings is ruled out by the condition $a>0$ so the only nontrivial possibility is $a=4/3$ corresponding to the octet solution,

$$4m_K^2=3m_\eta^2+m_\pi^2 \quad ; \quad \frac{1}{\sqrt{3}}\,(m_{K^+}^2-m_{K^0}^2) = \Delta_{\pi\eta} \quad ; \quad m_{\pi^+}^2-m_{\pi^0}^2 = 0 \ . \qquad (4.6)$$

These are the tadpole model results.

If we include explicit electromagnetic forces by acknowledging, besides the Goldstone bosons, the existence of the zero mass photon then the two meson-one photon state also contributes to the threshold behavior (see fig. 2b). This we call the nontadpole piece and should be included in a more complete treatment.

B. Baryon Mass Differences[2]

It is clear that once octet dominance has been established for the ground state meson mass splittings our hypothesis of threshold dominance will imply other matrix elements of symmetry breaking will also exhibit octet behavior. As an application we have examined the baryon $\underline{8}$ mass splittings (see fig. 2c). Proceeding as before we find, using current algebra to compute the absorptive part, that the 7 baryon mass splittings can be parametrized in terms of the meson mass splittings, $(f/d)_A=(1-\alpha)/\alpha$, the axial vector-baryon-baryon f/d ratio, and an arbitrary multiplicative cutoff. Using the octet relations (4.6) for the meson splittings there are, upon elimination of the multiplicative factor, 6 relations for the baryon masses. Five of these correspond to the tadpole model results but in addition we find a new relation between $(f/d)_B= (2/3)(M_\Xi-M_N)/(M_\Lambda-M_\Sigma)=-3.3$ and $(f/d)_A$ which is

$$(3/10)(f/d)_B = (f/d)_A/(3(f/d)_A^2-1) \quad . \qquad (4.7)$$

From this relation we obtain $\alpha=0.69$ while the experimental value is $\alpha^{exp}=0.66\pm0.02$. This successful result gives us confidence that we have identified the tadpole as Goldstone-boson pair states.

V. Hadronic Corrections to the Goldberger-Treiman Relation[13] (with A. Zepeda)

We would like to propose a theoretical problem: Where are the corrections to the Goldberger-Treiman relation? Everything one can calculate for these corrections turns out to be too small to account for the experimental number by one or two orders of magnitude. We used to wonder why the Goldberger-Treiman relation was so well satisfied. Now we may ask why isn't it satisfied better?

The corrections are given by

$$\Delta = 1 - \frac{Mg_A}{gf_\pi} = 0.08\pm0.02. \tag{5.1}$$

If we assume that the matrix element of $\partial_\mu A_\mu^a(x)$ between nucleon states satisfies an unsubtracted dispersion relation then on just this assumption

$$\Delta = \frac{\mu^2}{g\pi} \int_{(3\mu)^2}^{\infty} \frac{dt \ \mathrm{Im}K(t)}{t(t-\mu^2)} , \tag{5.2}$$

where $K(t)$, $(K(\mu^2)=g)$ is the pion-nucleon form factor with the pion extrapolating field defined by $\mu^2 f_\pi \pi^a(x)=\partial_\mu A_\mu^a(x)$ (see Fig. 3a). Electromagnetic contributions to Δ have been estimated and they are too small to account for (5.1) by two orders of magnitude.

Using the Schwartz inequality on the unitarity condition for $\mathrm{Im}K(t)$ and assuming that the pion propagator $\Delta_\pi(t)$ is dominated by the pion pole at $t=0$ one can rigorously prove that the contribution of the integral (5.2) for $4M^2<t<\infty$ is less than 0.004 in magnitude. Hence the high frequency contribution can be ignored and

$$\Delta \simeq \frac{\mu^2}{g\pi} \int_{(3\mu)^2}^{(2M)^2} \frac{dt \ \mathrm{Im}K(t)}{t(t-\mu^2)} . \tag{5.3}$$

The contribution of the $\rho\pi$ and $\sigma\pi$ states to the dispersion

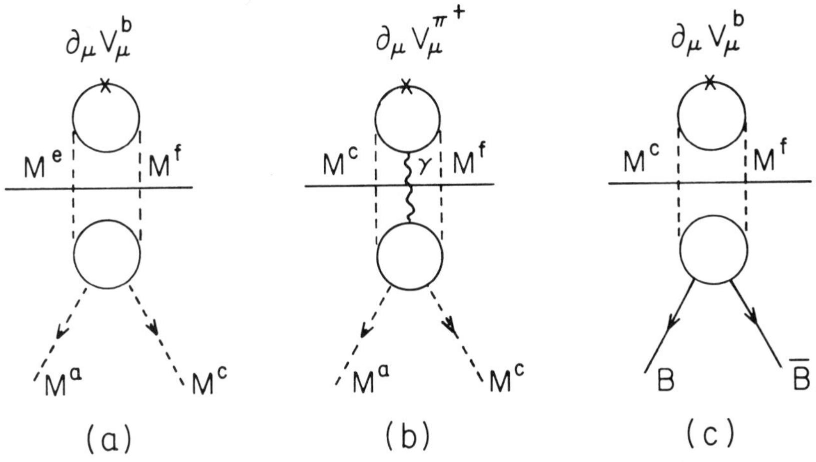

Figure 2. (a) Tadpole contribution to ground state meson mass splittings (b) Non-tadpole (c) Tadpole contribution to baryon $\underline{8}$ mass splittings.

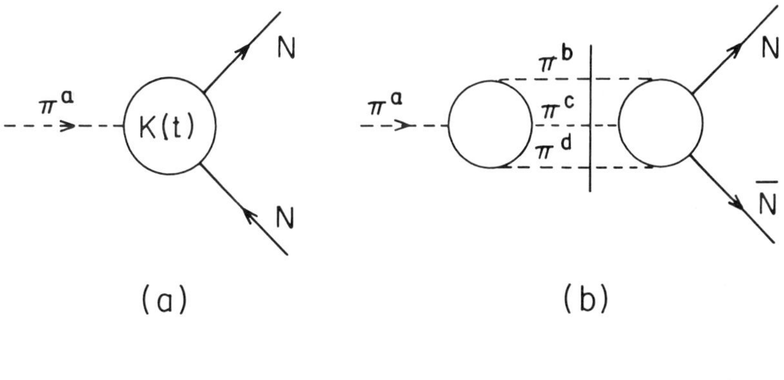

Figure 3. (a) Pion-nucleon form factor (b) Unitarity for 3π state.

integral (5.3) can be estimated. They are less than one order of magnitude too small and tend to cancel because they have opposite signs.

Finally there remains the 3π intermediate state (see fig. 3b). Computing ImK(t) in the chiral limit for which the pion mass vanishes and the threshold is at t=0 and doing the full 3 body angular integrations, using current algebra low energy theorems for the virtual processes $\pi\to3\pi$ and $3\pi\to NN$ we can establish the exact behavior of ImK(t) at threshold:[13]

$$\underset{t\to0}{\mathrm{ImK}(t)} = \frac{gt^2}{3(8\pi)^3 f_\pi^4}\left[\frac{g_A^2}{3}\left(\frac{5}{2}-\frac{17\pi^2}{35}\right)-\frac{7}{8}\right] . \qquad (5.4)$$

Substituting this exact theorem in the dispersion relation

$$\Delta \simeq \frac{\mu^2}{g\pi}\int_0^{(2M)^2}\frac{dt}{t^2}\,\mathrm{ImK}(t)\simeq -0.001 \quad , \qquad (5.5)$$

two orders of magnitude too small!

So where are the corrections to the Goldberger-Treiman relation? We consider three ways out (i) There is a subtraction. If this is true then our approach is invalid. But we also completely lose the rationale for thinking of SU(2)× SU(2) as a good symmetry for we have no way of justifying a small subtraction constant. (ii) The experimental numbers are wrong for g_A and g. If $g_A\simeq1.26$ instead of $g_A\simeq1.23$ and g was smaller by 5% then Δ would be consistent with our analysis. However quoted experimental errors do not allow for such a variation. (iii) The tripion. Perhaps there is a new state, a heavy pion. No such state has been seen but it ought to be looked for in $\pi+p\to\pi'+p\to3\pi+p$, or $\gamma+N\to\pi'+N\to3\pi+N$.

REFERENCES

1. Ling-Fong Li and Heinz Pagels, Phys. Rev. Letters, 26, 1204 (1971).
2. Ling-Fong Li and Heinz Pagels, Phys. Rev. Letters, 27, 1089 (1971).
3. Ling-Fong Li and Heinz Pagels, Phys. Rev., (to be published) (1972).
4. M. Gell-Mann, Phys. Rev. 125, 1067 (1962).
5. S. Coleman, Journ. of Math. Phys. 7, 787 (1966).
6. J. Goldstone, Nuovo Cimento, 19, 154 (1961).
7. T.P. Cheng and R. Dashen, Phys. Rev. Letters, 26, 594 (1971).
8. G. Holer, H.P. Jacob, and R. Strauss, Phys. Letters, 35B, 445 (1971).
9. H. Pagels and W.J. Pardee, Phys. Rev. D4, 3335 (1971).
10. R. Dashen, L.-F. Li, H. Pagels and M. Weinstein, (to be published).
11. See the review by M.K. Gaillard and L.M. Chounet (Cern 70-14) (1970). Also M.K. Gaillard private communication.
12. S. Coleman and S. Glashow, Phys. Rev., 134, B671 (1964).
13. H. Pagels and A. Zepeda (submitted to Phys. Rev.).

DISCUSSION

CARRUTHERS: I want to mention the possible existence of an additional singularity, noticed in the sigma model by Haymaker and me. In that case the existence of both spontaneously broken and normal solutions in the symmetry limit makes certain vacuum expectation values (VEV) of scalar fields multivalued functions of the symmetry breaking parameters. Hence these VEV are multivalued functions of the symmetry breaking parameters, characteristically having branch points which limit power series expansions from the various solutions in the symmetry limit. (This phenomenon occurs in the semi-classical, or tree approximation without taking into account closed loops.) It is not clear whether this phenomenon occurs in the real world, but is an effect which should be kept in mind and studied in greater generality since the numerical fit to the model in question indicates that the radius of convergence is exceeded by an order of magnitude by the physical values of the symmetry breaking parameters.

PAGELS: It certainly would be nice to know when this possible difficulty you refer to is going to occur and when it is not going to occur in actually calculating matrix elements. The problem is that we only have specific models to guide us and not general physical principles about symmetry realizations. What Li and I have shown is that in general for Goldstone symmetry realizations the leading order terms for symmetry breaking may not be analytic functions of the symmetry breaking parameters. As I understand it you are questioning whether the "leading order" is a good approximation and if the second order and higher are small relative to the leading order. This is an important question. Although one can devise special models in which the perturbation series diverges it is not clear to me whether this actually occurs in nature.

SCHNITZER: The difficulty with $K_{\ell 3}$ decay parameters may be resolved without abandoning the current algebra program by considering subtractions in the spectral representation for the two-point functions that appear in the Ward identities.

PAGELS: I agree that one of the principal assumptions in deriving all these bounds on $K_{\ell 3}$ form factors is the assumption of unsubtracted dispersion relations. It should also be pointed out that the negative result for any substantial continuum contributions to the Goldberger-Treiman relation also depends on the absence of subtractions. So maybe there are subtractions and the divergences of currents are not gentle operators.

SCHNITZER: The situation with respect to the Goldberger-Treiman relation may well be different. I would be prepared to consider unsubtracted dispersion relations for the three-point functions, while subtracting the spectral representations for the two-point functions. After all the two-point functions are quadratic divergent in the quark model.

PAGELS: Yes, I agree they are different.

ONEDA: I would like to make a comment about the perturbation approach to the symmetry breaking and I only discuss SU(3). There can be a non-perturbational approach. For example, I mention the approach based on the (dynamical) assumption of asymptotic SU(3). [S. Oneda, H. Umezawa and Seisaku Matsuda, Phys. Rev. Lett. 25, 71 (1970)]. In this approach, the creation and annihilation operators of physical ("in" or "out") particle, $a_\alpha(\vec{k})$ and $a_\beta(\vec{k})$, are assumed to obey linear SU(3) transformation (including mixing) in broken symmetry but only in the limit of $\underline{k}\to\infty$. This situation is realized in the model of free SU(3) multiplet with different masses, i.e., nature may read books on free field theory for SU(3). (Chiral SU(3) $\otimes$ SU(3) will not be so simple). Asymptotic SU(3), for example, relates the $K_{\ell 3}$-decay form factors to pion electromagnetic form factors (S. Oneda and H. Yabuki, Phys. Rev. D3, 2743 (1971). For reasonable choice of pion electromagnetic form factor, the scalar form factor of $K_{\ell 3}$-decay tends to behave differently from that given by smooth monotonic function. Namely the presence of a dip is indicated, thereby enabling us to obtain $\xi(0)\simeq -0.5$, although the value as small as $\xi(0)\simeq -1$ is not easy to obtain.

PAGELS: I think an important point in $K_{\ell 3}$ decays for the scalar form factor D(t) is that the cut at the $K+\pi$ threshold has a positive discontinuity at threshold as can be shown from current algebra. Consequently this function, if it begins to drop for small positive momentum transfer, must begin to rise again in response to that positive discontinuity. This is essentially the content of the formula for the slope of the scalar form factor.

OKUBO: I would like to make a comment on Professor Schnitzer's question on subtraction of two point functions. Even if we require one subtraction to two point functions, we can derive a bound for D'(0), provided that we may use a kind of the kappa dominance for the spectral weight function $\rho(t)$ with a rather slow t-dependence in the $t\to\infty$ limit. In this case, the bound is independent of any specific Hamiltonian such as GMORGW model, and it still has trouble with experiment if f_κ is reasonable. i.e. $|f_\kappa/f_\pi| \lesssim 0.5$. Therefore, I believe that the dilemma cannot be easily disposed of. The calculation has been done at Rochester with the collaboration of Mr. I-Fu Shih.

INTRODUCTION

Steven Weinberg
Massachusetts Institute of Technology

This is a session on cosmology. I think 10 years ago to announce that at a conference would create a certain sense of embarrassment, a feeling that cosmology, although not precisely a criminal activity, wasn't entirely respectable either. However, things have changed very much in the last decade. Before, let's say 1965, the primary concern of cosmologists (not the unique concern but the primary concern) was kinematic, the large scale structure of a very smoothed out universe -- a universe presumed to be homogeneous and isotropic. A certain amount of attention was paid to the mass density of the universe because the Einstein field equations imposed a relation between the apparent expansion rate and deceleration of the universe's expansion and the mass density of the cosmic matter. But the detailed structure of the contents of the universe was not something that was in the forefront of the minds of cosmologists. But things have been changing at a rapid rate, and I think that theorists really owe this to the experimentalists, and more than anything else to the discovery of the black body radiation in 1965, which is now generally regarded as a 2.7°K microwave background which appears to be extraordinarily isotropic and presumably fills the universe. The importance of this background radiation does not appear when you consider its energy content, which is 4 or 5 orders of magnitude less than the energy contained in the ponderable matter of the universe, but appears when you consider that there is roughly one baryon per 10^9 (give or take an order of magnitude) photons. That statement can be made a number of ways. You can say that the specific heat of the radiation in the universe is 10^9 times the specific heat of the matter in the universe, or you can say that the entropy in the radiation is 10^9, in dimensionless units, i.e. the entropy per baryon.

This has had a revolutionary effect on our thinking about the universe. For one thing it has changed our picture of nucleus-synthesis, or rather restored an earlier picture due to Gamow, Alpher and Herman in which nuclear reactions in the very early stage, when the temperature of the universe was 10^9 °K, were responsible for cooking a good deal of hydrogen into helium, but the cooking was prevented by the presence of this tremendous black body radiation flux from cooking all the matter of the universe into helium. The black body radiation has also had a very great effect on the topic that more particularly concerns us today: the evolution of irregularities. In searching for relics of this hot early condensed phase of the universe we find only a few to work with -- one is the black body radiation itself, another is the helium abundance that I just mentioned, but the most noticeable, that can be seen by anyone on a starry night, is the organization of matter into stars, into clusters of stars, into galaxies, into clusters of galaxies, perhaps into clusters of clusters of galaxies. And the theoretical elucidation of how this structure comes about has been a major preoccupation of theorists, particularly in the last five or so years.

The talks today deal with this subject rather than with the whole of cosmology and they are arranged essentially (although I haven't cleared this with the speakers and they may stray out of their limits) in inverse chronological order. That is, Alar Toomre will talk about honest-to-God galaxies as they really are on the photographs taken with those large telescopes, and what physical processes could result in some of the peculiar shapes that we see galaxies exhibit. George Field, as I understand it, will discuss the organization of matter on a somewhat grander scale of clusters of galaxies, and the intergalactic matter which may fill the space between the clusters, and all the physical effects this intergalactic matter might give rise to. And then Jim Peebles will carry us back, I believe, to an earlier period and look for the roots of this wonderful structure, this taxonomy of galaxies, in the early history of the universe. Without carrying this introduction any further I would like to call for the first speaker, Alar Toomre of Massachusetts Institute of Technology.

INTERACTING GALAXIES

Alar Toomre
Massachusetts Institute of Technology

This talk focused on i) photographs of real interacting galaxy pairs (notably Nos. 82, 85, 86, 87, 242, 243, 244 and 295 from Arp's (1966) Atlas of Peculiar Galaxies), ii) several related theoretical diagrams taken from a forthcoming Astrophysical Journal paper jointly with my brother Juri, and iii) a two-part computer-made movie already reported in part by Toomre and Toomre (1971).

Figures 1 and 2 are frames from that movie. They convey some of the graphic flavor of this presentation. Its essential message was that both some occasional spirals in which a major arm or bridge leads to a nearby companion, and the long narrow curving tails which extend from some other binary galaxies are -- contrary to what has often been conjectured -- simply relics of the violent but brief tidal forces to which two disk galaxies would have subjected each other during an almost interpenetrating close passage.

The computations which led to these pictues were deliberately naive: Except for a large mass (with inverse-square force) at the center of each "galaxy", all participants here were just massless test particles. By definition, they influenced the motions neither of each other nor of the central masses. Prior to the encounter, each disk composed of these test particles was flat and circular.

References

Arp, H. 1966, Atlas of Peculiar Galaxies (Pasadena: California Institute of Technology).

Toomre, A., and Toomre, J. 1971, Bull. American Astron. Soc., 3, 390.

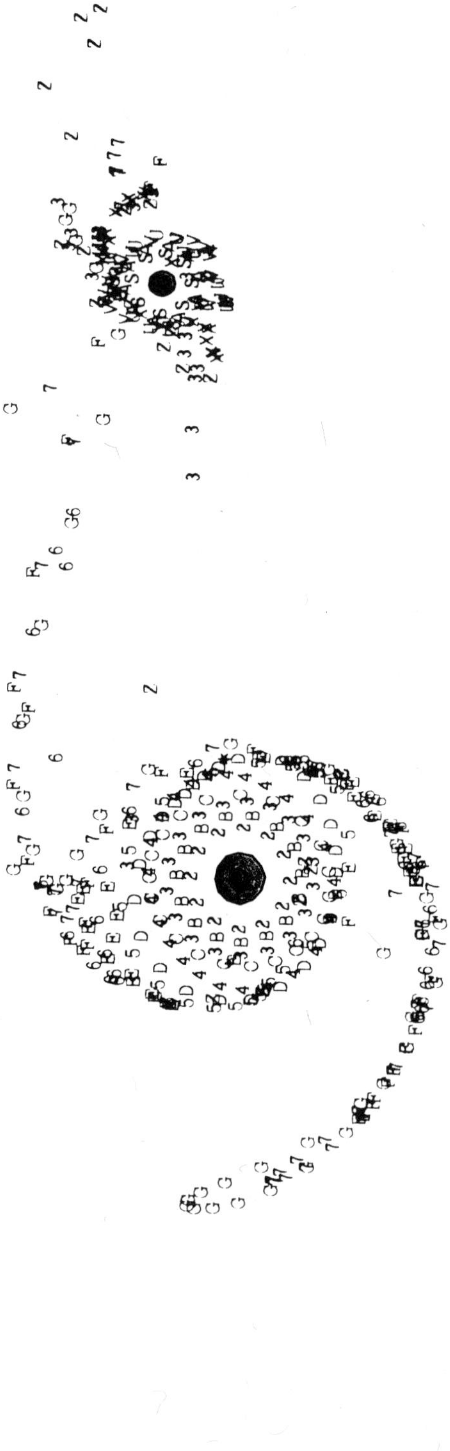

Figure 1. A bridge and counterarm, reminiscent of NGC 3808, which resulted from a parabolic, 45°-inclined close passage of a quarter-mass companion.

Figure 2. A pair of NGC 4038/9-like tails remnant from a close encounter of two equal, 60°-inclined disks in a relative orbit eccentricity $e \cong 0.5$. That orbit plane is viewed edge-on here.

DISCUSSION

FIELD: Are these galaxies in closed elliptical orbits around one another? And are they interacting at this time?

TOOMRE: Yes, to your first question. It is too much of a needle-in-the-haystack problem to imagine that enough totally unbound galaxies out in space just happen to be passing at close range. This must happen once in a while -- but any estimate you make misses by two orders of magnitude. Nor do stray galaxies pass that slowly, I would think. We now know that to make these good bridges and tails, one must have relatively slow passages, preferably sub-parabolic. And so indeed, what I envisage here are galaxies which have been in very elongated elliptic orbits. They just suffered an encounter. They are no longer strongly interacting -- but they sure interacted.

MISSING MASS IN THE UNIVERSE

George Field
University of California

The total mass content of the universe; this is a thorny and controversial topic. I think Jim Peebles may place it in a cosmological context as we go along here. I recommend to you a recent book by Jim Peebles called Physical Cosmology, of which one chapter is very close to what I will be saying today.

What is the mass content of the universe and what form is it in? The key notion is that when we see bodies in motion we are dealing with gravitational forces, and because those gravitational forces must originate in mass, we can infer the mass from the observed motions and distances of these bodies from each other. Let me write down an equation for the escape velocity from a system of mass M and write the mass in terms of the mean density $\bar{\rho}$ within the system:

$$V = \left(\frac{2GM}{R} \right)^{\frac{1}{2}} = \left(\frac{8\pi G\bar{\rho}}{3} \right)^{\frac{1}{2}} R \quad . \qquad (1)$$

The escape velocity divided by the scale of the system is a function of the density alone. This quantity is the inverse of the dynamical time, defined as the time for the system to fly apart if gravitation were to suddenly turn off. It is also of the order of the time for the system to execute one orbit.

Let's run through the kinds of objects that are familiar in astronomy. Let's start with the system that is most familiar to us, the solar system. Here the main body is the Sun. The mean density in equation (1) here refers to the average density obtained by spreading the matter in the Sun throughout the solar system. The dynamical time is then of the order of one year. When we move out to neighboring stars we find that many of them -- up to half -- are systems in which one star is orbitting another in a binary system. The orbital

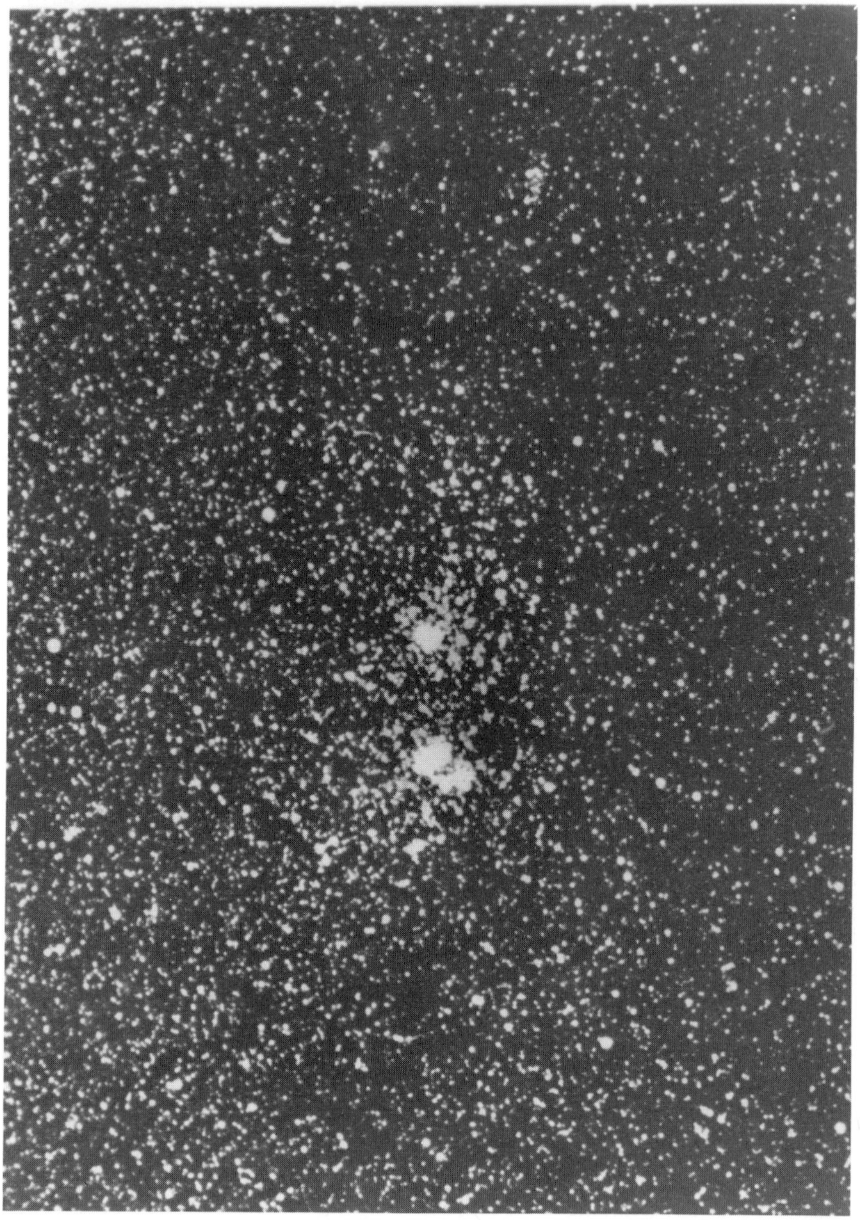

Fig. 1 A typical cluster of stars in our galaxy, the double cluster in Perseus, situated in a rich star field of the Milky Way. Such systems, having $\sim 10^3$ solar masses, are quite common in the Galaxy. (Yerkes Observatory, 10 inch Bruce telescope).

Fig. 2 The nearest massive galaxy, M 31 in Andromeda with satellite galaxies NGC 205 and NGC 221. Its mass can be found from its size and rotation velocity. The latter is deduced from the doppler shift of spectral lines. (Mount Wilson and Palomar Observatories, 48 inch telescope).

times in this case vary from something like from one day up to a thousand years. Figure 1 shows the next dynamical system in terms of increasing scale both in time and in distance; this is a cluster of stars, with an orbital time of the order of 10^6 years and a scale of the order of one parsec. It has 10^3-10^4 stars in it. The next figure (Fig. 2) is familiar: M 31 the Great Nebula in Andromeda, the nearest large galaxy to us. Notice the spiral arms of the galaxy, which are rotating around the nucleus of the galaxy, and notice the companions which are also galaxies. They form a multiple system like those Alar was just discussing and are bound to the main galaxy by gravitational forces. By studying something like this, using the Doppler effect to get the velocities in the main galaxy and measuring the distances by various methods, we can infer from equation (1) the dynamical time scale, the mean density, and the mass of the system, including the Great Nebula in Andromeda and its companions. That is how the masses of astronomical systems are measured. Figure 3 is a close-up of one of the galaxies which is a companion of the Great Nebula in Andromeda in the previous figure. In this picture it has been resolved into stars which are moving in the gravitational field of the whole group of stars to form a bound system. In this case one can infer the mass of the whole system by observing the angular radius, the distance, and the random motions of the stars (using the Doppler effect, which causes broadening of the spectral lines emitted by the galaxy). We can thus get the masses of the individual galaxies around the Great Nebula in Andromeda. Figure 4 shows a group of five galaxies, some of them interacting along the lines that Alar discussed. Such small groups are rather common in the sky so that not only are there clusters of stars imbedded in larger systems (galaxies) but the galaxies themselves cluster to form groups or clusters of galaxies. A larger cluster of galaxies in Hercules, having on the order of 1,000 galaxies bound into a single group, is shown in Figure 5.

Are there clusters of clusters of galaxies? We won't go into that, as it is a hotly debated question. The scale of the typical cluster of galaxies is of the order of one million parsecs or one megaparsec. These clusters lie at random in space and form the universe or the metagalaxy. What is the scale of the universe? It is on the order of 10^{10} parsecs (or 10^4 megaparsecs). It is believed that on scales between 100 megaparsecs and 10^4 megaparsecs, that the universe of clusters is relatively smooth. That is, a sample of space will contain roughly the same number of clusters of galaxies and of galaxies no matter where you take that sample.

When we apply dynamical reasoning to the universe itself, we might expect a relation between the mean density and the time scale of the universe. The actual recession velocity of a cluster of galaxies associated with the cosmological expansion obeys the observed Hubble relationship: that the redshift or the velocity of recession of the cluster is proportional to the distance of the cluster,

$$v = HR \quad . \tag{2}$$

When that is combined with equation (1) I get

$$\frac{v}{V} = \left(\frac{8\pi G\bar{\rho}}{3H^2} \right)^{\frac{1}{2}} \quad . \tag{3}$$

Let me define a critical density ρ_c by the equation

$$\rho_c = \frac{3H^2}{8\pi G} \quad . \tag{4}$$

From the known value of the Hubble constant ($H = 50$ km sec^{-1} mpc^{-1}), which is an observed fact, $\rho_c = 5 \times 10^{-30}$ grams per cubic centimeter. If I now define Ω by

$$\Omega \equiv \frac{\bar{\rho}}{\rho_c} \quad , \tag{5}$$

then I obtain

$$v = \Omega^{\frac{1}{2}} V \quad . \tag{6}$$

from equations (3) and (4). According to this result, whether the recession velocity exceeds the escape velocity depends

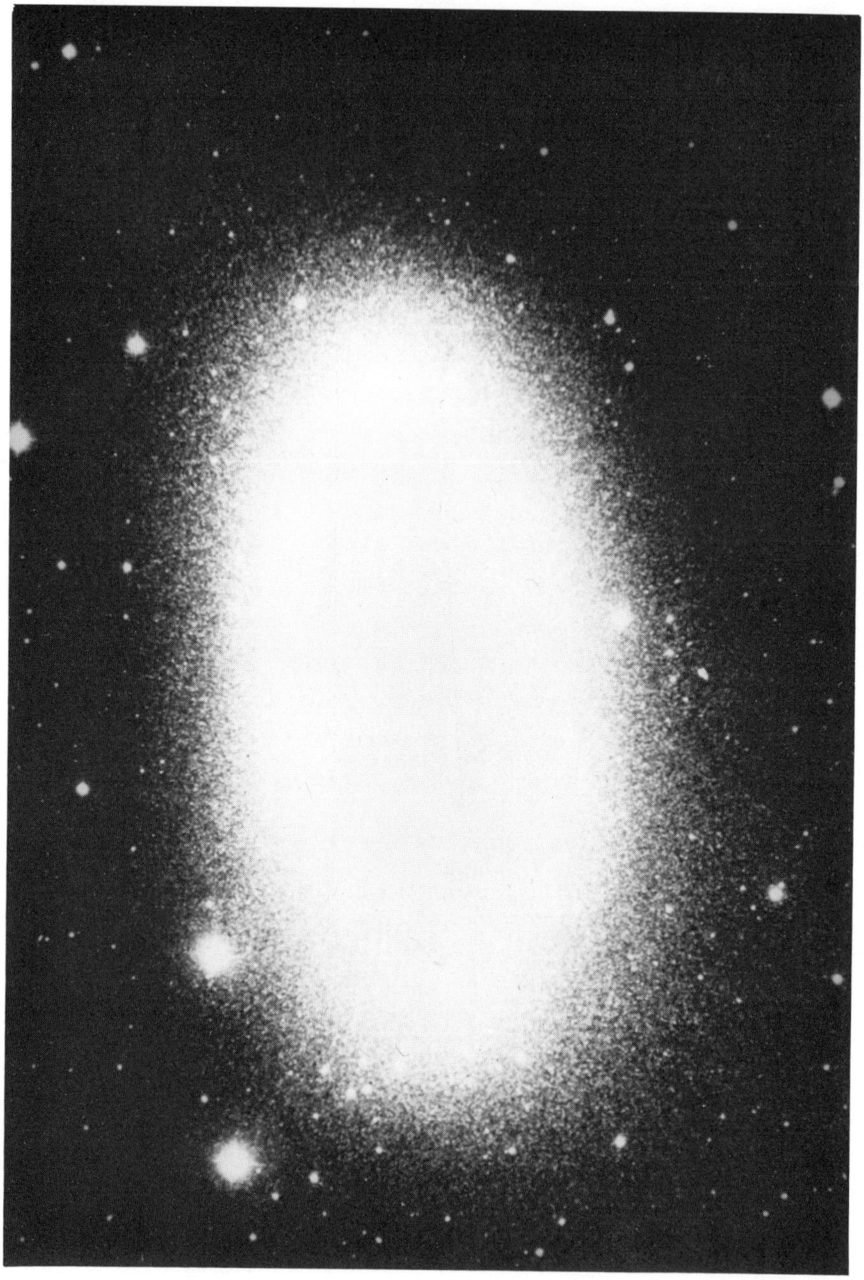

Fig. 3 NGC 205, an elliptical companion of M 31. In this case, rotational motions are small, so the random velocities are deduced from the width of spectral features. (Mt. Wilson and Palomar Observatories, 200 inch telescope).

Fig. 4 Stephan's Quintet, a remarkable system of five galaxies presumably orbitting in their mutual gravitational field. It is puzzling that systems containing so few members persist over cosmic time intervals, in view of the tendency, observed in computer simulations, for members to be ejected. (Mount Wilson Observatory, 60 inch reflector).

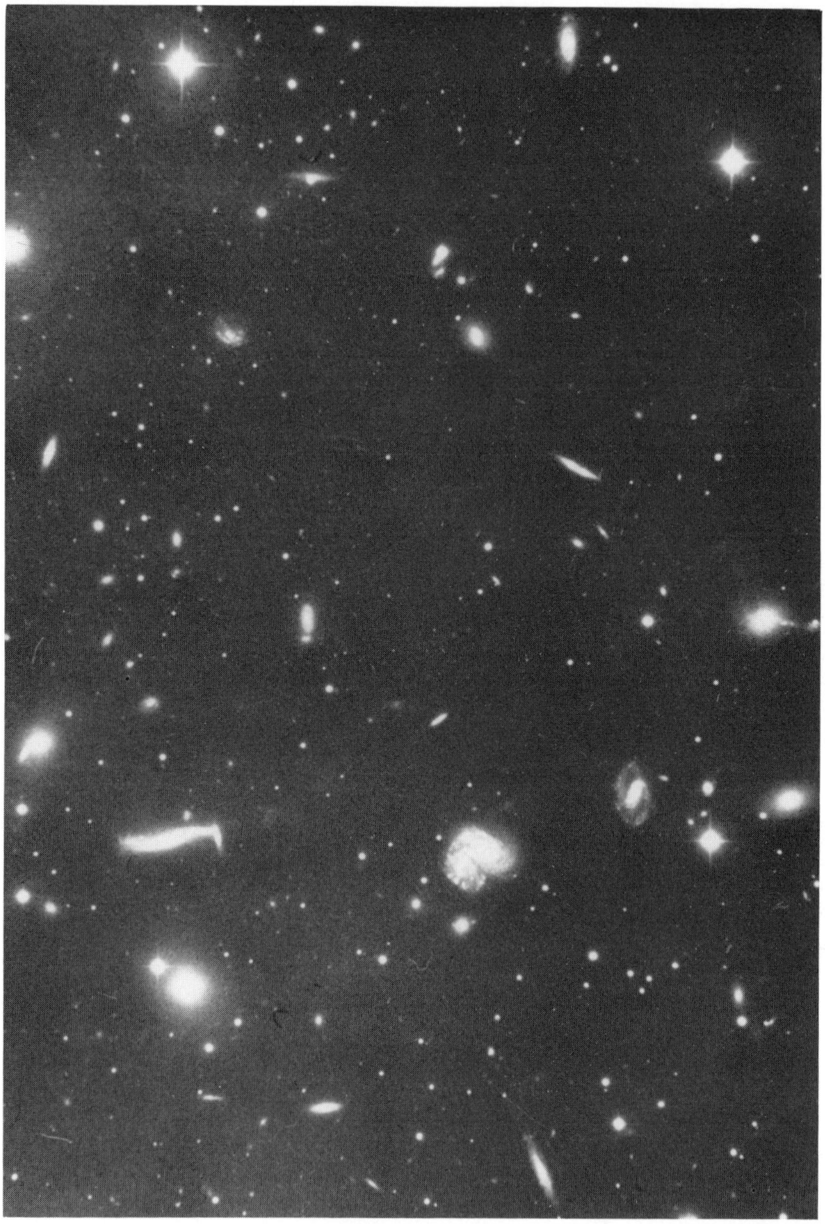

Fig. 5 A cluster of galaxies in the constellation Hercules. Many such clusters, containing $10\text{-}10^3$ galaxies are known. It is believed that their dynamics is qualitatively the same as that of star clusters like that shown in Figure 1. (Mount Wilson and Palomar Observatories. 200 inch telescope).

on Ω. This is a Newtonian argument, and therefore one cannot believe it for a system like the universe, where we have observed velocities up to something like 90% of the speed of light. We certainly have to use a relativistic theory of gravitation, which is just Einstein's theory of general relativity. The results of that theory are surprisingly similar to the Newtonian ones. Let me define the scale factor R to be the distance between two galaxies, which is a function of time in an expanding universe, and let me normalize that so it is unity at the present time. Then the present slope $\dot{R}$ is just the Hubble constant H, which relates velocity and distance in equation (2). There is one model which is essentially a straight line, an expansion out of a singularity at some point in the past, which goes out to infinity without any deceleration. In another possible model with the same slope at the present (H), the scale factor comes plunging back into a singularity in the future. There is also a solution which corresponds to just making it to infinity with zero velocity. Roughly speaking the first corresponds to a hyperbolic orbit, with matter rushing apart faster than the escape velocity, and reaching infinity with finite velocity. The others are like elliptical and parabolic orbits. In terms of Ω, the hyperbolic case has $\Omega < 1$, while the parabolic case has $\Omega = 1$, and the elliptical case is $\Omega > 1$. The velocity of recession is >, =, or < the velocity of escape in the three cases. Remarkably, the constants appearing in the relativistic solutions are the same as in the Newtonian ones (equations 1-6). The applicable model depends upon the value of Ω.

Our problem today is: what is the value of Ω? I can't give you a definitive answer now but I can point out how some of the new techniques in astronomy are giving us information about this problem we wouldn't have even hoped for 10 years ago. What about radiation? We observe 2.7°K blackbody radiation, but the corresponding Ω is 10^{-3}, so the contribution of radiation to closing the universe (as the case $v < V$ is called), is negligible. What about galaxies? Let's take a definite redshift, corresponding to a definite distance, count all the galaxies in that region and multiply in their masses.

This has been done several times, independently, with the result that Ω(galaxies) = 0.02. Galaxies themselves appear to be inadequate to close the universe. What about stars which might not be inside galaxies? Can we place any limits on these? Jim Peebles has pointed out that if you integrate up the brightness of such stars, it ought to give a diffuse glow around the earth which could be detected by the appropriate kind of experiments. From the fact that one does not see such a glow, Ω(stars) comes out to be less than 25%. That unfortunately is not a very tight limit and we would really have to improve the accuracy before we believe it very much. It is possible that stars close the universe, but it is not likely, judging by this number.

There is another argument which goes along a completely different line. Everything I'm going to be talking about is based on a big-bang model and as you know that has long been disputed; on the other hand the discovery of blackbody radiation which was predicted by big-bang cosmologists has given it an enormous boost. Figure 6 gives the cosmic time going from roughly a second after the big-bang up to 10^{10} years and the associated temperature. At about ten seconds or so, when the temperature was of the order of 10^{10}-10^{9} degrees, nuclear reactions produced helium and other elements. The results of calculations of that sort, due to Wagoner, Fowler, and Hoyle, are shown in Figure 7. There are protons and neutrons going into this phase and there is production of deuterium which comes out of the fireball with certain abundance. There is also production of helium-3, way down in abundance (10^{-4} or 10^{-5}). Helium-4 is very abundant and turns out to be about 30% by mass. That much is observed, approximately. What I want to focus on here are the very low abundances of deuterium and helium-3. Figure 8 gives the final abundances of deuterium and helium-3 as a function of the density of the universe at the present time. Let me explain why that comes in. We know that these reactions go on when the temperature is about 10^{9} degrees. We know the present temperature. We know how temperature scales with density. Therefore we can calculate the density at the time that the reactions were going on in terms of the present density. If the present density is low -- that

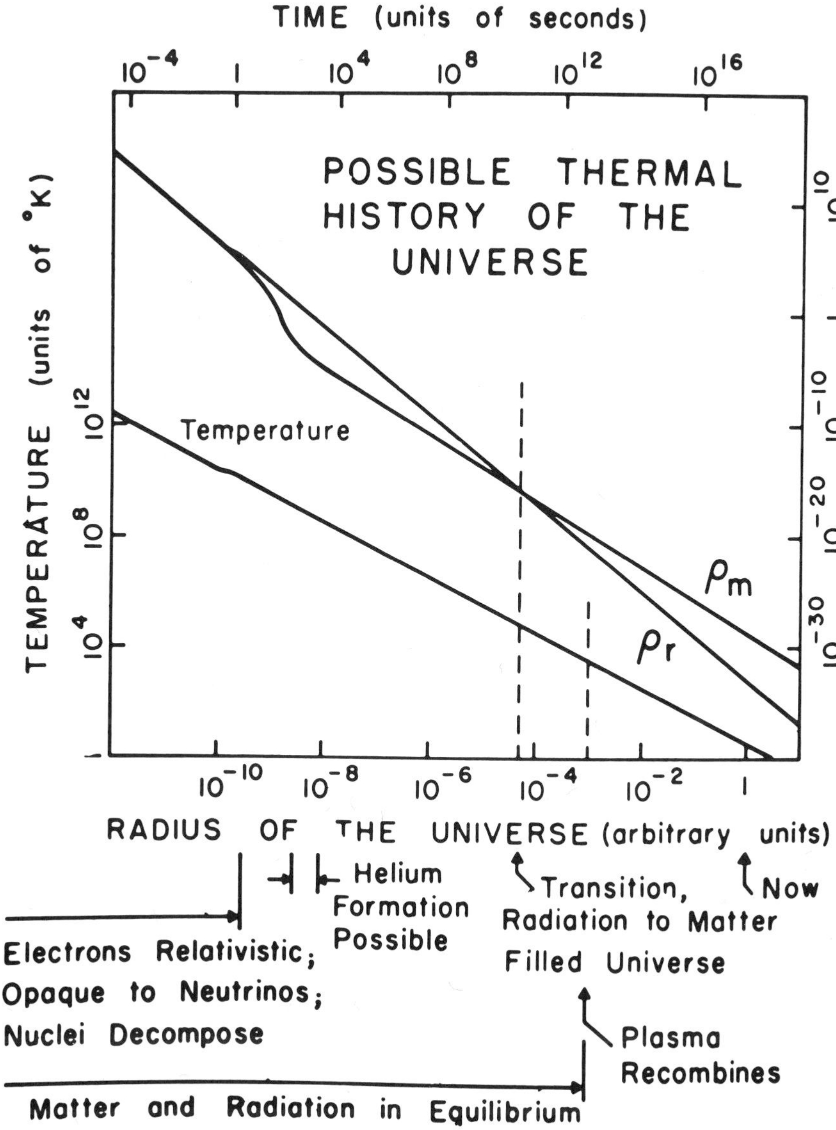

Fig. 6 Temperature against time in a big-bang universe. The present temperature is 2.7 °K, based on the spectrum of the observed blackbody background. Also shown is the density of radiation ($\rho_r = aT^4/c^2$) and the density of matter, ρ_m, assuming $\Omega \simeq 1$. The critical parameter for formation of nuclei heavier than hydrogen is the value of ρ_m when $T \simeq 10^9$ °K. (R. H. Dicke, P. J. E. Peebles, P. G. Roll, D. T. Wilkinson, Ap. J., 142, 418, 1965; reproduced with permission of the Astrophysical Journal.

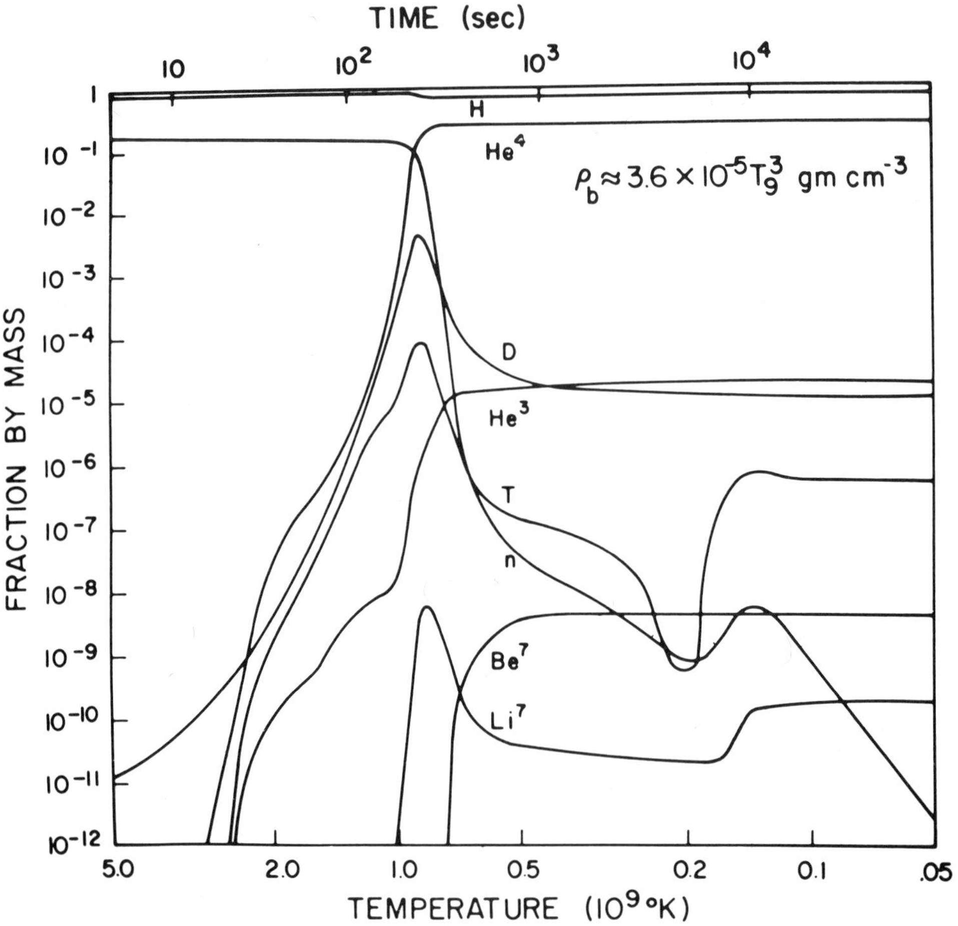

Fig. 7 Formation of various nuclei in the early universe, for a specific choice of matter density, ρ_b. While most of the D and He^3 burns completely to He^4, a small amount is left, and how much depends on ρ_b. (R. Wagoner, W. Fowler, and F. Hoyle, Ap. J., 148, 1, 1967; reproduced with the permission of Astrophysical Journal).

is, if Ω is small -- we get a low density, while if Ω is high, we get a high density at the time the temperature was 10^9 degrees. The abundances of the reaction products we predict are functions of the density at the time the reactions were going on, simply because the more particles there are, the more reactions can occur. A recent paper by Reeves and his collaborators argues, quite effectively, that the deuterium and helium-3 which are found in the Sun, in the solar wind, and on the earth, were created in the cosmic fireball a few seconds after the creation of the universe. There are some correction factors which I haven't time to go into and which are debatable, but if you follow this argument through, it leads to the conclusion that the present total density of the universe must be 3×10^{-31} g cm^{-3}, because that gives us the right deuterium and helium-3 abundances. From this argument Ω is equal to 0.06. Perhaps the universe isn't closed at all.

Figure 9 shows an attempt by Peach to use the Hubble relation itself to discuss the question of Ω. What is plotted here is the logarithmic brightness of galaxies in clusters and the redshift on a logarithmic scale. The fact that this line is nearly straight is a proof of the Hubble relationship equation (2). As the redshift is of the order of 0.4 at the maximum, the curves one predicts deviate there a little bit, depending upon the geometry of space-time. Depending upon the value of Ω, one gets either more or less curvature in this diagram. From this argument, $\Omega = 3(\pm 1.8$ probable error). There is not necessarily a contradiction among these various tests. From this measurement alone, Ω could be one, and on the other hand it could be very small, as the probable error is large. Let's stop and take stock. The indications are that from nucleosynthesis Ω is small. On the other hand the indications from the redshifts are that it may be large. This suggests that we continue to search for other forms of matter in the universe.

I'd like to turn to a different subject where somewhat the same problem has arisen, and that is within the clusters of galaxies themselves. The Coma cluster of galaxies, (not shown here) has a few very bright galaxies, but it also has roughly 1,000 ordinary galaxies. It is rather like the

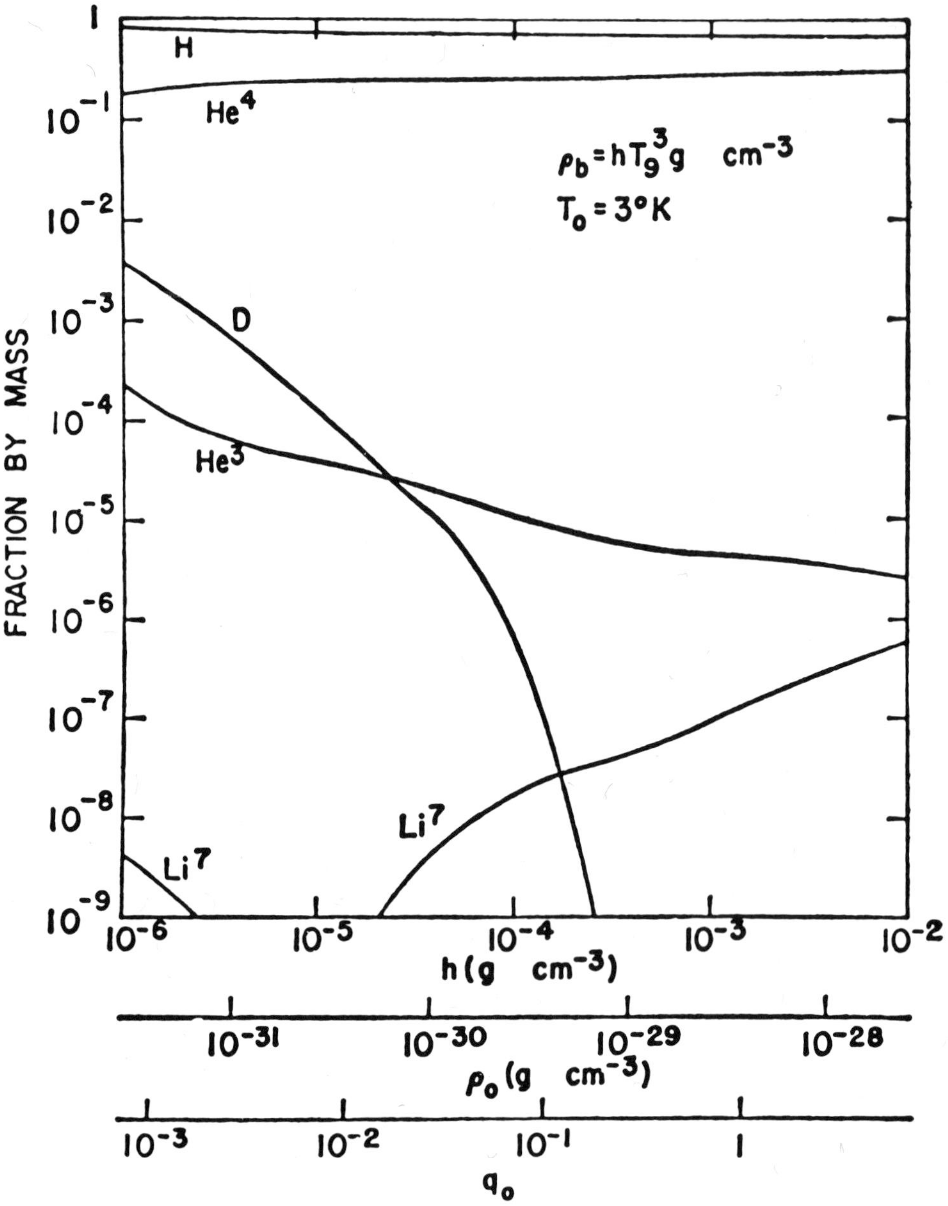

Fig.8 The final amounts of various nuclei produced in the early universe, as they depend on the matter density, ρ_b. Agreement with observation is obtained if $\rho_b \simeq 3 \times 10^{-31}$ gcm^{-3} at present. A high-density universe ($\rho_b = 5 \times 10^{-30}$ gcm^{-3}, $\Omega = 1$) would have negligible D, contrary to observation. (R. Wagoner, W. Fowler, and F. Hoyle, Ap. J., 148, 1, 1967; reproduced with the permission of Astrophysical Journal.

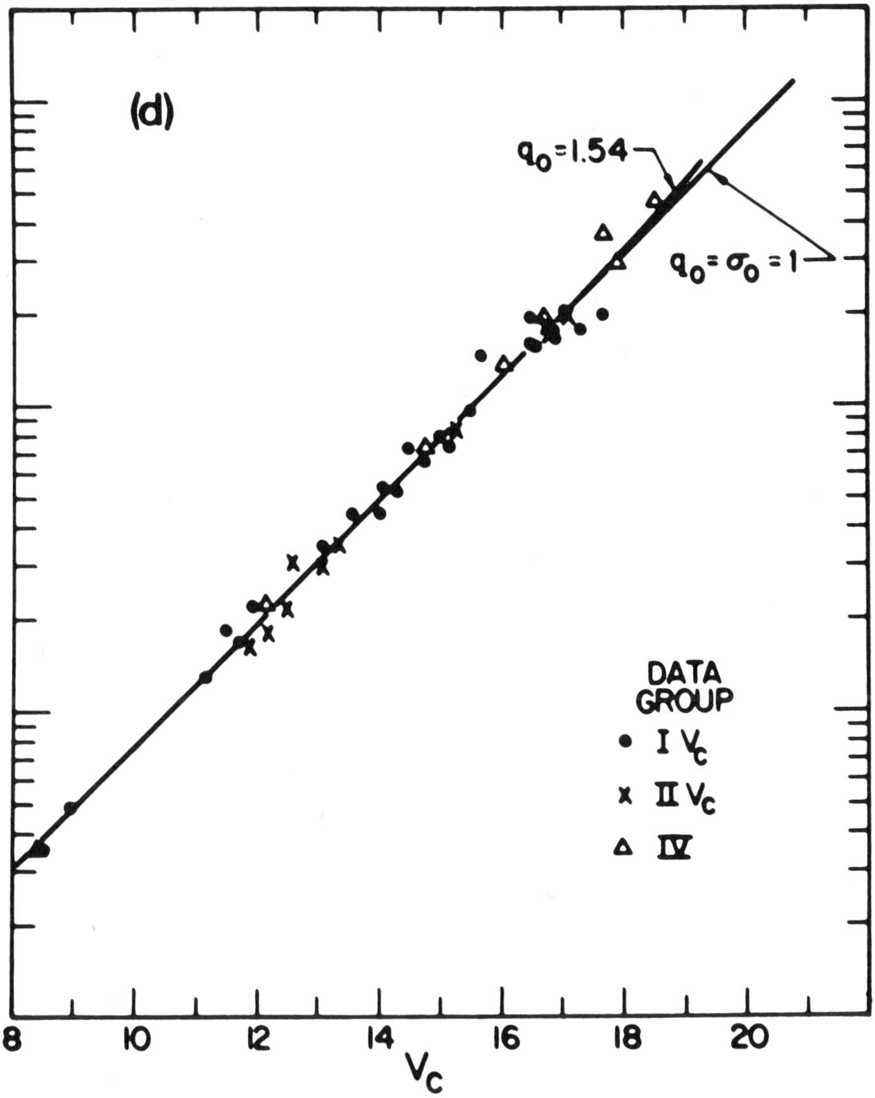

Fig. 9 A Hubble diagram for galaxies in clusters, logarithm of redshift against corrected apparent magnitude. The fit to a straight line of slope 0.2 is proof of the Hubble Law, equation (2). Different relativistic models of the universe predict different deviations from straight-line behavior as the redshift approaches and exceeds unity. The best fit gives $\Omega/2 = 1.5 \pm 0.9$ (J. Peach, Ap. J., 159, 753, 1970; reproduced with the permission of Astrophysical Journal).

cluster of stars I showed earlier, nicely elliptical in outline, quite closely packed on the sky, and fairly smoothly distributed. The average density of galaxies as a function of the distance from the center of the cluster falls off in a smooth manner. The distance of the system is 140 megaparsecs and, it subtends roughly 1° on the sky. One can apply equation (1) in which ρ is given by the ratio of the velocities to the distances. As one can observe the random velocities of the galaxies, and one knows their mutual distances, one can estimate the density and therefore the mass. It turns out that the mass that is obtained in this manner (the "virial mass") exceeds the total mass one would assign to the individual galaxies in the cluster. The virial mass divided by the mass of the galaxies turns out to be 8. The mass required to provide the gravitational field to keep the galaxies in orbit is 8 times the sum of the masses of the individual galaxies. The latter is obtained from studies of other galaxies and by simply adding up the total number of galaxies. Again, many different people have studied this question and they always get a discrepancy of this order. Some people argue that we can push things a factor or two here or there, but I don't think that is possible. What could be the possible explanation for this "missing mass problem"? Just as in the case of the closed universe, (if indeed it _is_ closed), the mass of material that we see is smaller than the total gravitational mass. What is the nature of this other mass? Before answering that question, let me make one other point. Many clusters have been studied, in at least an approximate manner, and one always finds that mass is missing. The Coma cluster is by far the best studied, so this number is a precise one, but inevitably, one finds discrepancies of a factor of 10, 100, or even sometimes 1,000. There is some kind of missing mass that has not yet been identified. What are the possibilities? First of all, you might say, the galaxies in clusters are different from galaxies which are not in clusters, so it is not correct to estimate the mass of galaxies in clusters the way I have. Well, people have gone into that -- they have tried to compare galaxies inside and outside of clusters and, as far as can be seen, they are exactly the same kind. Alright, maybe there are stars which are lying between the galaxies -- maybe the galaxies have collided, à la Toomre, have spun out stars, and the stars are there, not in the galaxies, but between them. Well, if that were true you would again apply the Peebles argument and you would see a diffuse glow of light coming from between the galaxies. In another cluster (Virgo) there is some indication of that, and there may be something like 40% of the mass of the galaxies between the galaxies. But a factor of 1.4 is too small to be of use.

Another possibility is that Newtonian physics is wrong. Of course, Newtonian gravitation is not exactly correct, for there is general relativity. But the general relativistic corrections for a system like this are of the order of the square of the ratio of the escape velocity to the speed of light. This is only 10^{-5}, which is completely negligible. Zwicky has suggested that the graviton has a finite rest mass. This would imply a finite range for gravitational forces of the order of one megaparsec (which is the scale of these systems), if the rest mass is 10^{-38}GeV. One trouble with this argument is that the gravitation would saturate at large distances, but that is exactly the wrong direction for solving our problem. There is something about these galaxies which gives them more gravitation rather than less. However, you may be led to think of other possibilities along that line. Another possibility that has been discussed very much recently is the black hole. We know that from stellar evolution calculations that a star can get into a problem in which its central density regions can begin collapsing -- in some cases the collapse is stopped either by degenerate electron pressure or neutron pressure -- to form a white dwarf or a neutron star. But in other cases there is no pressure yet known which is strong enough to resist the gravitational compression and there is an unlimited free fall to form a singularity -- a black hole. Maybe clusters of galaxies contain black holes. But the distribution of black holes would have to be like that of the galaxies, because from the structure of the cluster we can infer that the gravitational field must have the same variation as if it were due to the galaxies themselves, but larger by a factor of 8. Hence the missing mass cannot be in a central point mass, but must be distributed like the galaxies themselves. Suppose black holes occur within the galaxies. That won't help, because if they exist within cluster galaxies, they should also exist in other galaxies, and they would be counted when we reckon the masses of galaxies. Suppose they are in the cluster, but between the galaxies. It is hard to see why black holes would form in large numbers between the galaxies and not within them. That doesn't seem too plausible.

There is one suggestion that is really wild and may just be right. Most of the calculations in cosmology have been performed on a universe which is almost smooth, taken in the large. The reason is obvious, because that is the simplest case and one can actually do something about it. The zero-order state (where it is completely smooth), and the first-order state (perturbations) have been studied. But what if the universe was never smooth but rather was once extremely chaotic, in the sense that there are huge density fluctuations and associated large random velocities. Martin Rees has suggested that what may happen at that point early in the history of the universe, if it were chaotic, is that much of the energy of expansion may be converted into gravitational waves. We know that this can happen when bodies approach each other at the speed of light. This gravitational radiation would have a wavelength in this picture of the order of megaparsec and might be extremely powerful. Rees suggests its energy density might be enough to make Ω equal to 1. Such a hypothesis also offers a chance of explaining why the clusters of galaxies appear not to be bound. As a wave of gravitational radiation goes through a medium of galaxies, it of course accelerates them with respect to one another. Two galaxies separated by one wave length will be accelerated in opposite directions. Every few million years every galaxy would receive an impulse of 300 kilometers per second. In the next few million years it is going back the other way, and so on. This kind of jiggling of galaxies is very nice because it says that although in these big clusters there are slow orbits of the order of a billion years' period there would be a rapid jiggling with 300 kilometers per second

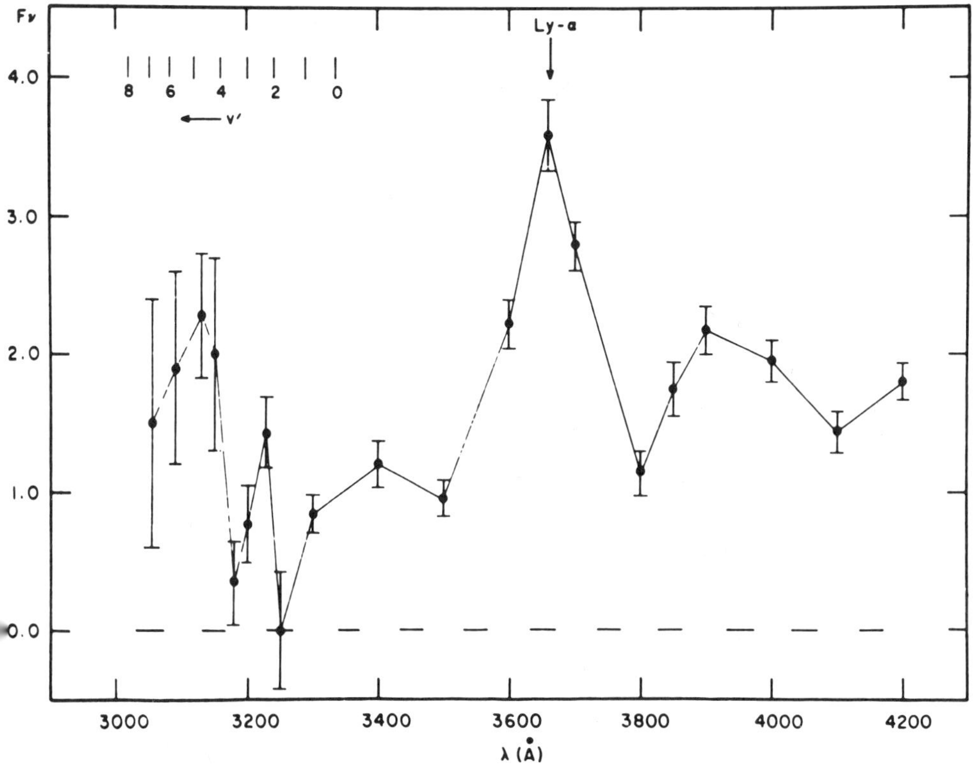

Fig. 10 A photoelectric spectrum of 3C9, a quasar with a red-shift of 2.01. Lyman α from the quasar appears in emission at 3650 Å. The absences of strong absorption at shorter wavelengths proves that hydrogen atoms with velocities between zero and that of the quasar are absent. The expected positions of lines in the Lyman band of H_2 are also shown. Their absence proves the absence of H_2. (G. Field, P. Solomon, and J. Wampler, Ap. J., 145, 351, 1966; reproduced with the permission of Astrophysical Journal).

amplitude. Because the frequency is high, there is never a large displacement, but there would be a measurable velocity shift. What we need for this case, it turns out, is 1,000 kilometers per second to make the orbital velocities appear bigger than they really are, and, by equation (1), the density to appear larger than it really is. The velocities of galaxies may be due partly to the acceleration by passing gravitational radiation, radiation which binds the universe. There are experimental tests which can be performed to see if Rees' suggestion is correct, which may be done in the next five years or so.

Another suggestion that might work out is intergalactic matter. You may have noticed that there was some black stuff in our picture of the Great Nebula in Andromeda. That black stuff is interstellar matter. It's diffuse matter lying between the stars; in our own galaxy interstellar matter is 3% of the stellar matter. One might speculate that when the galaxies formed they left behind much matter in the form of gas -- intergalactic matter. Then Ω(intergalactic matter) may be large. That's a matter that one can go into experimentally.

First of all, we know that most of the matter in the universe is hydrogen, so the very first thing one tries to do is look for intergalactic hydrogen. There are two possibilities: neutral hydrogen (HI) or ionized hydrogen (HII). HI can be seen by means of its 20-centimeter line. That experiment has been done in the Coma cluster, with the result that Ω(HI) (defined as the density of the material in question divided by the total gravitational mass density) turns out to be less than 0.02. Perhaps the hydrogen is ionized. A discovery was made very recently at the American Science and Engineering Company, (A. S. & E.) where various people, including Giaconni and Gursky, were involved in this experiment. What they did was scan over the Coma cluster with an x-ray detector. They found x-rays coming from the cluster; their energy spectrum is such that it is consistent with thermal bremsstrahlung with a gas temperature of 70 million degrees or about 6 kilovolts. That is a very exciting observation because the root-mean-square velocity of a particle at 70 million degrees would be about 1,100 kilometers per second, in one dimension, just about equal to the root-mean-square velocity that is necessary to support the gas in the gravitational field of the cluster. Furthermore, the diameter observed as A.S.& E. is about 0.5°, roughly what is seen for the galaxies. So we may be dealing here with a hot gas which is distributed throughout the cluster. However, there are other processes that can give x-rays, so the results do not definitely prove that gas is present.

Let's leave other possibilities aside and talk about gas. The intensity of radiation that you would get from a gas is a constant times the integral of the square of the particle density, times exp(-E/kT), integrated along the path. One does not measure directly the density of particles, but rather the mean square density of the particles. So there is an unknown "clumping factor", which is the mean square density divided by the square mean density. Then the mass of gas inferred from the observations is 4 x 10^{14} solar masses divided by $\sqrt{C}$. If C is greater than 1, so that the stuff is clumped into patches of higher density, then the gas is more efficient in emitting x-rays, and one can get by with a smaller amount of gas. The larger C is, the smaller the amount of gas required. As C is greater than 1, the mass is less than 4 x 10^{14}. That's a lot of mass in anybody's book, but the mass of the galaxies alone is somewhat larger. Moreover, the virial mass is 6 x 10^{15}. So the regrettable conclusion is that although this gas has the right properties and it is the kind of thing we are looking for, its mass is less than that in the galaxies, and less than a few percent of the total mass. The Coma cluster cannot be bound by gas of this kind.

The same ideas apply to the universe as a whole: there is a possibility that the universe is bound by gas. Figure 10 represents an attempt to find neutral gas in the universe. This is a spectrum of a quasar whose redshift is two; the wavelengths of the emission lines in the quasar are stretched by a factor of three. The strong emission line here is Lyman α-- the resonance line of hydrogen -- which is ordinarily at 1216 Angstroms. Here it appears at 3600 Angstroms, because of the redshift. If there were intergalactic matter we would

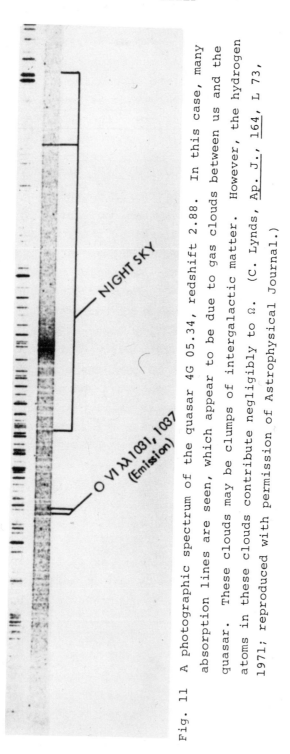

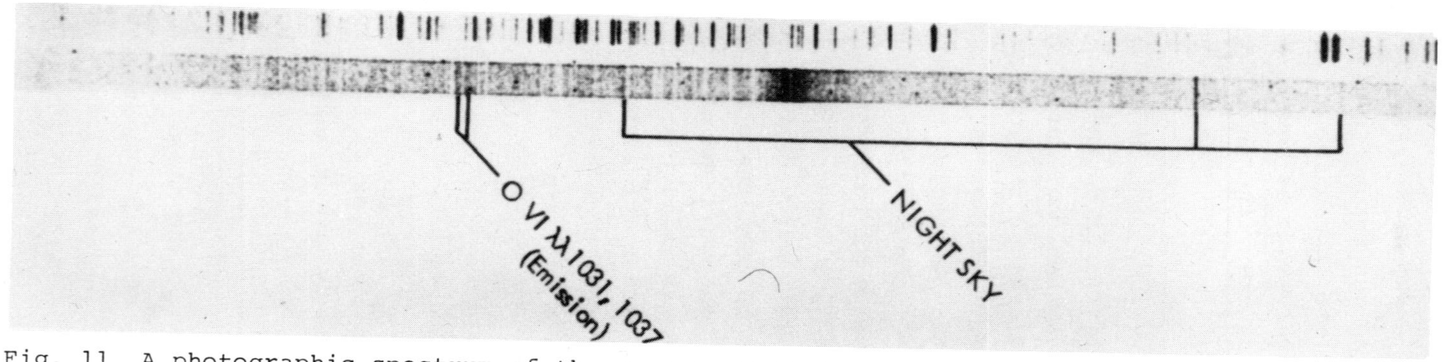

Fig. 11 A photographic spectrum of the quasar 4G 05.34, redshift 2.88. In this case, many absorption lines are seen, which appear to be due to gas clouds between us and the quasar. These clouds may be clumps of intergalactic matter. However, the hydrogen atoms in these clouds contribute negligibly to Ω. (C. Lynds, Ap. J., 164, L 73, 1971; reproduced with permission of Astrophysical Journal.)

expect to see it in absorption at velocities less than that of the quasar. We do not see any effect. The spectrum is very irregular, but there is definitely no effect which causes the quasar light to plunge down to zero and stay there. This experiment has been done on many quasars. Figure 11 is the spectrum of another quasar with Lyman α showing up prominently at 4700 Angstroms. This quasar has the largest redshift known, 2.88, so the wavelengths are stretched by a factor of four. There is a whole series of absorption lines in the spectrum. There are 93 absorption lines concentrated wavelengths shorter than Lyman α. John Bahcall has interpreted about half of those as due to several discrete redshift systems, as if there were patches of matter between us and the quasar through which the quasar is shining. He cannot identify the other half, and they could well be Lyman α. It may be that this is what we have been looking for -- intergalactic gas. If so, it is very patchy. If you add up the total amount of matter in these patches, it is far less than the amount required to bind the universe. That is, Ω(HI) for the universe is less than 3×10^{-7}. If Geoff Burbidge were here he would argue that these experiments are meaningless because the quasar redshifts are not cosmological. There have therefore been attempts to find HI by means of 21-centimeter absorption in objects whose redshifts are definitely cosmological. Figure 12 is a spectrum of the 21-centimeter line as seen against a powerful radio source. Here is the dip that you would expect to get if the universe were closed by neutral hydrogen, while this is what you actually get from the experiment. Therefore, Ω(HI) is less than 0.16, even if the quasar data is not valid. The search for HI is extremely negative if you believe the quasar story, and it is marginally negative if you do not believe the quasar story.

Now let me come to the one positive indication that the universe is in fact closed by gas. The x-ray astronomers discovered early in the game not only individual x-ray sources (which are now thought, some of them, to be binary stars or neutron stars) but also a diffuse background which glows over the entire sky. Figure 13 shows the spectrum of the diffuse x-ray background. On a plot of log (intensity) against log

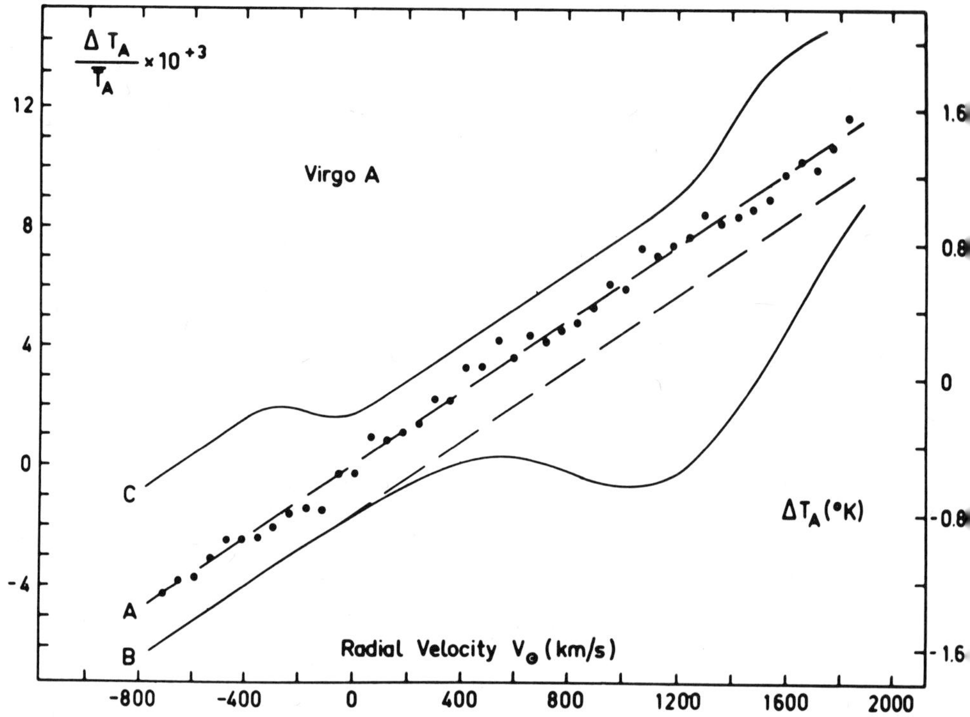

Fig. 12 A spectrum of the radio galaxy M87 in the vicinity of the 21-cm lines. The upper line shows the absorption trough expected if intergalactic HI contributes significantly to Ω. The data points indicate that it does not. (R. Allen, Astronomy and Astrophysics 3, 316, 1969; reproduced with permission of Astronomy and Astrophysics).

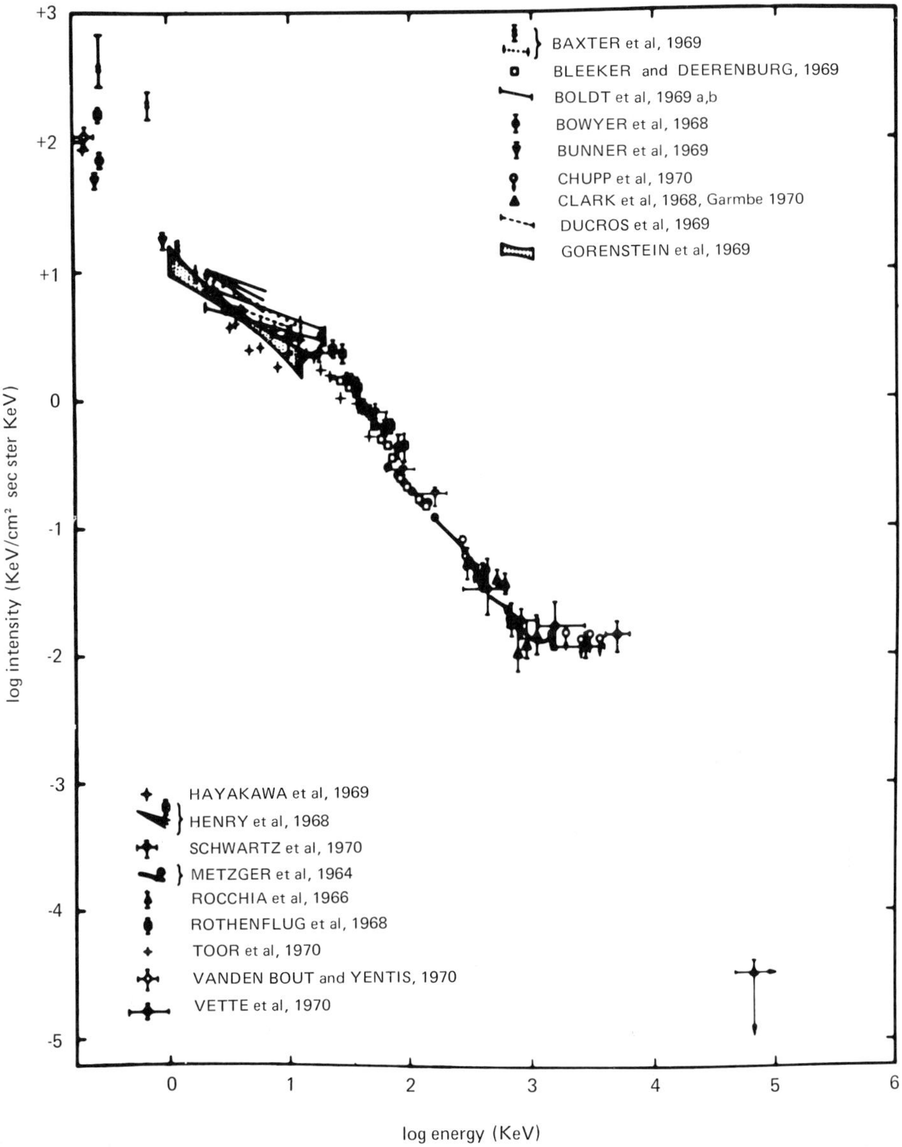

Fig. 13 The observed spectrum of the x-ray background. Data above about 1 MeV are controversial. (J. Silk, Lectures at the Institute of Scientific Studies, Cargese, Italy, July, 1971; reproduced with the permission of the author).

(energy) it has a slope of approximately -1. It does have a number of curious features, however. In the subkilovolt region it goes above the power law by approximately a factor of 5. In the region around 40 kilovolts there seems to be a change of slope from a small one at low energies to a steeper slope at high energies. These features may be telling us something about intergalactic matter. In Figure 14 we show some calculations which are put on to the same diagram with the observations. The dark line is an indication of what you would get if the universe were closed by gas with a temperature, in this case, of 2×10^8 degrees. You see the rather amaxing thing that there is a rough agreement between the intensity of the radiation that is observed and what is predicated on that basis. The spectrum does not fit very well, however. Another line indicates that if the universe is closed by hot gas, the radiation from it must exceed this line at some point, whatever its temperature. From that one can conclude that the temperature of the gas cannot exceed a certain amount, which turns out to be about 3×10^8 degrees. You see that in fact the radiation exceeds this line by small amounts, so the conclusion is that maybe the universe is bound by hot gas. From all this we find that Ω(HII) is less than or equal to $C^{-\frac{1}{2}}$. If the clumping factor were unity, Ω must be less than one, but only barely. It may be just equal to one. Ram Cowsik, a cosmic-ray physicist at Berkeley, has worked out a composite model of the x-ray background in which he explains these observational points. He works out the inverse Comptom mechanism that explains the low-energy points, he subtracts that off from the others and is left with a residue which is plotted in Figure 15. One finds that the residue fits very well on a bremsstrahlung curve for 3×10^8 °K. We must bear in mind the possibility that $C > 1$ and $\Omega < 1$, however.

Let's explore the possibility that the universe is closed, $\Omega = 1$, and $T = 3 \times 10^8$°K. This corresponds to an energy density, 3/2 NKT, equal to 0.25 eVcm^{-3}. The energy density of the cosmic blackbody radiation (T = 2.7°K) is also equal to 0.25 eVcm^{-3} -- an interesting coincidence. The trouble with this model, as I see it, is the tremendous energy that it

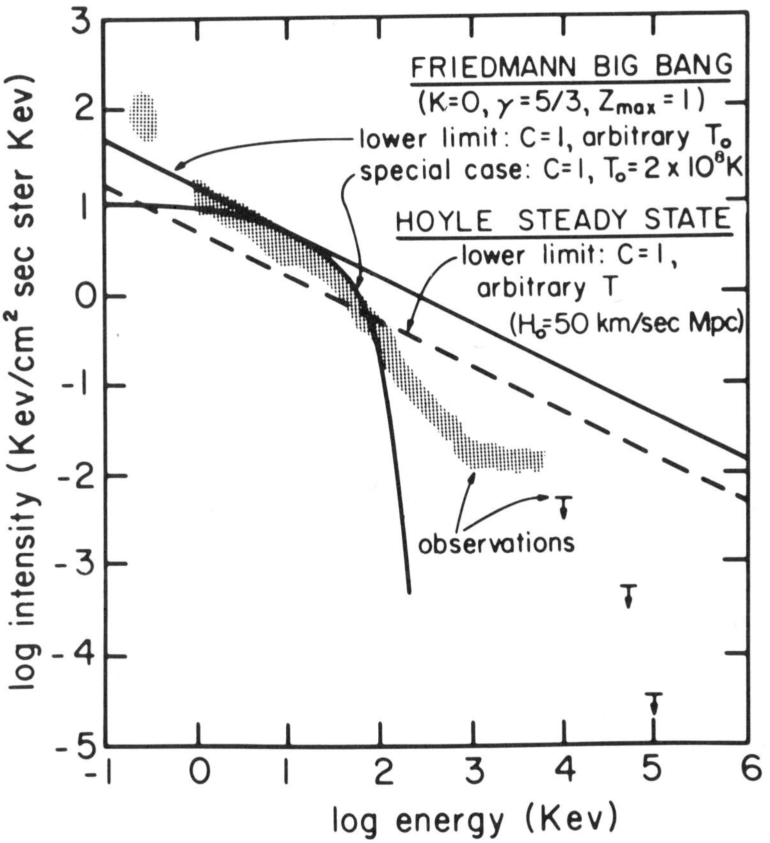

Fig. 14 Comparison of calculated thermal bremsstrahlung spectrum for $T = 2 \times 10^8$ °K, $\Omega = 1$ with the observational data copied from figure 13. Note the order-of-magnitude agreement in intensity, and the possible explanation of the break in the observed spectrum at 30 keV. Whatever the temperature, the observations must reach or exceed the straight line labelled "upper limit", if Ω is ≥ 1. (G. Field, Annual Reviews of Astronomy and Astrophysics, Vol. 10, 1972; reproduced with the permission of Annual Reviews).

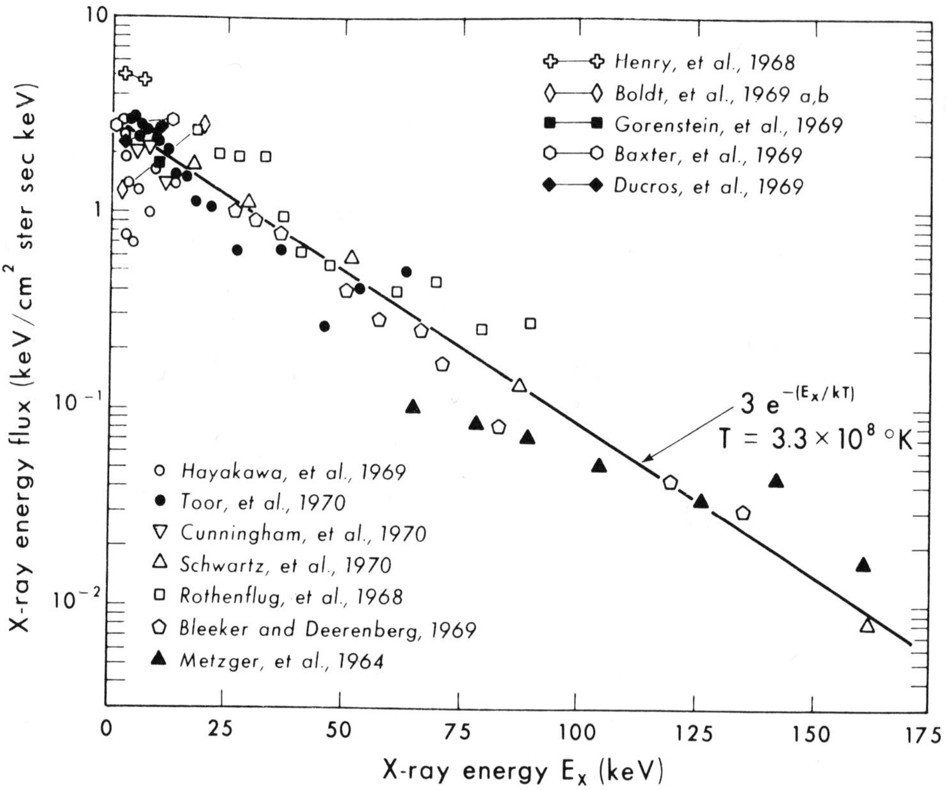

Fig. 15 Comparison of observed x-ray intensitites, corrected for contributions by the Galaxy and by radio galaxies, with a thermal bremsstrahlung spectrum of 3.3×10^8 °K. The intensity implies $\Omega \simeq 1$, in agreement with figure 14, if the gas is smoothly distributed. (R. Cowsik and E. Kobetich, to appear in the Astrophysical Journal, November, 1972; reproduced with the permission of the authors).

take to heat the intergalactic matter to such a high temperature. Because galactic matter is the only source available for accelerating particles and heating the gas, we are led to compute the energy density in the gas, divided by the rest-mass density of galaxies. We find that in order to heat the gas to a very high temperature, we would have to take 1% of the rest mass of the galaxies, put it into relativistic particles, have those relativistic particles go into space and heat the gas to the high termperature. Is that possible? I don't know, but there is now indication that every galaxy may go through an explosive phase something like a quasar. The magnitude of energies that are involved in a quasar explosion are not that different from what we need to heat the intergalactic gas.

In summary, we know that galaxies and neutral gas are inadequate to bind the clusters or the universe. Hot, ionized gas seems like a better bet, and in fact may occur within clusters of galaxies according to recent x-ray background observations, but whether it is enough to bind the universe depends upon the clumping factor, which is unknown. Both in clusters and outside of them, these would have to be tremendous energy sources like quasar explosions to heat the gas to the very high temperature required by the x-ray observations.

EVOLUTION OF IRREGULARITIES IN AN EXPANDING UNIVERSE*

P. J. E. Peebles
Princeton University

In the standard cosmology the Universe is described by a highly symmetric model - homogeneous, isotropic, and uniformly expanding. This picture has enjoyed a modest success[1], but it does neglect the obvious detail that the observed matter is not uniformly distributed; it is concentrated in galaxies, and the galaxies themselves are distributed in a highly organized fashion.[2] It seems clear that this irregular organization of the matter has something fundamental to say about the nature of the Universe, and it is an important challenge to cosmology to account for it. I will be describing below a few aspects of the phenomenon and a possible phenomenological picture for the development of irregularity in an expanding universe.

The positions in the sky of the brighter galaxies (apparent photographic magnitude ≤ 13) for most of one hemisphere are shown in Fig. 1. This graph was plotted from the data in the Zwicky catalogue of galaxies and clusters of galaxies. The graph is centered on the North pole of the Galaxy, the circles corresponding to angles of 30°, 60° and 90° from the pole. The galactic latitude is plotted so that equal areas in the sky correspond to equal areas on the graph. There are no galaxies in a strip on the lower right hand part because the catalogue does not cover that part of the sky. There are few galaxies near the largest circle, galactic latitude zero, because in this direction we must look through the obscuring interstellar dust in the disc of the Galaxy. These complications aside, it is apparent that the galaxies are far from uniformly distributed. There is a strong concentration

*This research was supported in part by the National Science Foundation.

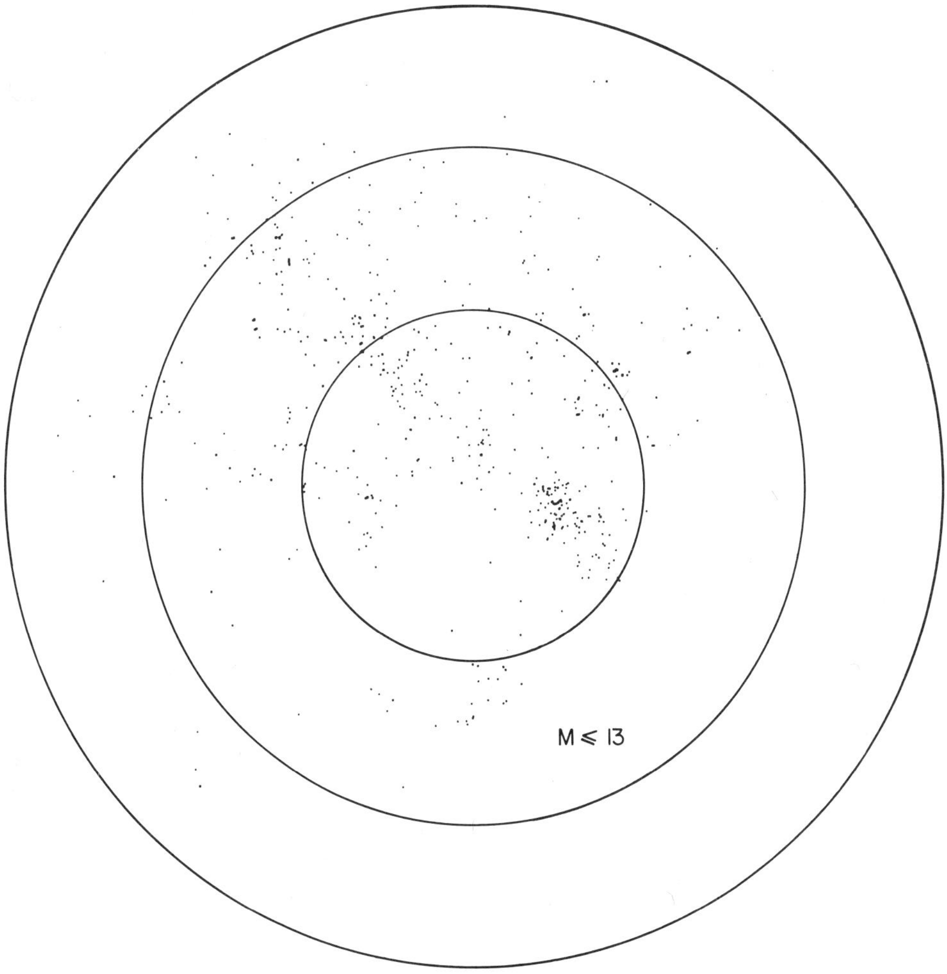

Figure 1. Positions of the Bright Galaxies ($m \leq 13$) from the Zwicky Catalogue. The graph is centered on the North Pole of the Galaxy and the circles are at galactic latitudes 30°, 60°, 90°.

of galaxies toward the right hand side of the inner circle. This concentration is called the Virgo cluster of galaxies. It is roughly 40 million light years away, and its mass is perhaps 10^{14} Solar masses. One can pick out many smaller groups of galaxies. There is a broad band or cloud of galaxies running up and to the left from the Virgo cluster and extending across a good fraction of the sky. There also appears to be a strong (and real) gradient in the mean density of galaxies going away from the Virgo cluster, for the abundance of bright galaxies in the other hemisphere, opposite to the Virgo cluster, is lower by a factor of two.

Figure 2 shows the view out to a larger distance. This graph shows all galaxies down to a limiting brightness a factor of ten lower than the galaxies in Fig. 1. The galaxies in Fig. 2 are typically at distance $\sim$ 200 million light years.

The Virgo cluster still is a prominent feature in Fig. 2, and one can see in the plot several other more distant massive and compact clusters of galaxies. Very near the center of the graph is the Coma cluster, perhaps 300 million light years away. The mass of this cluster is thought to be about 10^{15} Solar masses.

Irregularities on scales larger than the Coma cluster seem not to be so prominent. One receives the impression that on distance scales larger than about 10^{8} light years, encompassing masses substantially greater than 10^{15} Solar masses, the strong irregularities wash out and the standard symmetric cosmological picture may be a reasonable approximation.

The challenge to cosmology is to find an account of the origin of this structure. One line of thought is that the structure is a natural consequence of the gravitational instability of an expanding universe -- the inevitable small irregularities in the mass distribution in time grow into big irregularities like what we observe. Interestingly enough, one can also find quite the opposite view, that the universe began in a state of primeval chaos and grew more uniform. There is no scarcity of intermediate views.[3]

I presume that this question, whether an evolving expanding general relativistic cosmological model grows more or less irregular as times goes on, has a definite answer, and

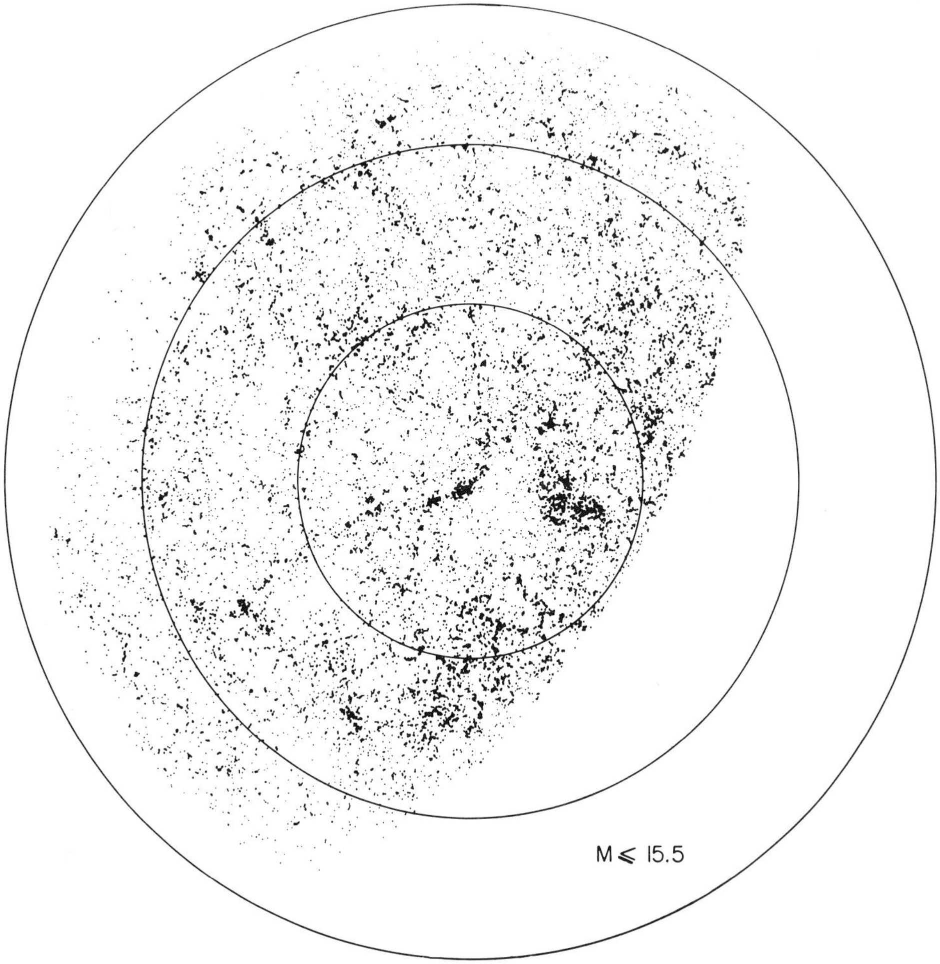

Figure 2. Positions of Galaxies ($m \leq 15.5$) from the Zwicky Catalogue.

that someday we will learn it. Meanwhile we can find gainful employment by attempting to construct naturalistic phenomenological models of evolution of irregularities within the framework of some chosen world view. By phenomenological I mean that the model should be constructed to evolve toward a state approximating what we observe about us. By naturalistic I mean that the model should commence from initial conditions that do not seem artificial or contrived, and should evolve according to a realistic picture for the properties of matter and the laws of physics. The degree of success of this program is one measure of the merit of the world view on which it is based.

I think the gravitational instability picture is the most reasonable first choice for an attack along these lines because it is simplest (fewest parameters, least detailed or involved chain of events). What is more this picture does admit some interesting models of the evolution of irregularities. I have in mind numerical computation of the motion of N bodies (N a few hundred) under the action of their mutual gravitational attraction. Examples are shown in Figures 3 and 4.

The model in Figure 3 is designed to mimic a possible course of evolution of the Coma cluster of galaxies (the dense knot close to the center of Fig. 2).[4] Figure 3-a represents an imagined time when protogalaxies had formed but the cluster had not. The 300 points (each with the same mass in this model) are distributed at random within a three-dimensional sphere. To represent the general expansion of the universe each point is given velocity pointing away from the center of mass, speed directly proportional to the distance from the center of mass. This is Hubble's law,

$$v = HR \quad . \tag{1}$$

One must imagine that the space outside the sphere is similarly populated with point masses. In computing the motion of the 300 points all this exterior mass is ignored. This is a realistic approximation, for to the extent that the exterior world is disposed in a spherically symmetric fashion around

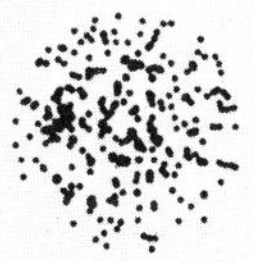

g. 3(a) Numerical Model for the Formation of the Coma cluster. When model is scaled to the observed dimension and velocity dispersion in the Coma cluster the model age is $t = 7 \times 10^8$ y.

Fig. 3(b) Numerical Model for the Formation of the Coma Cluster. When the model is scaled to the observed dimension and velocity dispersion in the Coma cluster the model age is $t = 1 \times 10^9$ y.

Fig. 3(c) Numerical Model for the Formation of the Coma cluster. When the model is scaled to the observed dimension and velocity dispersion in the Coma cluster the model age is $t = 6 \times 10^9$ y.

the sphere it can have no influence on the interior points (this is the relativistic analogue of Newton's theorem about the gravitational field within a hollow iron sphere).

To make a gravitationally bound cluster of galaxies, I assume the 300 points represent a growing density irregularity. Accordingly, I have fixed the constant of proportionality in equation (1) such that the system has negative total energy, and such that, if the system expanded in a homogeneous fashion, it would expand by a factor of about 10 in radius, stop, and then collapse. The figure shows that the actual course of evolution is a good deal more complicated than this. What is more, one finds that all the complicated detail shown in Figure 3-b is quite model-dependent. With a different set of initial random positions the picture looks quite different. However, the models always seem to end up looking like Fig. 3-c, with a compact core of galaxies surrounded by a diffuse halo, rather similar to an isothermal gas sphere but with particle orbits that tend to be radial so that the observed (line of sight) velocity dispersion decreases with increasing distance from the cluster center. The interesting thing is that this final state is a reasonable approximation to the observed features of the Coma cluster (for more details see ref. 4).

What are we to conclude from this? Certainly not that we have a theory of the origin of clusters of galaxies, because I adjusted the model by hand to make a bound system, and the physical parameters, mass and linear dimension, were scaled to fit the observations. However, the model does end up looking like the Coma cluster of galaxies, and this does not require nice adjustment of initial conditions. We do have a naturalistic phenomenological model.

The model in Figure 4 is designed to evolve into something that mimics what we see in Fig. 1. In making the model I have been guided by an opinion on what the distribution of brighter galaxies looks like. The two suggestive features are:

1) there is a prominent band of galaxies running across the Virgo cluster and across a good fraction of the sky;

2) the abundance of bright galaxies is lower (by a fac-

Model for the Formation of the Local Supercluster. The law of attraction among mass points is given in reference 4.

Fig. 4a (see text)

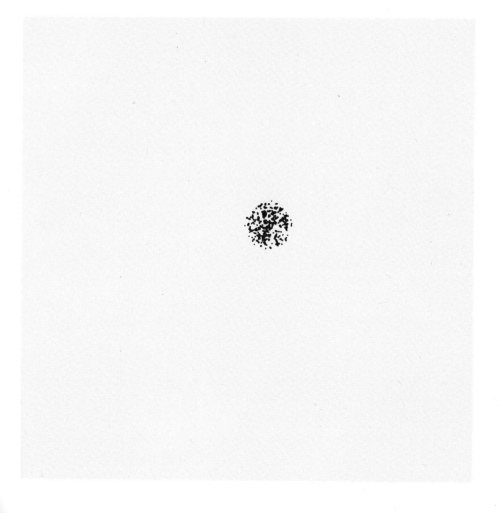

Model for the Formation of the Local Supercluster. The law of attraction among mass points is given in reference 4.

Fig. 4b (see text)

Model for the Formation of the Local Supercluster. The law of attraction among mass points is given in reference 4.
Fig. 4c (see text)

tor of two) in the hemisphere opposite the Virgo cluster. Following de Vaucouleurs[5], I assume this means that there is a Local Supercluster of galaxies, a local concentration of galaxies very roughly spherical and centered on the Virgo cluster, and extending out to our position near the edge of the concentration. This local concentration of galaxies apparently is expanding, for one observes that the line of sight velocity component is away from us for most of the galaxies in Figure 1, and that the line of sight velocity increases roughly in proportion to distance from us.

Now let us consider a further speculative point of interpretation. Let us assume that the Local Supercluster is a proto-cluster, a density irregularity that has grown and become prominent, but has not yet reached the point where the irregularity starts to collapse toward a steady state like that shown in Figure 3-c. That is, we imagine that the Local Supercluster might look something like Figure 3-b.

Since the nearby galaxies are moving apart at roughly the rate of the general expansion of the Universe, it appears that the kinetic energy of expansion of the Local Supercluster is not much less than the potential energy. Therefore, I choose the initial value of H in equation (1) such that the kinetic energy of expansion is just equal to the potential energy. There are again 300 points, each with the same mass, and the points are distributed at random within a sphere (different random numbers from Fig. 3).

Figure 5 gives an instant replay of the evolution of this same model, but this time with the scale on each plot adjusted so that, if the distribution expanded in a homogeneous and isotropic fashion, the points would not change position from one plot to the next. The growth of the initial chance irregularities in the random distribution is very apparent. By the time we get to Fig. 4-c we have indeed something that looks reminiscent of Figure 1. There happens to be a very prominent band of points across the cluster (although in views from other directions neither band nor plane is apparent), and we can pick out numerous large and small groups of points. I suppose that in a more realistic model some of the tight groups would coalesce into single galaxies. When the

Model for the Local Supercluster. This is the same as Fig. 4 except the scale has been adjusted to take out the general expansion.

Fig. 5a (see text)

Model for the Local Supercluster. This is the same as Fig. 4 except the scale has been adjusted to take out the general expansion.

Fig. 5b (see text)

Model for the Local Supercluster. This is the same as Fig. 4 except the scale has been adjusted to take out the general expansion.

Fig. 5c (see text)

evolution is followed well beyond the point shown here one finds typically that part of the original system collapses to a fairly massive cluster, the rest leaves with positive energy as individual points or smaller groups. Models based on different randomly chosen initial positions look quite different in detail as they evolve, but the general features remain.

One might entertain two very different objections to this model approach. First, the model in Figure 4 has expanded by a factor of 15 in radius in going from graph a to c, and in the process it has become very irregular. Since the Universe has expanded by a factor $\sim$ 1000 in radius since the Primeval Fireball released its grip on the matter, would one not expect the distribution of galaxies to be even more disorderly than what is observed in Figure 1? The answer is no, because nothing forces us to make the initial distribution random on the mass scale of galaxies, as was done in the model. With a smoother initial distribution things would take longer to happen.

A few years ago it was the custom to carry this latter point to the bitter extreme and start with the individual atoms randomly distributed, so that the fractional density contrast on the scale of big galaxies, containing perhaps 10^{68} atoms, would be very small, $(\delta\rho/\rho)_i \sim N^{-1/2} \sim 10^{-34}$, and it would be a long time indeed before anything interesting happened.[6] This is a valid point but to my mind not a very profitable one. The fractional density irregularity generally grows as a power of time (when $\delta\rho/\rho < 1$),

$$\delta\rho/\rho \sim (\delta\rho/\rho)_i \ (t/t_i)^n \quad ,$$

with n on the order of unity. If $(\delta\rho/\rho)_i$ is small all we have to do is choose the starting time t_i small enough. I think the real embarrassment is that there is nothing to prevent us from taking t_i smaller yet, thereby quite messing up the Universe before it can expand to the present epoch.

Until we can come to grips with this problem it is hard to see how we will be able to make an <u>a priori</u> theory of the origin sturcture. But I am encouraged by the models presented here to think that we may be able to construct a naturalistic

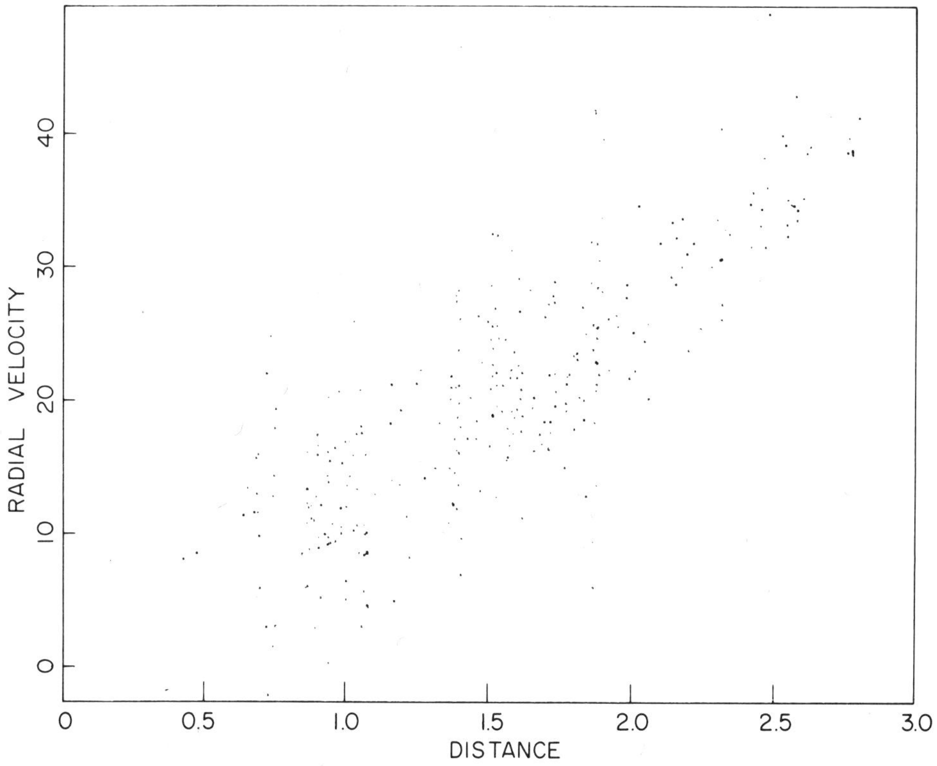

Fig. 6 "Hubble Plot" of Line of Sight Velocity vs. Distance. This graph applies to model 4c, and the distances and velocities are measured relative to an observer near the right hand edge of the distribution in Fig. 4c.

phenomenological model for the evolution of objects like the Coma cluster or the Local Supercluster based on the simple gravitational instability picture. Of course one can make more quantitative tests of the model, and M. Geller and I are working on this. I might report here on one interesting point, the development of peculiar velocities.

As irregularities develop, the points plotted in Figure 5 move from a random distribution into groups, and this motion represents a growing deviation from the simple law of general recession given by equation (1). Figure 6 shows a plot of line of sight velocity vs. distance of points as measured by an observer situated near the right hand edge of the distribution shown in Figure 4-c. In the initial state this would have been a smooth straight line (eq. 1). The scatter in the graph represents ihe peculiar velocities that developed along with the irregularities. One aspect of the peculiar velocities is that each separate group of points ends up with non-zero angular momentum about its center of mass. This angular momentum about its center of mass. This angular momentum is at least the right order of magnitude to account for the rotation of the galaxies, if one assumes that the galaxies formed by the same instability process that formed the groups of points in the numerical model.[7]

What I have been describing here is the relatively modest goal of deciding how large-scale irregularities may have been evolving in the fairly recent past. This should be an easy problem because the dominant effect is gravitation and it can be described by the Newtonian approximation. The results so far are encouraging because the models look rather like the real world. The simple-minded gravitational instability picture seems naturally to operate in an expanding universe and to produce natural-looking results.

References

1. For a recent survey see P. J. E. Peebles, Physical Cosmology (Princeton University Press) 1971.
2. This is very well described by J. A. Oort, Solvay Conference on the Structure and Evolution of the Universe, 1958, p. 1.
3. For a summary and references see P. J. E. Peebles, Comments on Astrophysics and Space Physics, 4, 1972, and B. J. T. Jones and P. J. E. Peebles, Comments on Astrophysics and Spa e Physics, 4, 1972.
4. P. J. E. Peebles, Astron. J., 75, 13, 1970.
5. G. de Vaucouleurs, Astrophys. J. Suppl., 6, 213, 1958 and earlier references therein.
6. This view is described by D. Layzer, Annual Review of Astronomy and Astrophysics, 2, 341, 1964.
7. P. J. E. Peebles, Astron. and Astrophys., 11, 377, 1971 and earlier references therein.

PANEL DISCUSSION

WEINBERG: I would like to begin by asking the panelists for any comments or questions they may want to direct to each other , and I will then ask for questions of the audience. George?

FIELD: Looking at the two guys sitting there, I would say that we do represent a conservative school, putting it bluntly. There are things in cosmology that haven't been mentioned here because they do not fit the picture that we have been discussing. One is of course the quasars, those objects with very large red shifts. If the red shifts are cosmological, they imply huge distances, and therefore energy productions which are truly fantastic, calling for the most extreme physical conditions, such as relativistic collapse or some new forces that we don't know about. If, on the other hand, the red shifts are not cosmological, as is claimed by Burbidge and Arp, then we have an even greater puzzle which would throw into doubt the whole question of expansion of the universe, which is of course based on the notion that the red shift is essentially a Doppler shift. I think Jim Peebles and I tend to take a similar line on some of these things and Alar clearly is working within the framework of conventional big-bang cosmology. But we have to point out that some things have not been explained at all within that framework. Jim, do you want to comment?

PEEBLES: I think that one of the things that Geoff Burbidge more than anyone else would emphasize is how tentative most of our interpretations are in astronomy. How hard it is to know what is going on, how rich are all the elementary classical theories like Newtonian mechanics plus hydrodynamics plus magnetic fields. How hard it is to interpret any phenomenon from those measely photographic plates that we have at our disposal, how easy it is to romanticize anything you see into something very complicated. I feel that I am much too young to become a reactionary, but it seems to me that it's very strange to make unconventional interpretations of phenomena until you have exhausted the conventional interpretations, and

I have never understood why my respected colleague Geoff Burbidge makes these unconventional interpretations. I might just give one example. George Field described the Coma-cluster of galaxies. This beautiful compact system of galaxies is gravitationally bound, and in dynamic equilibrium, here it is. And yet if you work up the mass needed for stability you get a number in units of solar masses per solar luminosity of around 300. Then you look at the individual objects and you find a number around 30. Something is wrong here. Now it might be that we have got the dynamics wrong, or we have the gravity theory wrong, or that the thing is perhaps really freely expanding. But also what may be wrong is that we have simply underestimated the mass to light ratio of an individual galaxy. The fact of the matter is that most of our understanding of masses of galaxies comes from the spirals, while the objects in the Coma-cluster of galaxies are all ellipticals and S-0's. What information we have on masses of S-0 galaxies is close to nil, on elliptical galaxies is just a couple of cases studied, and there only the mass to light ratio of the central part of the object. We do not know what that stuff is out on the outer halo; I presume it is stars, but I do not think we know what its mass to light ratio is.

<u>TOOMRE</u>: Having been told to ask wise questions, it seems that Field and Peebles have just made some wise statements instead. But I will here ask a <u>stupid</u> question: Whatever happened to Einstein's cosmological constant? I mean, is that really passé? Could one of you who tries to estimate the density out there again rationalize why this extra fudge factor isn't interesting any more?

<u>PEEBLES</u>: I think that the cosmological constant is an abomination for the simple reason that already without it we have more unknowns than measured quantities, so that you can add more parameters to the theory; it does not do you a bit of good.

<u>WEINBERG</u>: I have an overwhelming theoretical prejudice against the cosmological constant, just because Einstein's theory makes sense to me as the classical version of a quantum theory of particles of mass zero and spin 2. You see it is just barely possible that the fields that represent the particles couple

to conserved sources, and the conserved sources by well-known arguments have to include the fields themselves, and you build up the full nonlinearity of Einstein's equations. I don't say that there is any rigor to this argument, but one feels comfortable with this interpretation of gravitation; but the cosmological constant just throws the whole thing out -- the cosmological constant, as far as I know has no interpretation in terms of a quantum theory of mass zero, spin 2.

FIELD: I was going to ask you something which you are an expert on, concerning neutrinos. It has been argued that the universe is filled with degenerate neutrinos which make $\Omega = 1$. I know that at one time you worked on that and maybe you would like to comment on it.

WEINBERG: When I made that argument, those were in the good old days, of a cold universe, before the black body radiation had been discovered. It seemed natural then not to take into account the fact that the neutrinos would thermalize; and as you know it was first pointed out by Zel'dovich, if the neutrinos reached thermal equilibrium and then the universe expanded and all sorts of things happened, they would now have a temperature of 1.9° Kelvin, which is $(4/11)^{1/3}$ of the 2.7° Kelvin. I think that this is a definite prediction which may be true, and it is one of the great challenges to bright people to try to think of some way of detecting these. There is about one neutrino per photon in the universe and no one can think of any way of detecting it. There are strange things about neutrinos, and most recently there has arisen the strange problem of the solar neutrinos. Bahcall, Cabbibo and Yahil have suggested that perhaps the neutrinos have a little mass and that they decay into other particles and that that is why we don't see any neutrinos from the sun. Now, while a 1.9° Kelvin neutrino background would not help at all in closing the universe, if neutrinos had a little mass of the order of 20 electron volts then that would provide plenty of energy to stabilize the universe. You would have about as many neutrinos as you have photons (that wouldn't change), but the photons would go on being redshifted down to their present mean energy which is about 10^{-4} electron volts. The neutrinos, when the energy dropped to about 20 electron volts, could no longer be redshifted because the rest mass would take over, so that there would be an energy density of neutrinos of the order of 10^5 times the energy density of the black body radiation, which would be more than enough then to stabilize the universe. So I think if you take the neutrino masses somewhat less than the best limit on the neutrino mass, then the expected population of neutrinos provides more than enough energy to stabilize the universe. But we simply don't know what the mass of the neutrino is.

I would like to ask a question which is related to things that almost all of you said. There are very peculiar objects-- Arp has written recently a very interesting article in Science about them -- objects in which you have radio sources like beads on a string, galaxies with very strange nuclei, exploding galaxies. I'd like to ask whether or not, leaving the quasars aside, you think that these objects can be interpreted in some rather 18th century way that Alar was discussing, or whether these also require a new kind of physics?

TOOMRE: I suppose I am a reactionary. If put to the test, I would guess that at least 90 per cent of all the odd fellows in Arp's Atlas of Peculiar Galaxies, for instance, must have quite mundane explanations. And yet, there also exist pictures and situations that make one feel quite humble. One such example is the jet of M87. Another is the chain of five galaxies which Sargent studied a few years ago. The redshift of one member of this chain differs by some 20,000 km/sec, as I recall, from that of the others. Now this could be a projection effect, due to a lucky line of sight. But as Arp is especially fond of pointing out, there are other such "coincidences" as well. I would have to say that collectively they are worrisome.

FIELD: I would be surprised that anyone could come up with a rational explanation within the framwork of Newtonian physics on such objects. Even relativity seems in question, with factors of ten or even a hundred in redshifts. The object is either moving away, out of the region at a very high speed, in which case we don't understand why it is still there, or, on the other hand, if it is a gravitational redshift associated with the object itself. When we go into the physics of this, with the magnitude of gravitational redshift that one would need, one can't explain the other properties of the object. So we keep running into these brick walls in trying to understand how an object of low redshift is associated with an object of high redshift. It is just a very, very, puzzling phenomenon.

TOOMRE: Things at very different distances could be overlapping.

FIELD: That's right, but the number of these surprising events is increasing to the point that it is becoming hard to believe that.

PEEBLES: You showed a photograph earlier of the compact group of galaxies called the Stephan Sextet. Perhaps I could start with the remark: statistics is a subtle subject. I can tell you what my license number is and then I can say, "what a coincidence it turned out to be that number." You scan large areas of sky and look for the most peculiar thing in that area and then say "My God, that could never be an accident!" But the trouble is that if you look hard enough you will find just about anything you care to see. The full difficulty in assessing the reality of these aberrant redshifts is to know which area of the sky has been searched thoroughly down to some well-defined a priori limit of brightness and compactness of objects, and what, in that whole area of sky surveyed, is the probability of getting one aberration. We don't know any of these numbers and they are awfully difficult to estimate.

Now this is a beautiful example of many of these problems that face us. Five of the members of the sextet have redshifts that were measured by Sargent and one is so faint that one can't get its spectrum. Four have very nearly the same redshift, and three are compact elliptical galaxies of a type that is not too common, but don't look too badly off, one looks like a loose spiral galaxy, and one has a redshift that is larger by some 10,000 km/sec. So if this last one is indeed a part of this dynamic system, we are really embarrassed.

So that is a good example of the sort of thing that Dr. Arp has been talking about. How did it come about that this guy so close to the others has such a very different redshift? It surely cannot be dynamic -- it has to be some bad news in the laws of physics or it's got to be an accident. The problem is that we don't know whether it's an accident or not.

I'd like to make some remarks that perhaps Dr. Toomre could answer. Suppose we take away the maverick. Then let's ask ourselves a question that was first asked, I believe, by Ambartsumian, namely, can you imagine that a system of four or five objects like this can remain in a tightly bound system for 10^{10} years, given that the crossing time for one to zip back and forth in the system is $\sim 3 \times 10^8$ years, so that the system is well-mixed? Would this system stay looking the way it is for 10^{10} years? The naive answer is, "No, it would not." The whole system would relax. Two of the systems would form a tight binary and would kick out the others. In fact, numerical models suggest that this is about what happens. Can we understand why these systems have remained so close and uniformly distributed, not relaxing according to the ordinary laws of Newtonian mechanics, for 10^{10} years? And can we understand why they didn't amalgamate, because they are in rather lossy systems? Now maybe this is a type of interaction that happened only once. But it does worry me that this system, even ignoring the maverick, could have lasted for 10^{10} years. I would not like to make each of the galaxies a maverick. I would hope that this is really a compact system in dynamic equilibrium, but I just don't see how it could be.

TOOMRE: Once Peebles removes his maverick from the Stephan Sextet, I think it just possible that one can weasel out of the time scale problem as well. For one thing, the remaining five could be a rather young bunch of galaxies -- that's a common enough excuse. And it would also help if the cluster itself were more disklike than spherical. In such a revolving system, I have the impression that galaxies don't get tossed out quite as fast as they would in the thoroughly three-dimensional case. But I am only offering alibis, not solid excuses.

WEINBERG: I'd just like to say first a word about the reactionary flavor of this little symposium. I am just going to repeat what Jim Peebles said in different words. I think there is a certain conservatism which is a function of the amount of experimental data which is appropriate. When you have very little experimental data then you take the most simple assumptions and try to work them out with the known laws of physics and see how far you can go. As the data improves, the very simplest ideas don't prove satisfactory and you can go to slightly less simple ideas. It seems to me that the level of conservatism which is represented by the talks today is highly appropriate for the kind of data which is now available and as our knowledge of the mass of the universe and structure of the universe improves, undoubtedly less conservative ideas will begin to be appropriate because they will begin to be things that can be confronted with data. Now I think it is about time to ask if there are any questions from the audience. Wally?

GREENBERG: Weber's experiment to detect gravitational radiation, which might be relevant to cosmic evolution, has not been mentioned so far. Has Weber detected 1600 Hertz gravitational radiation? If so, where does it come from? What can we learn from Weber's observations?

FIELD: Well, I think that most of you have seen Weber's publications in Physical Review Letters. It's very difficult to see how the error was made, if there was an error, so at least on face value it seems that there may be some kind of external radiation source that varies with sidereal time. As you remember he gets activity particularly in one direction; he has a directionality which is awfully hard to explain away. On the other hand, one can argue along this line: that if the radiation originates within the galaxy -- and you can argue that it probably would from the sidereal time variation -- then, knowing something about the sensitivity of his detectors, one can then conclude something about the flux of radiation and thus the total power (given a distance to the source). For example, if the energy is coming from the center of the galaxy, which is what seems to fit the data best, it turns out that one has to annihilate of the order of a hundred solar masses per year. Now the total mass of the galaxy is roughly 10^{11} solar masses, so this says that the time scale for a significant fraction of the entire mass of the galaxy to go into gravitational radiation is only 10^9 years. That is only a tenth of the age of the galaxy. One finds that the dynamics of the galaxy is affected by mass loss of this kind. If the gravitating mass in the center of the galaxy is decreasing, this will have an effect on the stellar orbits in the outer parts of the galaxy. The galaxy expands, and from the observations one can put an upper limit on such an expansion. It turns out that the upper limit is about equal to the value that Weber would derive from his observations, about 100 solar masses a year. It is a very good question, Wally, because if one follows this chain of reasoning to its conclusion, one would suggest that gravitational radiation is a significant factor in the universe, so much so that there may be more energy in gravitational waves than there is in ordinary matter. Such a phenomenon is then related to the omega question. Is the Universe perhaps bound by these gravitational waves? It is very difficult to understand how one gets a hundred different bursts per year, as Weber does. This is totally contrary to ordinary ideas of stellar evolution, because the number of stars per year that are getting into a fix where they might emit gravitational radiation is down a factor of a hundred, probably a thousand. So it is another one of these cases where experiments, which are tentative because they have not yet been confirmed by other groups, point to rather drastic changes in the picture that we have been discussing today.

WEINBERG: I'd like to see Weber give an automatic definition of what his coincidences are. If he has data on automated coincidences, I await that with interest.

ROHRLICH: Listening, as an outsider, to this morning's talks on cosmology I was struck with the paucity of solid input data. At the risk of making things worse I want to raise a question which has had fluctuating popularity in the recent history of physics: if the "universal constants" have in fact been changing at early times, how can one unravel the early history of the universe?

PEEBLES: We have some information. We know we have samples from the time the solar system formed, for example. And we have radioactive decay of meteorites from uranium decay, potassium decay, argon decay, others, going by different processes, giving concordant ages. Over the last four and half billion years, the atomic physics parameters and subatomic physics parameters haven't been grossly changing. Also one can look at spectra of very distant quasars from distant galaxies and see fine structure lines in the spectra at the right spacing. So things for the atomic parameters don't seem to be grossly varying.

FIELD: But Jim, what about the results of your colleagues, Bob Dicke and others, who have been working on scalar-tensor gravitation? Such theories would have the gravitational constant changing, and that would again change much of what we have been saying this morning. As far as I know, the experimental results from tracking spacecraft near the sun and looking for retardation of light effects which are different in the scalar-tensor theory of gravitation and ordinary Einstein relativity are not really definitive yet. However, there are a number of secondary theories of gravitation -- one of them being Whitehead's theory -- which were in fact put to the test and were found lacking because of some of these recent experiments. In fact, there is a preprint circulating which eliminates half of last year's viable theories of relativity on the basis of such experiments.

WEINBERG: Well, there have been observations by Shapiro of the present rate of increase or decrease of the gravitational constant, not inferred from tests of the Brans-Dicke theory but referred directly, and they put an upper limit on $\dot{G}/G$ which is comparable to the value that would be present in Dirac's cosmological theory in which, as I recall, G changes like $t^{1/3}$. However, it is no where near good enough to rule out the rate of change of G that is expected in Brans-Dicke theory. As to Whitehead, I haven't even heard of that theory, so I can't answer that.

FELD: Are galaxy interactions capable of accounting for most of all of the observed spiral galaxies, or is there some other mechanism operating on isolated galaxies that can account for the spirals in a convincing fashion?

TOOMRE: Here is a lovely instance of a physicist seeking simplicity, but nature does not oblige! I have almost no doubt that several spectacular galaxies got their spiral structures from outside. You saw a few of them this morning. Yet it seems equally clear that almost every respectable barred galaxy, and many a raggedly spiral as well, has somehow managed to do it from within. Many of the latter seem to have no companions, at least none big enough nearby. So I am afraid life is just complicated.

FIELD: Do people understand how isolated galaxies get their spiral structure?

TOOMRE: Well, yes and no. For close to a decade now, my distinguished colleague C. C. Lin has been saying how attractive it is to imagine that self-made spirals are wavelike -- that is, one has a density wave running round and round a disk, and the gas goes through it and forms stars in the right places. In principle this seems far more sensible than any other approach I know of. And yet, we still have only the foggiest notions of whatever caused such waves, and why they haven't disappeared by now.

WEINBERG: Just to force you to the wall Alar, isn't it true that spirals are somewhat more numerous among field galaxies than in rich clusters? Proportionally speaking?

TOOMRE: I think that is true, but there are a number of complicating reasons. One is that in rich clusters the galaxies tend to be gas poor. We know that the very pretty spiral structure is definitely correlated with the abundance of the interstellar gas and junk, from which you create new stars. Many cluster galaxies lack this raw material.

BREIT: I don't remember hearing anything about the complications with the effect of the parts of the galaxy charging up electrically. For one thing, are there estimates on the effects of such charges that might account for secondary phenomena, e.g. photoelectric effect, ejecting electrons, complications of cosmic rays? When you have a body with electric charge that moves such that it produces a magnetic field in a purely statistical way, there would be some probable fluctuation of this magnetic field. Is it not likely that one could explain why the magnetic field has this and this value or average value? It would be interesting to know whether the statistical fluctuation of such a field is involved with what is observed. Another thing is the perturbing effect of such charges on the calculations regarding gravity effects, depending of course on the masses of the objects.

WEINBERG: I think that Field has an answer to that and also a comment on the last question.

FIELD: Many people have looked into this kind of a question and the conclusion always seems to be simple: that for your charges there are always discharges and the reason is that the conductivity of the gases in space is approximately equal to that of copper. In trying to separate charges within a body made of copper you get into trouble. That is basically the problem. The conductivity simply is so high that you can't build up any significant charge separation in the medium.

BREIT: My question is entirely motivated by a past experience. At one time I was at the department of Carnegie Institution of Washington which was concerned with terrestrial magnetism. As you doubtlessly know, at the surface of the earth there is an electric gradient of about 200 volts/meter.

FIELD: Are you talking about the atmospheric electric field?

BREIT: Yes, but the fact that the field remained unexplained for years makes me wonder.

FIELD: Right. I'm not sure what the explanation is. However, it is certainly quite possible to have such a field in view of the fact that the lower atmosphere is not a good conductor, whereas the gases in space are partially ionized and therefore are a good conductor.

I was going to comment on something that Alar said. It is perhaps an indication of the way we think that, he said "well, I don't think I can attribute the spiral arms to companions because in many of the cases that we have looked at there are no companions." But all that one can really say is that there are no visible companions. This does not rule out massive black holes. A star with a mass of the order of that of the sun has no trouble in settling down into an equilibrium configuation which can last for billions of years. Even after that, degenerate electron pressure may keep it from collapsing and if that doesn't work the degenerate neutron pressure can. However, for burned-out objects having a mass greater than some critical value which is of the order of a few solar masses there is no known equilibrium configuation, so a singularity seems inevitable, the reason being that once the gravitational forces get beyond a certain point in strength there is no force that can oppose them. The masses of galaxies are 10^{11} solar masses. If Jim Peebles had continued his calculations to very high densities, it seems to me he would have ended up with assemblages of particles each having 10^{11} solar masses; they would have free fallen into a central body which would in fact be a black hole of 10^{11} solar masses. I challenge the other speakers on this panel to come up with some argument why such black holes do not exist and why they may not in fact be prevalent throughout the universe. In particular, they could be orbitting normal galaxies. I think it is the other way; it is hard to understand how matter always manages to fragment into small droplets, which have bust less than the critical mass for making a black hole, just less than a few solar masses, namely stars. I would argue that maybe that isn't the case, that we are just being misled, and that the universe may be full of black holes which obviously we do not know about for obvious reasons. Maybe Alar's spiral arms are telling us something.

Participants in the Panel Discussion on Astrophysics at the 1972 Coral Gables Conference on Fundamental Interactions in Physics and Astrophysics.

THE DIRAC HYPOTHESIS

Edward Teller
University of California
Berkeley, California

All of us imagine that simple explanations of important and interesting problems in physics really exist: they do not have to be constructed, they have to be found. Past experience bears out this belief to which we are dedicated.

P.A.M. Dirac, our great friend, uses words that are more sharp and more clear on the subject. I should like to attempt to formulate what I like to call the Dirac Hypothesis

Each important question in physics will eventually have a solution.

One of my ingenious and skeptical friends said to me: "Yes, when a solution is found, then and no sooner shall we know that the question was really important." I tend to disagree. The taste and consensus of the physicists probably can decide ahead of time what the really interesting questions are.

Half a century ago, when Professor Dirac made his first magnificent contributions to physics, Dirac's Hypothesis was almost fulfilled in a strong sense. It seemed we still should explain the proton, the electron, and the light quantum. How these interact we knew; what they are and why seemed the only problems left (apart from the mysteries of the nucleus).

Professor Dirac was the first one to open up new problems and to point the way towards a more complex physics which today looks to us almost hopelessly involved. He did that by the relativistic generalization of the equation for the electron which in turn predicted the positron. This started us on our present Odyssey which lasted longer than the time it took Ulysses to get back from Troy to Ithaca (and we have not arrived yet). We are now talking about much more than the protons, electrons and light quanta. We have hadrons, leptons, gravitons and probably other entities.

Professor Dirac himself introduced in 1959 the somewhat surprising commutation relations between the components of the energy momentum tension tensor. Thus he has brought quantum mechanics and the gravitational puzzles closer to one another. This may be one important step towards fulfilling the promise of the Dirac Hypothesis.

We should remember that physics is merely that portion of the interesting part of the world that we have (or imagine we have) really understood. It is Archimedes' fixed point from which we hope to move the world. Indeed, by understanding physics we have understood the principles of chemistry as well. Biology, not yet. Astrophysics, maybe to some extent.

The Dirac hypothesis, I believe should apply to physics and to everything that physics should become. In the usual modest way of a physicist, I want to say that this indeed should include everything that is interesting.

Every great discovery leaves a "poison" that tends to inhibit further development. After Newton everything had to be explained in terms of $\frac{1}{r^2}$. The wave-theory of light gave us the concept of ether. The success of relativity and quantum mechanics fixed in the minds of contemporary physicists the purpose that a universal explanation of all remaining problems must be found. Dirac does not participate in this trend. It was always his style to answer questions one by one with lucidity and with logic.

Participants at the 1972 Coral Gables Conference on Fundamental Interactions at High Energy of the Center for Theoretical Studies, University of Miami.

ZITTERBEWEGUNG OF THE NEW POSITIVE-ENERGY PARTICLE

P.A.M. Dirac
Florida State University
Tallahassee, Florida

I spoke about a new wave equation last year and I would like now to talk about some recent developments of it. At present we don't know how to apply this equation. I am not sure if it applies at all, but it is so closely connected with the successful equations that I feel there must be some application for it.

The equation reads like this:

$$\left(\frac{\partial}{\partial x_o} + \alpha_r \frac{\partial}{\partial x_r} + \beta\right) q\psi = 0 \quad . \qquad (1)$$

Here the three alphas and beta are 4 x 4 matrices which anti-commute with one another; the α's have their squares equal to unity and β has its square equal to minus one. So you see this is exactly the usual relativistic equation for the electron if we dissolve this letter q.

Now the letter q makes a very big difference; q means a column matrix of four elements $(q_1\ q_2\ q_3\ q_4)$. The above mentioned 4 x 4 matrices act on this column matrix in the usual way according to matrix multiplication, and the four q's act upon ψ. ψ has just one component instead of the four components of the usual electron wave equation.

The four operators q that act on ψ don't commute with one another. They satisfy

$$q_a\, q_b - q_b\, q_a = i\beta_{ab}\ ;\ a,\ b = 1,2,3,4.$$

With a suitable choice of β these conditions would be $q_1q_3 - q_3q_1 = i$, $q_2q_4 - q_4q_2 = i$, all other q's commute. The four q's can be looked upon as the dynamical variables describing two harmonic oscillators, so we can look upon our particle as having internal degrees of freedom consisting of two harmonic oscillators.

There is just one ψ in this theory. It is a function of the four x's, x_0, x_1, x_2, x_3 and of any two commuting q's, say q_1 and q_2. We have just one wave function but four equations which it has to satisfy. In general you cannot have one function satisfying four differential equations. It's necessary for certain consistency conditions to be fulfilled.

Well it turns out that of these four equations only three are independent. Still, some consistency conditions must be fulfilled. We find that they are fulfilled provided ψ satisfies

$$\left(\frac{\partial^2}{\partial x_\mu \partial x^\mu} + 1\right) \psi = 0$$

which is the de Broglie equation corresponding to the units $\hbar = 1, m = 1$. So we have a particle of a mass 1 with some internal motion and satisfying the usual de Broglie equation for all values of the internal variables.

Then one finds that the equation (1) is really a relativistic equation. The proof is similar to the proof that the ordinary wave equation for the electron is relativistic. However, there is a further condition, in that the alpha and beta matrices cannot be completely general matrices satisfying the algebraic conditions. They must satisfy the further condition that all their matrix elements are real. It is quite easy to satisfy this further condition.

We may examine the eigenstates of momentum and energy and we find that the energy is always positive. We can make further developments in the theory similar to those of the ordinary electron wave equation. We can introduce a charge-current density expression. One finds that the density is always positive. I dealt with all this last year. The result is that we have an equation which is satisfactory from the mathematical point of view, but we don't have any physical image of what it describes.

To get a physical image it is best to pass to the Heisenberg picture. In the Heisenberg picture we have dynamical equations of motion. Of course in the Heisenberg theory the dynamical variables are noncommuting, but one can ignore non-commutation to a certain degree of approximation and then one has something which one can visualize.

Let us now try passing over to the Heisenberg picture and getting the Heisenberg equation of motion for our new wave equations. Schrödinger did the corresponding piece of work for the ordinary electron wave equation a long time ago, and it led him to the discovery of what he called the Zitterbewegung. He found that the motion of an electron in the Heisenberg picture consists of the classical motion of a particle, moving with the velocity connected with the momentum by the classical formula, plus a certain extra trembling motion. This further motion of the electron results in the velocity of the electron always being plus or minus the velocity of light. It is only the average velocity throughout a number of periods which is the usual observed velocity, connected with the momentum by the classical formula.

What I want to do today is to work out the corresponding Zitterbewegung for the new electron wave equation. Here is a definite problem, and it's just a question of following through well known rules and seeing where they lead to.

The passage to the Heisenberg picture is not as straightforward as it usually is because we have a wave function satisfying supplementary conditions. We have first to understand how to pass over to the Heisenberg picture from a Schrödinger theory involving supplementary conditions.

There are general rules for doing that. I'll just remind you what they are. Let us take first the ordinary case of one Schrödinger equation, which we can write like this: $(H-W)\psi = 0$, where W is the operator $i\partial/\partial t$. Now to pass to the Heisenberg picture we take this operator H-W and use it as a Hamiltonian to give us the Heisenberg equations of motion. Suppose g is any dynamical variable. It will vary with respect to some independent variable, which for the time being I just call τ, according to the Heisenberg equation

$$\frac{dg}{d\tau} = [g, H-W] .$$

You see that if we take g equal to the time t, we get $dt/d\tau = 1$, so t and τ are just the same thing if we choose the origin suitably. Thus the independent variable is now just the ordinary time.

Now let us pass to a Schrödinger theory where there are supplementary conditions. This means we start off with several conditions for ψ:

$$\phi_1 \psi = 0 , \phi_2 \psi = 0 , \ldots\ldots .$$

Here the ϕ's are something like H-W, but there is the possible difference that they might not be hermitian.

We can multiply the ϕ's by any coefficients on the left and we get equations which are still valid. We can form an equation like

$$(\lambda_1\phi_1 + \lambda_2\phi_2 + \ldots.) \psi = 0 ,$$

and this is true for any λ's. We choose these λ's so as to get operators which are hermitian. We take as many independent choices of λ as possible which lead to hermitian operators on ψ. Let us call these equations $\phi_m \psi = 0$, the ϕ_m being hermitian. Then the ϕ_m can be used as Hamiltonians in the Heisenberg picture. In fact we have to use them. This means that, for a general dynamical variable g, we have equations of motion in the Heisenberg picture like this:

$$dg/d\tau_m = [g,\phi_m] .$$

We now have several independent variables τ_m, one for each of the hermitian ϕ_m's. The Heisenberg equations of motion give us the dynamical variables depending on all the independent variables τ_m. The equations which we get will be integrable. This follows from the fact that the ϕ_m's have to satisfy certain consistency conditions in order that the equations for ψ may have solutions simultaneously. The consistency conditions just lead to the result that the Heisenberg equations are integrable, so that we can get all our dynamical variables expressed as functions over the τ space.

Now what is the physical meaning of these Heisenberg equations? One of them might be just a usual equation telling us how the dynamical variables vary with the time, but the others will give us variations of the dynamical variables of a different character. These variations are of a type which the variables can be subjected to mathematically but which are not of physical importance. They are of the nature of gauge transformations. Whenever we have a Schrödinger theory involving supplementary conditions, the physical situation is that the Heisenberg dynamical variables are subject to gauge transformations, and the Heisenberg equations tell us how the dynamical variables vary with the gauge transformations as well as how they vary with time. The variations with respect to the gauge transformations and with respect to the time may be mixed up, but that doesn't matter. The equations contain the time variation and they also contain all the gauge transformations.

In the Heisenberg picture, the hermitian ϕ's, besides giving us the Hamiltonians for the equations of motion, also provide us with constraints $\phi_m = 0$ between the Heisenberg dynamical variables. Well, that is the general theory of how to pass over to the Heisenberg picture when one starts with a Schrödinger theory containing supplementary conditions.

Let us now apply this general method to the new wave equation and see what we get. We have four equations here, in equation (1), and everyone of them corresponds to a non-hermitian ϕ. We have to bring in suitable λ's as multiplying factors in order to get hermitian ϕ's. These λ's we take to be linear functions of the operators q, so the operators ϕ become quadratic in the q's.

Thus we have four primary equations (1) which are linear in the q's. We get from them secondary equations which are quadratic in the q's by multiplying them by q's and adding. We get altogether sixteen of these secondary equations. We look through the list of them and we see that some of them are hermitian. Those are the important ones, which will give our ϕ_m's for the Heisenberg picture. They are

$$\phi_0 \, \psi \equiv (s_{23}p_1 + s_{31}p_2 + s_{12}p_3) \, \psi = 0 \; ,$$

$$\phi_1 \, \psi \equiv (s_{23}p_0 + s_{30}p_2 + s_{02}p_3) \, \psi = 0 \; ,$$

two similar equations $\phi_2\ \psi = 0$, $\phi_3\ \psi = 0$, obtained from the preceding one by cyclic permutation of the suffixes 1,2,3, and finally

$$\phi_5\ \psi \equiv (1 - 2s_{\mu 5}\ p^{\mu})\ \psi = 0 \quad .$$

The quantities s_{uv} here (u, v = 0,1,2,3,5), are antisymmetric between u and v, and are defined by

$$s_{\mu\nu} = -\frac{1}{8}\, q^{\sim}(\alpha_{\mu}\beta\ \alpha_{\nu} - \alpha_{\nu}\beta\ \alpha_{\mu})\ q\ ;\ \mu,\nu=0,1,2,3,$$

$$s_{\mu 5} = \frac{1}{4}\, q^{\sim}\ \ \alpha_{\mu}\ q\ ,$$

with $\alpha_0 = 1$, and $q^{\sim}$ denoting the four q's written as a row matrix. These s_{uv}'s are just the infinitesimal operators of the 3 + 2 de Sitter group, satisfying the appropriate commutation relations.

Well here we have the ϕ's which are to be used in our Heisenberg theory, and we must examine them to see what equations of motions they give. This is a straightforward piece of analysis. We see immediately that, for all the Hamiltonians, the momentum-energy variables p_{μ} are constants.

Let us see how our dynamical variables vary with respect to the independent variable τ_5. We have

$$dx_{\mu}/d\tau_5 = [x_{\mu},\ \phi_5] = 2s_{\mu 5}\ ,$$

showing that the $s_{\mu 5}$ are the velocity variables with respect to the independent variable τ_5, apart from the factor 2. Now let us form the acceleration variables $ds_{\mu 5}/d\tau_5$. This we can work out from our knowledge of the commutation relations. We obtain

$$ds_{\mu 5}/d\tau_5 = 2\ s_{\mu\nu}\ p^{\nu} \quad .$$

The new variables $s_{\mu\nu}$ occur in the accelerations. We finally get $ds_{\mu\nu}/d\tau_5 = 2\ (s_{\nu 5}\ p_{\mu} - s_{\mu 5}\ p_{\nu})$.

We can proceed to integrate these equations and the result is that

$$x_{\mu} = p_{\mu}\ \tau_5 + b_{\mu}\ \sin 2\tau_5 + b'_{\mu}\ \cos\tau_5 + a_{\mu} \quad .$$

The quantities b_μ, b'_μ, a_μ appear as constants of integration. They are constants with respect to τ_5. They do not commute with each other. We are dealing all the time with noncommuting quantities, but even so we can apply ordinary methods of integration.

If you look at these equations for x_μ you see that, concentrating your attention on the first and last terms on the right, the x_μ's vary according to the classical formula with τ_5 as the proper time. So we are to picture τ_5 as being the proper time and our particle as moving with respect to τ_5 according to the usual formula, with the addition of the two middle terms, which are oscillatory and give us Zitterbewegung. b_μ, b'_μ are the amplitudes of this Zitterbewegung.

Let us now consider the other Heisenberg equations. Instead of proceeding to work out the integration of the equations of motion in the general case, let us restrict ourselves to a particle with zero momentum. That means that we put $\vec{p} = 0$, $p_0 = 1$. After we have worked out the equations of motion, we substitute the above values for the p's, and we may also make use of the constraint equations. You see that the $\phi_1 = 0$ constraint equation will just tell us that $s_{23} = 0$, and it similarly follows that $s_{31} = s_{12} = 0$.

The equations that follow with ϕ_0 as Hamiltonian make everything constant with respect to τ_0. Those that follow with ϕ_1 as Hamiltonian are

$$\frac{dx_0}{d\tau_1} = 0 , \quad \frac{dx_1}{d\tau_1} = 0 , \quad \frac{dx_2}{d\tau_1} = s_{30} , \quad \frac{dx_3}{d\tau_1} = - s_{20} .$$

Also

$$\frac{ds_{10}}{d\tau_1} = 0 , \quad \frac{ds_{20}}{d\tau_1} = - s_{30} , \quad \frac{ds_{30}}{d\tau_1} = s_{20} .$$

We can integrate these equations and we find that s_{10} is a constant c_1, and

$$s_{20} = c_2 \cos\tau_1 + c_3 \sin\tau_1 , \quad s_{30} = c_2 \sin\tau_1 - c_3 \cos\tau_1 ,$$

with c_2 and c_3 also constants.

Let us look at these equations and see what they mean. The time x_0 isn't changing at all, so they are entirely gauge transformations. If one completes the integration one gets $x_1 = a_1$,

$$x_2 = a_2 - c_2 \cos\tau_1 - c_3 \sin\tau_1 \quad ,$$

$$x_3 = a_3 - c_2 \sin\tau_1 + c_3 \cos\tau_1 \quad ,$$

with the a's constants.

You see that this just gives you the particle moving around in a circle in the plane of x_2 and x_3. The center of the circle has the coordinates $a_2 = x_2 + s_{20}$ and $a_3 = x_3 + s_{30}$.

Well there we have the effect of gauge transformations for a particle in the case when it has no momentum and we use the Hamiltonian ϕ_1. There will be similar equations that follow from the other two Hamiltonians ϕ_2 and ϕ_3 . Each of these gauge transformations makes the particle move round in a certain circle, and when they are all combined the particle will run about over a sphere. For any of the gauge transformations the particle keeps to the surface of the sphere. The center of the sphere has the coordinates $y_r = x_r + s_{r0}$.

We can put all these results together and get a Heisenberg picture consisting of a sphere with respect to the frame of reference in which the particle has zero momentum. When we set up a four dimensional picture for a particle moving with a general momentum, we shall have a tube in four dimensions and our particle will move about over the surface of this tube under the influence of the gauge transformations, with the passage of time corresponding to going along the axis of the tube. The Zitterbewegung consists now of an arbitrary motion on the surface of the tube as one goes along its axis.

The gauge transformations are not of physical importance. The quantities of physical importance are those that describe the tube itself -- the coordinates of its axis and its radius. Such quantities are unaffected by the gauge transformations.

Now you might say, if the y's describing the axis of the tube are the variables of physical importance, why not work entirely with them and forget all about the x's. Well you can't do that because the y's don't commute with each other. The variables which occur in the Schrödinger function must be a set of commuting x's, so we cannot set up a Schrödinger function in terms of the y's. The description with gauge transformations is the best we can do for setting up a Schrödinger function: it refers to a particle on the surface of the tube which can move about freely under the influence of the gauge transformations.

Now there is one final development that we can make, namely with regard to the discussion of the spin of this particle. For getting the spin of any particle, we can give a meaning to the total angular momentum, spin plus orbital angular momentum, because this total angular momentum is connected with the operators of the group of rotations in space, which are always well defined. The total angular momentum is thus a well defined quantity always. So if we can decide what we are to take as the orbital angular momentum, then the difference will be the spin.

Now the usual way of defining orbital angular momentum is to take the 1 component to be $x_2 p_3 - x_3 p_2$. If we use this in the present case, and then take its difference from the total angular momentum as the spin, we get a spin which depends on the momentum of the particle. That is not very useful. Experimental physicists never have a spin depending on the momentum the particle.

It would seem more reasonable to take a new definition for orbital angular momentum, namely $y_2 p_3 - y_3 p_2$, and subtract this from the total angular momentum to get the spin. We find then that the spin is zero. This spin is a physical quantity independent of gauge transformations, and it is zero for all values of the momentum.

Of course using this new definition of orbital angular momentum, we ought to check up the commutation relations of its components. When we do work them out we find that they come out correctly. This is not an obvious result at all, because we have a lot of noncommuting quantities occuring and we just have to carry out the calculations and see what they give. There is thus no physical objection to our using the new definition. So I propose that when one works with this wave equation one should use only the new definition for the orbital angular momentum, and then one has the spin of the particle always zero.

Well this is as far as I have got in the development of this theory. I hope I have succeeded in giving you some physical picture of what it means, instead of just a mathematical equation whose interpretation is obscure.

DISCUSSION

ROHRLICH: Years ago Newton and Wigner analyzed relativistic wave equations. They found that when one restricts oneself to positive energy solutions the particles are no longer localizable. Is there a similar difficulty with the present equation?

DIRAC: There is localizability here. I expect Wigner's analysis was incomplete through not taking into account the possibility of supplementary conditions.

I might mention another piece of work that was done many years ago, namely, a relativistic wave equation was proposed by Majorana having only positive energies for the particle. Now Majorana's theory is closely connected with the present one. We can get Majorana's theory from the present one by retaining one of its equations and dropping all the others.

In Majorana's theory there is a whole succession of rest masses going from 1 down to 0 and you find that the rest mass 1 corresponds to a particle with zero spin, just like in the present theory. The rest masses which are less than 1 correspond to higher spins. To a physicist it's a bit obnoxious to have a decreasing value of rest mass with increasing spin and that is why Majorana's theory did not attract much attention.

If any of you are interested in Majorana's equation and don't read Italian, I would suggest the paper by Fradkin which gives an English version of Majorana's theory and a very good version too. The reference to Fradkin is American Journal of Physics 1966, 134, 314. There is also a complete translation of Majorana's paper into English given by Orzalesi, University of Maryland, Department of Physics and Astronomy, Technical Report 792 (1968).

A MASTER WAVE EQUATION

Behram Kursunoğlu
Center for Theoretical Studies
University of Miami
Coral Gables, Florida

"God used beautiful mathematics
in creating the world"
P.A.M. Dirac

I. INTRODUCTION

In the two previous papers[1] (to be referred to as I and II) a relativistic wave equation describing particles with different spins and masses has been proposed and discussed in detail. For the half-integral spin systems there are no subsidiary conditions on the wave function and for the integral spin systems the wave equation itself contains the required subsidiary conditions. The particle classification is based on the finite dimensional representations of the group $G \doteq [SO(3,2) \otimes U(3,1)]$ where N_o (dimension number of SO(3,2)) assumes only the values 4,5 and 10 while N (the dimension number of U(3,1)) ranges over the entire representation spectrum of U(3,1) viz. N=1, 4, 6, 10, 15, 20,... The wave functions (suppressing the SO(3,2) index) are of the form

$$\begin{array}{llllll} & \Psi, & \Psi_{\mu}, & \Psi_{[\mu\nu]}, & \Psi_{\{\mu\nu\}}, & \Psi_{[[\mu\nu],[\rho\sigma]]}, \\ N= & 1, & 4, & 6, & 10, & 15, \\ & & & & \Psi_{\{[\mu\nu],[\rho\sigma]\}}, & \Psi_{\{\mu\nu\rho\}} \\ & & & & 20, & 20, \ldots, \end{array} \tag{I.1}$$

where we observe that the trace operation, for example, in

$\Psi_{\{\mu\nu\}}$ and $\Psi_{\{\mu\nu\rho\}}$ by means of the metric $g^{\mu\nu}$ ($g_{44} = 1$, $g_{4j} = g_{j4} = 0$, $g_{jk} = -\delta_{jk}$, $j,k = 1,2,3$) does not commute with the U(3,1) transformations[2]. Thus the $\Psi_{\{\mu\nu\}}$, $\Psi_{\{\mu\nu\rho\}}$ which are reducible under O(3,1) correspond to irreducible representations of U(3,1). A square bracket around the indices implies antisymmetry while a curly bracket implies symmetry under permutations of the respective indices.

If we represent the vector index μ by the symbol $4 = \square$ $(=(0,\frac{1}{2}) \times (\frac{1}{2},0) = (\frac{1}{2},\frac{1}{2}))$ and the fully antisymmetric symbol $[\mu\nu\rho]$ (the dual of μ) by $\bar{4} = \begin{array}{|c|}\hline \ \\ \hline \ \\ \hline \ \\ \hline\end{array}$, then some of the Young patterns for the representations of the U(3,1) are of the form

$$4\times 4 = \square \times \square = \square\!\square + \begin{array}{|c|}\hline \ \\ \hline \ \\ \hline\end{array} \quad (=[1,1)+(0,0)]+[(0,1)+(1,0)]),$$

representing $\Psi_{\{\mu\nu\}}$ and $\Psi_{[\mu\nu]}$, respectively. Similarly we have

$$\bar{4}\times 4 = \begin{array}{|c|}\hline \ \\ \hline \ \\ \hline \ \\ \hline\end{array} \times \square = \begin{array}{|c|}\hline \ \\ \hline \ \\ \hline \ \\ \hline \ \\ \hline\end{array} + \begin{array}{|c|c|}\hline \ & \ \\ \hline \ & \multicolumn{1}{c}{} \\ \cline{1-1} \ & \multicolumn{1}{c}{} \\ \cline{1-1}\end{array} \quad (=[(1,1)+(0,1)+(1,0)]+[(0,0)]),$$

representing Ψ and $\Psi_{[[\mu\nu],[\rho\sigma]]}$. The remaining Young tableaux up to N = 84 are given in the appendix.

The fundamental wave equation

$$(B^{\mu}p_{\mu} - imc)\Psi = 0 \tag{I.2}$$

contains the generators for the finite dimensional representations of G, where the possible forms of the matrices B_{μ} are given by

$$B_{\mu} = (\frac{1}{\rho}\Gamma_{\mu\nu} + g_{\mu\nu})\gamma^{\nu}, \quad B_{\mu} = (\frac{1}{\rho}\Gamma_{\mu\nu} + g_{\mu\nu})\beta^{\nu}, \tag{I.3}$$

where the β-matrices correspond to $N_o = 5$ or $N_o = 10$ dimensional representation of SO(3,2). The representations (I.3) are to be used for the classification of the baryons and mesons. The parameter ρ as discussed in I and II is greater than 1 in all cases except the lepton classification. Furthermore the commutation relations of U(3,1) are invariant under the substitutions $\Gamma_{\mu\nu} \to \pm \Gamma_{\mu\nu}$, $g_{\mu\nu} \to \pm g_{\mu\nu}$ and therefore double the spectrum implied by the wave equation (I.2) and must be incorporated into it by replacing the generators $\frac{1}{\rho}\Gamma_{\mu\nu} + g_{\mu\nu}$ of U(3,1) by

$$\frac{1}{\rho} q \Gamma_{\mu\nu} + y\, g_{\mu\nu} \quad , \tag{I.4}$$

where q and y can be represented by matrices of the form

$$y \text{ or } q = \begin{bmatrix} 1 & 0 \\ 0 & -1 \end{bmatrix} , \quad \begin{array}{l} q^2 = 1,\ y^2 = 1,\ [q,y] = 0 \\ [q, \Gamma_{\mu\nu}] = [y, \Gamma_{\mu\nu}] = 0 \quad . \end{array} \tag{I.5}$$

Thus for the fixed sign of the parameter ρ the wave function Ψ must be tagged with two other indices q and y (=1,2) acted on by the matrices q and y alone and they remain unaffected by Lorentz transformations.

The q-symmetry of the wave equation (I.2) can be exhibited in the N = 4 representation of the U(3,1). Thus, if

$$(M_{\mu\nu})_{\rho}^{\ \sigma} = i(g_{\mu\rho}\delta^{\sigma}_{\nu} - g_{\nu\rho}\delta^{\sigma}_{\mu}) \tag{I.6}$$

are the usual generators for the Lorentz transformation of the coordinates then the N = 4 representation of U(3,1) can be constructed in the form

$$J_{\mu\nu} = \frac{1}{2} i\ (M_{\mu\rho}M_{\nu}^{\ \rho} - M_{\nu\rho}M_{\mu}^{\ \rho}) = M_{\mu\nu} \quad , \tag{I.7}$$

$$\Gamma_{\mu\nu} = \frac{1}{4} g_{\mu\nu}\ M_{\rho\sigma}M^{\rho\sigma} - \frac{1}{2}(M_{\mu\rho}M_{\nu}^{\ \rho} + M_{\nu\rho}M_{\mu}^{\ \rho}) = \Lambda_{\mu\nu} \ . \tag{I.8}$$

Now the N = $\bar{4}$ representation of U(3,1) follows from the dual matrices of $M_{\mu\nu}$

$$M'_{\mu\nu} = \frac{1}{2}\,\varepsilon_{\mu\nu\rho\sigma}\ M^{\rho\sigma} \quad , \tag{I.9}$$

where $\varepsilon_{\mu\nu\rho\sigma}$ is the usual fully antisymmetric fourth rank tensor. Hence using

$$M'_{\mu\rho}\ M'^{\ \rho}_{\nu} = \frac{1}{2} g_{\mu\nu}\ M_{\rho\sigma}\ M^{\rho\sigma} - M_{\nu\rho}\ M_{\mu}^{\ \rho}$$

we obtain from (I.7), (I.8) the results

$$J'_{\mu\nu} = J_{\mu\nu} = M_{\mu\nu} \ ,$$

$$\Gamma'_{\mu\nu} = -\,\Gamma_{\mu\nu} \ , \qquad (I.10)$$

which, of course, obey the commutation rules of U(3,1) as given by (A2.1) of reference 1.

The q and y invariance of the commutation relations for the Lie algebra of U(3,1) will be interpreted as an internal symmetry of the particle classification of this theory.

II. COMPOSITE STRUCTURE OF THE SUPERMULTIPLETS

All the representations of the U(3,1) can be constructed from its 4 and $\bar{4}$ representations. This fact is quite similar to the construction of all SU(3) representations from its 3 and $\bar{3}$ representations. However in the present case, the Lorentz invariance of the wave equation (I.2) implies clearly that all the representations of U(3,1) transform as we change from one inertial frame to another.

In order to see the composite structure of a member of a supermultiplet $[N_o, N]$ we observe that the matrices $M_{\mu\nu}$ can be decomposed according to

$$M_{\mu\nu} = \frac{1}{2}\left(\tau_{\mu\nu} + \tau'_{\mu\nu}\right) \ , \qquad (II.1)$$

where

$$\tau_{\mu\nu} = M_{\mu\nu} - i\, M'_{\mu\nu} \ , \quad \tau'_{\mu\nu} = M_{\mu\nu} + i\, M'_{\mu\nu} \qquad (II.2)$$

and where $M'_{\mu\nu}$ is defined by (I.9). The τ and τ'-matrices commute

$$[\tau_{\mu\nu} \ , \ \tau'_{\rho\sigma}] = 0 \qquad (II.3)$$

and they are constructed as

$$\tau_{jk} = \varepsilon_{jk\ell}\, \tau_\ell \ , \quad \tau_{4j} = i\, \tau_j \ ,$$

$$\tau'_{jk} = \varepsilon_{jk\ell}\, \tau'_\ell \ , \quad \tau'_{4j} = -i\, \tau'_j \ , \qquad (II.4)$$

where the hermitian matrices τ_j and τ'_j are given by

$$\tau_j = M_j - iN_j \ , \ \tau'_j = M_j + iN_j, \ N_j = M_{4j} \ , \ M_j = \frac{1}{2} \varepsilon_{jk\ell} M_{k\ell} \ . \tag{II.5}$$

More explicitly we have

$$\tau_1 = \begin{bmatrix} 0 & 0 & 0 & 1 \\ 0 & 0 & -i & 0 \\ 0 & i & 0 & 0 \\ 1 & 0 & 0 & 0 \end{bmatrix}, \quad \tau_2 = \begin{bmatrix} 0 & 0 & i & 0 \\ 0 & 0 & 0 & 1 \\ -i & 0 & 0 & 0 \\ 0 & 1 & 0 & 0 \end{bmatrix}, \quad \tau_3 = \begin{bmatrix} 0 & -i & 0 & 0 \\ i & 0 & 0 & 0 \\ 0 & 0 & 0 & 1 \\ 0 & 0 & 1 & 0 \end{bmatrix} \tag{II.6}$$

$$\tau'_1 = \begin{bmatrix} 0 & 0 & 0 & -1 \\ 0 & 0 & -i & 0 \\ 0 & i & 0 & 0 \\ -1 & 0 & 0 & 0 \end{bmatrix}, \quad \tau'_2 = \begin{bmatrix} 0 & 0 & i & 0 \\ 0 & 0 & 0 & -1 \\ -i & 0 & 0 & 0 \\ 0 & -1 & 0 & 0 \end{bmatrix}, \quad \tau'_3 = \begin{bmatrix} 0 & -i & 0 & 0 \\ i & 0 & 0 & 0 \\ 0 & 0 & 0 & -1 \\ 0 & 0 & -1 & 0 \end{bmatrix}. \tag{II.7}$$

They obey the relations

$$[\tau_j, \tau_\ell] = 2i \ \varepsilon_{j\ell s} \ \tau_s \ , \ [\tau'_j, \tau'_\ell] = 2i \ \varepsilon_{j\ell s} \ \tau'_s \ , \tag{II.8}$$

$$\{\tau_j, \tau_\ell\} = 2\delta_{j\ell} \ , \ \{\tau'_j, \tau'_\ell\} = 2\delta_{j\ell} \tag{II.9}$$

and

$$[\tau_j, \tau'_\ell] = 0 \quad . \tag{II.10}$$

More generally,

$$[\frac{1}{2} \tau_{\mu\nu}, \frac{1}{2} \tau_{\rho\sigma}] = i \ (g_{\rho\nu} \frac{1}{2} \tau_{\mu\sigma} + g_{\sigma\nu} \frac{1}{2} \tau_{\rho\mu} - g_{\mu\rho} \frac{1}{2} \tau_{\nu\sigma} - g_{\mu\sigma} \frac{1}{2} \tau_{\rho\nu}), \tag{II.11}$$

and

$$\frac{1}{2} \{\tau_{\mu\nu}, \tau_{\rho\sigma}\} = - i \ \varepsilon_{\mu\nu\rho\sigma} + g_{\mu\rho} g_{\nu\sigma} - g_{\mu\sigma} g_{\nu\rho} \quad . \tag{II.12}$$

The above relations are obeyed by $\tau'_{\mu\nu}$ also. Hence we see that the matrices $\tau_{\mu\nu}$ and $\tau'_{\mu\nu}$ generate 4-dimensional SL(2,C) transformations. Further relations are obtained by Hermitian conjugation operation (†) as

$$F\tau^{\dagger}_{\mu\nu}F = \tau'_{\mu\nu}, \quad \tau^{*}_{\mu\nu} = -\tau'_{\mu\nu} \tag{II.13}$$

and

$$\tau_{\mu}\,\tau_{\rho\sigma} - \tau^{\dagger}_{\rho\sigma}\,\tau_{\mu} = 2i\,(g_{\mu\rho}\,\tau_{\sigma} - g_{\sigma\mu}\,\tau_{\rho}) \,, \tag{II.14}$$

$$\tau'_{\mu}\,\tau'^{\dagger}_{\rho\sigma} - \tau'_{\rho\sigma}\,\tau'_{\mu} = 2i\,(g_{\mu\rho}\,\tau'_{\sigma} - g_{\sigma\mu}\,\tau'_{\rho}) \,, \tag{II.15}$$

where F is the matrix form of the 4-dimensional parity tensor (as defined by III.11) and where $\tau_4 = I_4$ (4-dimensional unit matrix) together with τ_j $(j = 1,2,3)$ transform as 4-vector operator according to

$$L^{\dagger}\,\tau_{\mu}\,L = \Lambda^{\nu}_{\mu}\,\tau_{\nu} \,, \tag{II.16}$$

where

$$\Lambda = LL^{*} = L^{*}L \tag{II.17}$$

is, because of (II.1) and (II.3), a Lorentz matrix defined by

$$\Lambda = \exp\,[\tfrac{1}{2}\,i\lambda f^{\rho\sigma}\,M_{\rho\sigma}] \,, \tag{II.18}$$

$$L = \exp\,[\tfrac{1}{4}\,i\lambda f^{\rho\sigma}\,\tau_{\rho\sigma}] \,, \tag{II.19}$$

$$L^{*} = \exp\,[\tfrac{1}{4}\,i\lambda f^{\rho\sigma}\,\tau'_{\rho\sigma}] \,. \tag{II.20}$$

The spin matrices of the supermultiplet [4,4] are given by

$$S_{\mu\nu} = M_{\mu\nu} + \tfrac{1}{2}\,\sigma_{\mu\nu} = \tfrac{1}{2}\,(\tau_{\mu\nu} + \tau'_{\mu\nu} + \sigma_{\mu\nu}) \,. \tag{II.21}$$

Hence we see that for fixed eigenstates of q and y as defined in (I.4) and (I.5) the two $s = \frac{1}{2}$ and one $s = \frac{3}{2}$

states of [4,4] are composite states of three individual spin $\frac{1}{2}$ states whose spin matrices are $\frac{1}{2}\underline{\tau}$, $\frac{1}{2}\underline{\tau}'$ and $\frac{1}{2}\underline{\sigma}$, where

$$\sigma_{\mu\nu} = -\frac{1}{2} i \, [\gamma_\mu, \gamma_\nu]$$

and γ_μ are the usual Dirac matrices.

In terms of τ and τ'-matrices the N = 4 generators of SU(3,1) can be written as

$$J_{\mu\nu} = M_{\mu\nu} = \frac{1}{2}(\tau_{\mu\nu} + \tau'_{\mu\nu}) \, ,$$
$$\Lambda_{\mu\nu} = q\,\Gamma_{\mu\nu} = -\frac{1}{4} q\,(\tau_{\mu\rho}\,\tau'^{\rho}_{\nu} + \tau_{\nu\rho}\,\tau'^{\rho}_{\mu}) \, , \qquad \text{(II.22)}$$

where $q = \pm 1$ and $\tau_{\mu\nu}\,\tau'^{\mu\nu} = 0$ so that $g^{\mu\nu}\,\Lambda_{\mu\nu} = 0$.

The fact that all representations of U(3,1) are constructed out of its 4 and $\bar{4}$ representations implies that all of the supermultiplets [4,N] are composite states of the three sub-nuclear elements introduced above. The role of charge conjugation operation in the τ-formalism consists of the complex conjugation operation alone. If we include the Pauli matrices σ_j then the charge conjugation operation on the spin operators

$$\underline{S} = \frac{1}{2}(\underline{\tau} + \underline{\tau}' + \underline{\sigma}) \qquad \text{(II.23)}$$

of the [4,4] is given by $\sigma_2\bar{C}$ where $\bar{C}$ is the complex conjugation operation and the Pauli matrix σ_2 commutes with τ and τ'. Hence the charge conjugate total spin operators are

$$\underline{S}_c = \sigma_2\bar{C}\,\underline{S}(\sigma_2\bar{C})^{-1} = -\underline{S} \quad . \qquad \text{(II.24)}$$

Finally we note that the spin matrices $\tau_{\mu\nu}$ and $\tau'_{\mu\nu}$ can also be written in terms of τ_μ and τ'_μ as

$$\tau_{\mu\nu} = -\frac{1}{4}\varepsilon_{\mu\nu\rho\sigma}[\tau^\rho, \tau^\sigma] - \frac{1}{2} i\,[\tau_\mu, \tau_\nu] \, , \qquad \text{(II.25)}$$

$$\tau'_{\mu\nu} = \frac{1}{4}\varepsilon_{\mu\nu\rho\sigma}[\tau'^\rho, \tau'^\sigma] - \frac{1}{2} i\,[\tau'_\mu, \tau'_\nu] \quad . \qquad \text{(II.26)}$$

The τ_μ and τ'_μ matrices have the following trace property

$$g_{\mu\nu} = -\frac{1}{4}\,\text{trace}\,(\tau_\mu\tau_\nu) = -\frac{1}{4}\,\text{trace}\,(\tau'_\mu\tau'_\nu)\ . \qquad \text{(II.27)}$$

Furthermore, we have explicitly the relations

$$(\tau_j)_4{}^\ell = (\tau_j)_\ell{}^4 = \delta_{j\ell},\ (\tau_j)_k{}^\ell = -i\varepsilon_{jk\ell},\ (\tau'_j)_4{}^\ell = (\tau'_j)_\ell{}^4 = -\delta_{j\ell},$$

$$(\tau'_j)_k{}^\ell = -\,i\,\varepsilon_{jk\ell} \qquad \text{(II.28)}$$

and

$$\tau^*_\mu = g^\nu_\mu\,\tau'_\nu\ . \qquad \text{(II.29)}$$

III. ELECTRODYNAMICS WITHOUT POTENTIALS

Instead of the usual minimal electromagnetic interaction described in terms of the vector potential A_μ and the massive current J_μ, we would like to introduce an equivalent but more general procedure where the usual gauge transformations of the second kind (i.e. $A_\mu \rightarrow A_\mu + \frac{\partial u}{\partial x^\mu}$) are eliminated and only the gauge transformations of the first kind (i.e. $\Psi \rightarrow \exp(i\alpha)\Psi$) are retained. The electromagnetic interactions of a supermultiplet are described by the wave equation (I.2) provided the operators p_μ obey the commutation relations

$$[p_\mu,\ p_\nu] = -\frac{ie\hbar}{c}\,f_{\mu\nu}\ , \qquad \text{(III.1)}$$

where $f_{\mu\nu}$ is an external electromagnetic field.

The relativistic equations of motion for a point electric charge can be written as

$$\frac{dp}{ds} = -\frac{e}{mc^2}\,f\,p\ , \qquad \text{(III.2)}$$

where p represents a 4-dimensional column vector and where f ($= f_\mu{}^\nu$) is the electromagnetic matrix with $f_{4j} = E_j$, $f_{jk} = \varepsilon_{jk\ell}\,H_\ell$. The invariant parameter s represents, in the classical case, the proper-time. Formally, the equations (III.2) can be solved by

$$p(s) = \Lambda(s)\,p(0)\ , \qquad \text{(III.3)}$$

where the proper-time dependent Lorentz matrix $\Lambda(s)$ is given by

$$\Lambda(s) = \exp\left[-\frac{e}{mc^2}\int_0^s f(u)\,du\right] \quad . \qquad \text{(III.4)}$$

The $\Lambda(s)$ because of the relations

$$f = -\frac{1}{2}\, i\, f^{\mu\nu} M_{\mu\nu} \quad , \quad F\tilde{f}F = -f \ ,$$

satisfy the condition

$$\tilde{\Lambda}\, F\, \Lambda = F$$

and also the group property

$$\Lambda(s_1)\,\Lambda(s_2) = \Lambda(s_1 + s_2) \quad . \qquad \text{(III.5)}$$

Furthermore, from (III.4) we see that $\Lambda(s)$ obeys the equation

$$\frac{d\Lambda}{ds} = -\frac{e}{mc^2}\, f\, \Lambda \ . \qquad \text{(III.6)}$$

Hence the momentum of a classical point charge in an external electromagnetic field is obtained by an unfolding of a continuous Lorentz transformation at every space-time point as described by (III.3). For example, for a constant electromagnetic field the coordinates of the trajectory are given by integrating (III.3) in the interval (o,s),

$$x(s) - x(0) = -\frac{c}{ef}\,[\Lambda(s) - 1]\, p(0) \ , \qquad \text{(III.7)}$$

where now

$$\Lambda(s) = \exp\left(-\frac{es}{mc^2}\, f\right) \quad . \qquad \text{(III.8)}$$

Now let us consider the factorized form of a Lorentz matrix Λ as given by (II.17). The SL(2,C) transformations L and L* act on complex vectors η and η^*, respectively. In fact L and L* result, up to a unitary transformation, from the decomposition $D(\frac{1}{2}, \frac{1}{2}) = D(0, \frac{1}{2}) \times D(\frac{1}{2}, 0)$ of a Lorentz matrix Λ. Thus the vector p_μ can be written as

$$p_\mu = \bar{\eta}\, \tau_\mu\, \eta \quad , \tag{III.9}$$

where

$$\bar{\eta}_\mu = g_\mu^\nu\, \eta_\nu^* \tag{III.10}$$

and the parity tensor g_μ^ν is defined by $g_4^4 = 1$, $g_j^4 = g_4^j = 0$, $g_\ell^j = -\delta_{j\ell}$ and is an improper Lorentz matrix satisfying the relations

$$g_{\rho\sigma}\, g_\mu^\rho\, g_\nu^\sigma = g_{\mu\nu} \;;\; g_\mu^\nu = \{-1 + \tfrac{1}{2}\, \tau^\rho \tau'_\rho\}_\mu^\nu \quad . \tag{III.11}$$

The same are satisfied by L and L*, as

$$g_{\rho\sigma}\, L_\mu^\rho\, L_\nu^\sigma = g_{\mu\nu} \;,\; g_{\rho\sigma}\, L^{*\rho}_\mu\, L^{*\sigma}_\nu = g_{\mu\nu} \;.$$

The matrices L, L* and the complex vectors η and η^* satisfy the equations

$$\frac{dL}{ds} = \frac{ie}{4mc^2}\, f^{\mu\nu}\, \tau_{\mu\nu}\, L \;,\; \frac{dL^*}{ds} = \frac{ie}{4mc^2}\, f^{\mu\nu}\, \tau'_{\mu\nu}\, L^* \tag{III.12}$$

and

$$\frac{d\eta}{ds} = \frac{ie}{4mc^2}\, f^{\mu\nu}\, \tau_{\mu\nu}\, \eta \;,\; \frac{d\eta^*}{ds} = \frac{ie}{4mc^2}\, f^{\mu\nu}\, \tau'_{\mu\nu}\, \eta^* \;, \tag{III.13}$$

where

$$-\tfrac{1}{2}\, i\, (f^{\rho\sigma}\, \tau_{\rho\sigma})_\mu{}^\nu = f_\mu{}^\nu - i\, \phi_\mu{}^\nu \;,\; -\tfrac{1}{2}\, i\, (f^{\rho\sigma}\, \tau'_{\rho\sigma})_\mu{}^\nu =$$

$$= f_\mu{}^\nu + i\, \phi_\mu{}^\nu \;,\; \phi_{\mu\nu} = \tfrac{1}{2}\, \varepsilon_{\mu\nu\rho\sigma}\, f^{\rho\sigma} \;.$$

From (III.13) it follows that p_μ as defined by (III.9) obey the equations of motion (III.2). Furthermore using the definitions (II.6) of the τ_μ matrices and (III.9) we obtain

$$p_\mu\, p^\mu = m^*\, m \;, \tag{III.14}$$

where

$$m = \eta_\mu \eta^\mu \quad . \tag{III.15}$$

Also the definition (III.9) yields

$$p_4 = |\eta_4|^2 + |\underline{\eta}|^2 \tag{III.16}$$

so that the time like classical vector p_μ lies in the future light cone.

For the quantum theory the complex vectors η_μ, η^*_μ are subject to the commutation relations

$$[\eta_\mu, \eta_\nu] = -\frac{e\hbar}{2cm^*} (f^{\rho\sigma} \tau_{\rho\sigma})_{\mu\nu} = -\frac{ie\hbar}{2cm^*} (f - i\phi)_{\mu\nu} \,, \tag{III.17}$$

$$[\eta^*_\mu, \eta^*_\nu] = -\frac{e\hbar}{2cm} (f^{\rho\sigma} \tau'_{\rho\sigma})_{\mu\nu} = -\frac{ie\hbar}{2cm} (f + i\phi)_{\mu\nu} \tag{III.18}$$

$$[\eta_\mu, \eta^*_\nu] = 0 \ . \tag{III.19}$$

The use of the commutation relations (III.17)-(III.19) together with the definition (III.9) yield the commutation relations (III.1). The definition (III.1) and the commutation relations (III.17)-(III.19) are invariant under the gauge transformations of the form $\eta \rightarrow \exp\,(i\alpha)\eta$, where α is real. From the commutation relations

$$[[\eta_\mu, \eta_\nu],\ m^*] = [[\eta^*_\mu, \eta^*_\nu],\ m] = 0 \tag{III.20}$$

it follows that

$$[f - i\phi,\ m^*] = 0 \quad , \quad [f + i\phi,\ m] = 0 \ . \tag{III.21}$$

For the spinless case we may use the "Hamiltonian" $H = \frac{1}{2m} mm^*$ and record the equations of motion,

$$i\hbar \frac{d\eta}{ds} = \frac{1}{c} [\eta, H], \ i\hbar \frac{d\eta^*}{ds} = \frac{1}{c} [\eta^*, H] \ , \tag{III.22}$$

which yield the results

$$i\hbar \frac{dp}{ds} = \frac{1}{c} [p,H], \quad \frac{dm}{ds} = \frac{dm^*}{ds} = 0 . \qquad \text{(III.23)}$$

For the case of spin we shall consider the supermultiplet [4,4] in the presence of an electromagnetic interaction. The wave equation (I.2) can be squared in the form

$$[\frac{1}{2} \{B_\mu,B_\nu\}p^\mu p^\nu + \frac{ie\hbar}{4c} f^{\mu\nu} [B_\mu,B_\nu] + m^2c^2]\Psi = 0 , \qquad \text{(III.24)}$$

where we used the commutation relations (III.1) for p_μ and where

$$\frac{1}{2} \{B_\mu,B_\nu\}p^\mu p^\nu = \frac{1}{\rho^2} [\sigma^{\mu\nu}M_{\mu\nu}-s(s+1)- \frac{5}{4} - (\rho- \frac{q}{2})^2]p^2 + \frac{(3-2\rho q)}{\rho^2} (\Gamma_{\mu\nu} + \frac{1}{2} g_{\mu\nu})p^\mu p^\nu , \qquad \text{(III.25)}$$

$$\frac{ie\hbar}{4c} [B_\mu,B_\nu]f^{\mu\nu} = \frac{e\hbar}{2c} f^{\mu\nu} \sigma_{\mu\nu} (\frac{3}{4\rho^2} - 1) - \frac{3e\hbar}{4c\rho^2} f^{\mu\nu} (\tau_{\mu\nu} + \tau'_{\mu\nu}) - \frac{e\hbar}{c\rho^2} (q\rho+1) f^{\mu\nu} \sigma_{\mu\rho} \Gamma_\nu{}^\rho . \qquad \text{(III.26)}$$

In the above derivation we used the relations for N = 4:

$$\{\Gamma^\rho_\mu,\Gamma_{\nu\rho}\} = - 4 \Gamma_{\mu\nu} + \frac{9}{2} g_{\mu\nu} ,$$

$$\{\Gamma_{\mu\nu},M_{\rho\sigma}\} = g_{\mu\nu}M_{\rho\sigma} + g_{\sigma\nu}M_{\mu\rho} - g_{\rho\nu}M_{\mu\sigma} + g_{\sigma\mu}M_{\nu\rho} - g_{\mu\rho}M_{\nu\sigma} , \qquad \text{(III.27)}$$

$$\frac{1}{2} (\sigma_{\mu\rho}M_\nu{}^\rho + \sigma_{\nu\rho}M_\mu{}^\rho)p^\mu p^\nu = \Gamma_{\mu\nu}p^\mu p^\nu + [\frac{9}{4} + \frac{1}{2} \sigma^{\mu\nu}M_{\mu\nu}-s(s+1)]p^2, \qquad \text{(III.28)}$$

$$\{\Gamma_{\mu\nu},\Gamma_{\rho\sigma}\} = - (\Gamma_{\mu\nu})_{\rho\sigma} + g_{\mu\nu}\Gamma_{\rho\sigma} + g_{\rho\sigma}\Gamma_{\mu\nu} - g_{\rho\nu}\Gamma_{\mu\sigma} - g_{\sigma\nu}\Gamma_{\mu\rho} - g_{\mu\rho}\Gamma_{\nu\sigma} - g_{\mu\sigma}\Gamma_{\rho\nu} , \qquad \text{(III.29)}$$

$$(\sigma^{\mu\nu}M_{\mu\nu})^2 = 12 - 4 \sigma^{\mu\nu}M_{\mu\nu} , \qquad \text{(III.30)}$$

$$- \frac{e\hbar}{4c\rho^2} f^{\mu\rho}\sigma^{\nu\sigma} \{\Gamma_{\mu\nu},\Gamma_{\rho\sigma}\} = - \frac{e\hbar}{c\rho^2} f^{\mu\nu}\sigma_{\mu\rho}\Gamma_{\nu}{}^{\rho} + \frac{3e\hbar}{c\rho^2} f^{\mu\nu}\sigma_{\mu\nu} . \quad \text{(III.31)}$$

A further relation refers to the inverse of the coefficient of p^2 in (III.25), given by

$$\frac{1}{\sigma^{\mu\nu}M_{\mu\nu}-s(s+1)-\frac{5}{4}-(\rho-\frac{q}{2})^2} = \frac{\sigma^{\mu\nu}M_{\mu\nu}+s(s+1)+\frac{21}{4}+(\rho-\frac{q}{2})^2}{[\frac{3}{4}-s(s+1)-(\rho-\frac{q}{2})^2][\frac{29}{4}+s(s+1)+(\rho-\frac{q}{2})^2]} . \quad \text{(III.32)}$$

From (III.25) and (III.26) it is clear that what we have obtained is a relation between magnetic moments and that the magnetic moments for various states are yet to be derived.

IV. MASS LEVELS AND PARITIES FOR N = 4, 6 and 10

The free particle wave equation (I.2) for N=6 can be written as

$$(\gamma^{\mu}p_{\mu} - i\, Mc)\, \Psi = 0 , \quad \text{(IV.1)}$$

where

$$Mc = \frac{2\tau_0 g p^2}{2mc\rho} (x^2 - \rho^2 - 3A) ,$$

$$x = \frac{mc\rho q}{p} , \quad A = \frac{1}{3}[\frac{15}{4} - S^2] , \quad A^2 = A ,$$

$$s(s+1) = S^2 = \frac{W}{p^2} , \quad W = W_{\mu}W^{\mu} , \quad W^{\mu} = \frac{1}{2}\varepsilon^{\mu\nu\rho\sigma}S_{\nu\rho}p_{\sigma} ,$$

$$S_{\mu\nu} = J_{\mu\nu} + \frac{1}{2}\sigma_{\mu\nu} .$$

From (IV.1) we obtain

$$\frac{mc\rho q}{p} = BZ + B\sqrt{(\rho^2 + 3A + 1)} . \quad \text{(IV.2)}$$

In the derivation of (IV.1) we used the projection operators (6.1)-(6.2) of I which act on $\Gamma_{\mu\nu}\gamma^{\mu}p^{\nu} = \underline{\Gamma}$ according to

$$\Gamma_{++}\Gamma_{-} = 4\tau_0 \underline{p}\, \Gamma_{++} , \quad \Gamma_{+-}\underline{\Gamma} = 0 , \quad \Gamma_{-+}\underline{\Gamma} = - 2\tau_0\underline{p}\, \Gamma_{-+} ,$$

$$\underline{\Gamma}\Gamma_{--} = \Gamma_{--}\underline{\Gamma} = -\frac{1}{p^2}\, \underline{p}\underline{p}\, \underline{\Gamma}\Gamma_{--} = -\frac{\underline{p}}{p^2}[-2\tau_0 p^2 + i\sigma^{\mu\rho}\Gamma_{\nu\rho}p_{\mu}p^{\nu}]\Gamma_{--} = 6\underline{p}\tau_0\Gamma_{--} ,$$

where

$$\underline{p} = \gamma^\mu p_\mu \ , \ \frac{1}{p^2}\sigma^{\mu\rho}\Gamma_{\nu\rho}p_\mu p^\nu \Gamma_{--} = 4i\tau_o\Gamma_{--} \ ,$$

$$2\tau_o = \frac{\Gamma_{\mu\nu}p^\mu p^\nu}{p^2} \ , \ (2\tau_o)^2 = 1 \ .$$

On substituting from (IV.2) in (IV.1) we obtain

$$[\gamma^\mu p_\mu - i2\tau_o BZ \frac{mc\rho q}{BZ+B\sqrt{(\rho^2+3A+1)}}] \ \Psi = 0 \ .$$

In the rest frame, using (IV.2) and $2\tau_o = \Gamma_{44}$, we obtain

$$\beta\Gamma_{44}\Psi = BZ\Psi \ . \qquad \text{(IV.3)}$$

Hence we see that the parity operation

$$\beta\Gamma_{44} \ I_s \ \Psi(0) = \beta\Gamma_{44} \ \Psi(0) = BZ\Psi(0)$$

shows that BZ are the eigenvalues of the parity operator

$$\mathcal{P} = \beta\Gamma_{44} \ I_s \ , \ \mathcal{P}' = BZ$$

of N = 6, where I_s acts on p_μ according to

$$I_s \ \underline{p} = - \underline{p} \ , \ I_s \ p_4 = p_4 \ ,$$

and where Z = ± 1 and B(=±1) is the baryon number.

The mass formulae for the N = 4 and N = 6 can be written in a single relation as

$$\frac{m\rho}{M} = BZ + B\sqrt{[(\rho + \frac{1}{2}q(Y-1))^2 + 3A+Y]} \ , \qquad \text{(IV.4)}$$

where Z, the parity quantum number of the particle, assumes the value + 1 when N = 4 and $s = \frac{3}{2}$ and ± 1 for all other states of N = 4 and N = 6. Furthermore Y = 0 for N = 4 and Y = 1 for N = 6. The "internal" quantum number q (= ± 1) is defined in the previous section. It must be noted that because of the relation

$$\{\Gamma_5 , \Gamma_{\mu\nu}\} = 0$$

for the N = 6 representation (see 4.6 and (A2.21) of I) the wave functions for the two internal states q=1 and q=-1 are given by Ψ and $\Gamma_5\Psi$, respectively.

Thus the mass spectrum for the supermultiplets [4,4] and [4,6] depend on the baryon (B), parity (Z), spin (s) quantum numbers as well as on q and Y defined above. The two parameters m and ρ play, of course, the most crucial role. For N = 1 the wave equation (1.2) becomes

$$(\gamma^{\mu} p_{\mu} -i\, mc)\, \Psi = 0 \tag{IV.5}$$

and therefore the parameter m will be interpreted as the mass of a spin $\frac{1}{2}$ particle corresponding to the "elementary" state of matter. All other states higher than N = 1 are composite structures of the N = 1 elementary states. A preliminary fit of the low lying levels (proton, neutron, Λ etc.) with (IV.4) implies the value of 1200 to 1300 Mev for m and 3.5 to 4 for ρ. The nucleon is placed in Y = 1, q = 1 (proton), q = - 1 (neutron) with A = 1 (i.e. s = $\frac{1}{2}$) and Z = 1. Thus the proton-neutron mass difference vanishes for free particle states. The mass difference may be revealed from the binding of the three elementary systems, each of mass m, in the presence of interaction. Furthermore, it must be pointed out that not all the members of a supermultiplet [4,N] (N = 4,6,10...) can be found in so far observed baryon resonances. It is also quite conceivable that the proton and neutron when bound inside the nuclei may each assume various mass levels differing from their free state masses.

For the supermultiplet [4,10] the wave equation (I.2), using the representation N = 10 of the appendix, can be written as

$$[(\rho+q)\gamma^{\sigma}p_{\sigma}-imc\rho]\Psi_{\{\mu\nu\}}-q[\gamma_{\mu}p^{\sigma}\Psi_{\{\sigma\nu\}}+\gamma_{\nu}p^{\sigma}\Psi_{\{\sigma\mu\}}+ p_{\mu}\gamma^{\sigma}\Psi_{\{\sigma\nu\}}+p_{\nu}\gamma^{\sigma}\Psi_{\{\sigma\mu\}}] = 0 , \tag{IV.6}$$

where the spinor index of the wave function $\Psi_{\{\mu\nu\}}$ has been suppressed. From the reduction $[(0,\frac{1}{2})+(\frac{1}{2},0)] \otimes [(1,1)+(0,0)] = [(0,\frac{1}{2})+(\frac{1}{2},0)] + [(1,\frac{1}{2})+(\frac{1}{2},1)] + [(1,\frac{3}{2})+(\frac{3}{2},1)]$, it follows that the supermultiplet [4,10] contains a pair ($q = \pm 1$) of $s = \frac{1}{2}$ triplets, a pair of $s = \frac{3}{2}$ doublets and a pair of $s = \frac{5}{2}$ singlets. The wave functions for the $s = \frac{1}{2}$ states are of the form

$$\Psi_1 = g^{\mu\nu}\Psi_{\{\mu\nu\}}, \quad \Psi_2 = \frac{1}{p}\gamma^\mu p^\nu \Psi_{\{\mu\nu\}}, \quad \Psi_3 = \frac{1}{p^2} p^\mu p^\nu \Psi_{\{\mu\nu\}},$$

and satisfy, as follows from (IV.6), the coupled set of equations

$$[(\rho+q)\gamma^\mu p_\mu - imc\rho]\Psi_1 - 4qp\Psi_2 = 0 , \qquad \text{(IV.7)}$$

$$- 2qp\Psi_2 + [(\rho-q)\gamma^\mu p_\mu - imc\rho]\Psi_3 = 0 ,$$

$$- qp\Psi_1 + [(\rho+q)\gamma^\mu p_\mu + imc\rho]\Psi_2 + 2(\rho-2q)p\Psi_3 = 0 .$$

Hence the mass spectrum for the pair of $s = \frac{1}{2}$ triplets is the solution of the cubic equation

$$x^3 - B(\rho-q)x^2 - [(\rho-q)^2 + 12]x + B[(\rho-q)^3 + 4(\rho+q)] = 0 , \qquad \text{(IV.8)}$$

where

$$x = \frac{m\rho}{M} .$$

Putting $x = t + \frac{1}{3}B(\rho-q)$, the equation (IV.8) becomes

$$t^3 - 12t\,[1 + (\frac{\rho-q}{3})^2] + 8\,B\,[2\,(\frac{\rho-q}{3})^3 + q] = 0 . \qquad \text{(IV.9)}$$

The discriminant of the equation is given by

$$\Delta = - 48 - \frac{64}{3}(\frac{\rho-q}{3})^2\,[(\rho-q - \frac{1}{2})^2 + \frac{35}{4}]$$

which is negative and hence the cubic equation (IV.8) has

three distinct real roots. The three roots x_1, x_2, x_3 satisfy the relation

$$x_1 + x_2 + x_3 = \rho - q \quad .$$

Hence we obtain the sum rule

$$\frac{1}{M_1} + \frac{1}{M_2} + \frac{1}{M_3} = \frac{1}{m} - \frac{q}{m\rho} \quad . \qquad \text{(IV.10)}$$

In order to complete the sum rule we must eliminate the parameters ρ and m from (IV.10). This can be done by calculating the spectrum for $s = \frac{3}{2}$ and $\frac{5}{2}$. The wave functions for $s = \frac{3}{2}$ states are of the form

$$u_\mu = (\Gamma_+)^\nu_\mu \, \phi_\nu \ , \ \upsilon_\mu = (\Gamma_+)^\nu_\mu \, \eta_\nu \ , \qquad \text{(IV.11)}$$

where we have the identities

$$\gamma^\mu u_\mu = p^\mu u_\mu = 0 \ , \ \gamma^\mu \upsilon_\mu = p^\mu \upsilon_\mu = 0$$

and where

$$\phi_\mu = \frac{1}{p} \, p^\nu \Psi_{\{\mu\nu\}} \ , \ \eta_\mu = \gamma^\nu \Psi_{\{\mu\nu\}} \quad .$$

The spin projection operator Γ_+ is given by (8.7) of I. In this case the corresponding mass spectrum is degenerate in q and is given by

$$\frac{m\rho}{M} = BZ + B\sqrt{(\rho^2 + 4)} \quad . \qquad \text{(IV.12)}$$

For the $s = \frac{5}{2}$ state the spin projection operator is given by

$$\Gamma = \frac{1}{40} \, (S^2 - \tfrac{3}{4}) \cdot (S^2 - \tfrac{15}{4})$$

and the corresponding wave function

$$t_{\{\mu\nu\}} = \Gamma^{\rho\sigma}_{\mu\nu} \, \Psi_{\{\rho\sigma\}}$$

satisfies the 28 identities

$$p^\mu p^\nu t_{\{\mu\nu\}} = 0, \ \gamma^\mu p^\nu t_{\{\mu\nu\}} = 0, \ g^{\mu\nu} t_{\{\mu\nu\}} = 0, \ p^\nu t_{\{\mu\nu\}} = 0 \ .$$

It obeys the wave equation

$$[(\rho+q)\gamma^\sigma p_\sigma - imc\rho]t_{\{\mu\nu\}} = 0 \quad . \tag{IV.13}$$

Hence the masses of the $s = \frac{5}{2}$ states are

$$\frac{m\rho}{M} = B(\rho+q) \quad . \tag{IV.14}$$

By using (IV.12) and (IV.14) in (IV.10) we obtain the sum rule for the supermultiplet [4,10] as

$$\frac{1}{M_1} + \frac{1}{M_2} + \frac{1}{M_3} = \frac{1}{M_5} - q\left(\frac{1}{M_3} - \frac{1}{M_4}\right) , \tag{IV.15}$$

where M_3 and M_4 correspond to $Z = 1$ and $Z = -1$, respectively, in (IV.12) and M_5 is given by (IV.14).

It is interesting to observe that the $N = 1, 4, 6, 10$ representations yield in the supermultiplets $[4,N]$ positive parity states of one singlet, two doublets and one triplet. It may therefore be tempting to associate all of these eight states into an SU(3) octet of baryons with $N = 1$ for Λ-particle, $N = 4$ for Ξ^-, Ξ^0, $N = 6$, for p and n, and finally $N = 10$ for Σ^+, Σ^0, Σ^-.

V. MESON MASS SPECTRA AND PARITIES

From the structure of β_μ and $J_{\mu\nu}$ matrices it is clear that the mesons are composite states of two elementary systems. For the singlet [5,1] this was discussed in A7 of I. The meson wave equation

$$\left[\frac{q}{\rho} \Gamma_{\mu\nu}\beta^\mu p^\nu + y\ \beta^\mu p_\mu - imc\right] \phi = 0 \tag{V.1}$$

for the supermultiplet [5,4] can be written as

$$[p^2((\rho q - \tfrac{1}{2} - 2\tau_0)^2 + 2 + 2\tau_0) - m^2c^2\rho^2]\ \phi = 0$$

where

$$2\tau_o = -\frac{1}{p^2}(\Lambda_{\mu\nu} + \frac{1}{2}g_{\mu\nu})p^\mu p^\nu, \quad (2\tau_o)^2 = 1$$

$$= -\frac{1}{p^2}\bar{\Lambda}_{\mu\nu}\,p^\mu p^\nu \quad ,$$

$$(\frac{1}{p^2}\bar{\Lambda}^\alpha_\mu\,\bar{\Lambda}_{\alpha\nu}p^\mu p^\nu)^\sigma_\rho = 2\,\delta^\sigma_\rho + 2\,\frac{p_\rho p^\sigma}{p^2} = 3 + 2\tau_o \quad .$$

We observe that the matrices

$$B^o_\mu = \frac{q}{\rho}\,\Gamma_{\mu\nu}\,\beta^\nu + y\,\beta_\mu$$

for the mesons, in contrast to B_μ for the fermions, are singular and therefore the equation (V.1) contains appropriate supplementary conditions since not all of the states are filled in this case. On the other hand B_μ for fermions being non-singular matrices, do not lead to any supplementary conditions and therefore all states are filled. Thus for the mass spectrum we have

$$(\frac{m\rho}{M})^2 = [\rho - (\frac{1}{2} + Z)q]^2 + 2 + Z \ , \qquad \text{(V.2)}$$

where

$$2\tau_o\ \phi = Z\ \phi \quad .$$

From

$$S^2 = \frac{3}{2}\Lambda_- + \frac{1}{2}\tau\cdot\tau'\,\Lambda_- \ , \quad \Lambda_- = \frac{1}{2}(1-\beta_5) \ ,$$

$$= \Lambda_- - 2\tau_o\Lambda_-$$

we obtain

$$2\tau_o = 1 - S^2 = 1 - s(s+1)$$

so that for $s = 0$ we have $Z = 1$ and for $s = 1$ we have $Z = -1$. Thus the spectrum (V.2) describes four masses of a pair of 0^+ and a pair of 1^- mesons.

A similar analysis for the supermultiplet [5,6] yields the equation

$$[(\rho^2 + 4\rho q\tau_o + 3)p^2 - m^2c^2\rho^2]\ \phi = 0 \qquad (V.3)$$

where now $2\ \tau_o = \frac{1}{p^2}\ \Gamma_{\mu\nu}p^\mu p^\nu$, $(2\tau_o)^2 = 1$. The corresponding mass spectrum is

$$(\frac{m\rho}{M})^2 = (\rho + qZ)^2 + 2 \quad , \qquad (V.4)$$

where Z as given by

$$2\tau_o\phi = \Gamma_{44}\phi = Z\phi \qquad (V.5)$$

is the parity quantum number. In this case we get a pair of 1^+ and a pair of 1^- mesons. For $\rho=0$ in [5,6] we obtain Maxwell's equations. Thus the [5,6] corresponds to a massive electrodynamics.

For the supermultiplet [5,10] the wave equation requires the calculation of the 10×10 matrices

$$(\Gamma_{\mu\gamma}\Gamma_\nu{}^\gamma)_{\{\rho\sigma\ \alpha\beta\}} = -2\Gamma_{\mu\nu} - 2iJ_{\mu\nu} + 5g_{\mu\nu}\ I_{\rho\sigma,\alpha\beta} + 2g_{\alpha\beta}\ I_{\mu\nu,\rho\sigma} + 2g_{\rho\sigma}\ I_{\mu\nu,\alpha\beta} \quad , \qquad (V.6)$$

where (for N=10)

$$(\Gamma_{\mu\nu})_{\rho\sigma,\alpha\beta} = g_{\mu\nu}\ I_{\rho\sigma,\alpha\beta} - g_{\alpha\sigma}\ I_{\mu\nu,\rho\beta} - g_{\beta\rho}\ I_{\mu\nu,\alpha\sigma} - g_{\alpha\rho}\ I_{\mu\nu,\sigma\beta} - g_{\sigma\beta}\ I_{\mu\nu,\rho\alpha} \quad ,$$

$$I_{\mu\nu,\rho\sigma} = \frac{1}{2}\ (g_{\mu\rho}\ g_{\nu\sigma} + g_{\mu\sigma}\ g_{\nu\rho}) \quad .$$

Also

$$-(J_{\mu\gamma}J_\nu{}^\gamma)_{\rho\sigma,\alpha\beta} = 2\Gamma_{\mu\nu} + iJ_{\mu\nu} + 2g_{\rho\sigma}\ I_{\mu\nu,\alpha\beta} + 2g_{\alpha\beta}\ I_{\mu\nu,\rho\sigma} - 4g_{\mu\nu}\ I_{\rho\sigma,\alpha\beta}, \qquad (V.7)$$

where

$$(J_{\mu\nu})_{[\{\rho\sigma\},\{\alpha\beta\}]} = i(g_{\nu\beta}\ I_{\rho\sigma,\mu\alpha} - g_{\mu\beta}\ I_{\rho\sigma,\nu\alpha} + g_{\nu\alpha}\ I_{\rho\sigma,\mu\beta} - g_{\mu\alpha}\ I_{\rho\sigma,\nu\beta}) \quad .$$

Hence

$$S^2=\frac{1}{2}\,J_{\mu\nu}J^{\mu\nu}-\frac{1}{p^2}(J_{\mu\rho}J_\nu{}^\rho)p^\mu p^\nu=4I_{\rho\sigma,\alpha\beta}-2g_{\rho\sigma}g_{\alpha\beta}+2\,\frac{\Gamma_{\mu\nu}p^\mu p^\nu}{p^2}+$$

$$\frac{2}{p^2}\,(g_{\rho\sigma}p_\alpha p_\beta+g_{\alpha\beta}p_\rho p_\sigma)\;.$$

In analogy with N = 4 we define a parity matrix by

$$\mathcal{p}=\frac{\Gamma_{\mu\nu}p^\mu p^\nu}{p^2}+4\,\frac{p_\rho p_\sigma p_\alpha p_\beta}{p^4}\quad, \tag{V.8}$$

and obtain the results

$$\mathcal{p}^2=1\;,$$

$$(S^2-4-2\mathcal{p})^2=-\,6(S^2-4-2\mathcal{p}) \tag{V.9}$$

or

$$S^2=4+2\mathcal{p}\quad,\quad S^2=2(\mathcal{p}-1) \tag{V.10}$$

so that for $\mathcal{p} = 1$ and $\mathcal{p} = -1$ we have the spins $S^2 = 6,(s=2)$, $S^2 = 0,(s=0)$ and $S^2 = 2,(s=1)$, respectively. The mass spectrum is given by

$$\frac{m^2\rho^2}{M^2}=\mathcal{p}^2+S^2+1+2\mathcal{p}(qy\rho-2)+4(3-qy\rho)\pm4\sqrt{[(3-qy\rho)^2}$$

$$+\,\frac{1}{2}\,(qy\rho-2)(4+2\mathcal{p}-S^2)]\,, \tag{V.11}$$

where for 1^- and 2^+ only the minus sign of the square root is to be taken.

Now, if we set $\rho = 0$ in (V.1) we obtain

$$\Gamma_{\mu\nu}\;\beta^\mu p^\nu\;\phi=0\quad, \tag{V.12}$$

or

$$-p_\mu\phi_{\{\rho\sigma\}}+p_\rho\phi_{\{\sigma\mu\}}+p_\sigma\phi_{\{\rho\mu\}}+g_{\mu\sigma}p^\nu\phi_{\{\rho\nu\}}+g_{\mu\rho}p^\nu\phi_{\{\sigma\nu\}}=0\;.$$

Contracting with respect to μ,σ and also ρ,σ we obtain

$$p_\mu \pi_1 = -5\, p^\rho \phi_{\mu\rho} \; , \; p_\mu \pi_1 = -4\, p^\rho \phi_{\mu\rho} \quad .$$

Hence the wave function of the graviton satisfy the 8 conditions

$$p_\mu \pi_1 = 0 \; , \; p^\rho \phi_{\{\mu\rho\}} = 0 \tag{V.13}$$

and it, therefore, has only two states of polarization. On setting $\rho = 0$ we get the wave equation

$$p^2 \phi_{\{\mu\nu\}} = 0 \quad . \tag{V.14}$$

Thus for $\rho \neq 0$ the [5,10] describes a "massive gravitational" field.

VI. CURRENT DENSITY

The conserved current vector (see I) is given by

$$J_\mu = -i\, \bar{\psi}\, B_\mu\, \Psi \tag{VI.1}$$

where the adjoint wave function $\bar{\psi}$ for the [4,6] is defined as

$$\bar{\psi} = \psi^\dagger\, \beta\Gamma_{44} \quad .$$

For the [4,6] the eigenvalues of the matrix $-iB_4$ are given by

$$(-iB_4)' = \frac{m}{M} = \frac{1}{\rho}\,[BZ+B\surd(\rho^2+(1+\lambda^2)(3A+1))] \quad . \tag{VI.2}$$

We can decompose the supermultiplet current (VI.1) into the sum of elementary currents of the individual members of a supermultiplet. Thus consider a transformation matrix V which acts on B_4 to diagonalize it according to

$$\bar{V}\,(-iB_4)V = \frac{m}{M} \quad .$$

The current (VI.1) reduces to

$$J_\mu = \sum_{zsqy} \left(\frac{m}{M}\wp\right)_{zsqy} [-i\,\bar{\phi}\,\gamma_\mu\,\phi]_{zsqy} \quad , \tag{VI.3}$$

where

$$\Psi = V\phi \ , \ \bar{\Psi} = \bar{\phi}\ \bar{V} \ ,$$

and where $\mathcal{P}$ is parity and the elementary currents are $(-i\ \bar{\phi}\ \gamma_\mu\ \phi)_{zsqy}$ for the various states described by the quantum number z(parity), s(spin), q and y. The coefficients $(\frac{m}{M}\mathcal{P})_{zsqy}$ for fixed spin s can, by choosing appropriate values of z, q and y, be made to assume positive values only. Thus in the total supermultiplet current J_μ the quantity J_4 is the sum of positive contributions from the elementary currents in the supermultiplet.

A different derivation of (VI.3) can be based on the reduced equation of the [4,6]

$$(\frac{m}{M}\ \mathcal{P}'\ 2\tau_0\ \gamma_\mu\ \frac{\partial\psi}{\partial x_\mu} - \frac{mc}{\hbar})\ \psi = 0 \qquad \text{(VI.4)}$$

(see IV.3), where $\mathcal{P}' = BZ$. Using a boost transformation we may write

$$2\tau_0 = L(p)\ \Gamma_{44}\ L^{-1}(p)$$

where

$$L(p) = \exp\ [i\ J_{4j}\ \hat{p}_j\ \text{Log}\ (\sqrt{(\frac{p_4-p}{p_4+p})})]\ ,$$

$$\hat{p} = \frac{p}{P}\ ,$$

is just a Lorentz transformation on index a of the wave function ψ_a. Thus the equation (VI.4) can be transformed into

$$\frac{m}{M}\ \mathcal{P}'\ \Gamma_{44}\ \gamma_\mu\ \frac{\partial\psi_o}{\partial x_\mu} - \frac{mc}{\hbar}\ \psi_o = 0 \ ,$$

where

$$\psi_o = L^{-1}\ \psi\ ,\ \bar{\psi}_o = \bar{\psi}\ L\ .$$

The corresponding conserved current density is

$$J_\mu = - i \Sigma \bar{\psi}_o \frac{m}{M} \mathcal{P} \Gamma_{44} \gamma_\mu \psi_o$$

or

$$J_\mu = - i \Sigma (\bar{\psi} \frac{m}{M} \mathcal{P}' 2\tau_o \gamma_\mu \psi) \quad . \tag{VI.5}$$

This is the general expression for the current density. For the [4,4] multiplet

$$2\tau_o = - \frac{\Gamma_{\mu\nu} p^\mu p^\nu}{p^2} - \frac{1}{2} , \; (2\tau_o)^2 = 1 .$$

* * * * * *

It is an honor to acknowledge the inspiration received during the development of this theory from Professor P.A.M. Dirac, both from the magnificent structure of his published work and from his personal encouragement and criticism during the past few years.

REFERENCES

1. B. Kursunoglu, Phys. Rev. D, Vol. 1, No. 4, 115, 1970 and Phys. Rev. D, Vol. 2, No. 4, 717, 1970.
2. M. Hamermesh, GROUP THEORY, Chapter 10, Addison-Wesley Publishing Company, 1962.

APPENDIX

The generators of U(3,1) for the irreducible nonunitary representations can easily be constructed. Thus the generators which act on Ψ_a (N=6) are, as was shown in I, given by

$$(J_{\mu\nu})_{ab} = -\frac{1}{2} i \, (Q_{\mu\rho a} Q_{\nu}{}^{\rho}{}_{b} - Q_{\nu\rho a} Q_{\mu}{}^{\rho}{}_{b}) , \tag{1}$$

$$(\Gamma_{\mu\nu})_{ab} = \frac{1}{4} g_{\mu\nu} Q_{\rho\sigma a} Q^{\rho\sigma}{}_{b} - \frac{1}{2} (Q_{\mu\rho a} Q_{\nu}{}^{\rho}{}_{b} + Q_{\nu\rho a} Q_{\mu}{}^{\rho}{}_{b}) , \tag{2}$$

where a,b = 1,2,..,6 and where the symbols $Q_{\mu\nu a}$ are defined by (A2.2)-(A2.3) of I. In terms of space-time indices the representations for N=6 are given by

$$(J_{\mu\nu})_{[[\rho\sigma],[\alpha\beta]]} = \frac{1}{2}[g_{\alpha\rho}(M_{\mu\nu})_{\sigma\beta} + g_{\sigma\beta}(M_{\mu\nu})_{\rho\alpha} - g_{\alpha\sigma}(M_{\mu\nu})_{\rho\beta} - g_{\rho\beta}(M_{\mu\nu})_{\sigma\alpha}] \tag{3}$$

$$(\Gamma_{\mu\nu})_{\{[\rho\sigma],[\alpha\beta]\}} = \frac{1}{2}[g_{\alpha\sigma}(\Lambda_{\mu\nu})_{\rho\beta} + g_{\beta\rho}(\Lambda_{\mu\nu})_{\alpha\sigma} - g_{\alpha\rho}(\Lambda_{\mu\nu})_{\sigma\beta} - g_{\sigma\beta}(\Lambda_{\mu\nu})_{\rho\alpha}] \tag{4}$$

where $M_{\mu\nu}$ and $\Lambda_{\mu\nu}$, as defined by (I.7) and (I.8), are the generators of U(3,1) for N=4. The expressions (3) and (4) follow from (1) and (2) by multiplying them by $-\frac{1}{4} Q_{\rho\sigma a} Q_{\alpha\beta b}$ and summing over a and b (see appendix A of I).

The metric δ_{ab} of N=6 vector space spanned by the six-vector Ψ_a corresponds to the metric

$$\delta^{[\rho\sigma]}_{[\mu\nu]} = -\frac{1}{4} Q_{\mu\nu a} Q^{\rho\sigma}{}_{b} \delta_{ab} = \frac{1}{2} (\delta^{\rho}_{\mu}\delta^{\sigma}_{\nu} - \delta^{\sigma}_{\mu}\delta^{\rho}_{\nu}) \tag{5}$$

of the vector space spanned by the tensors $\Psi_{[\mu\nu]}$. The Casimir operator for N=6 is given by

$$I^{[\rho\sigma]}_{[\mu\nu]} = \frac{1}{2} (J_{\alpha\beta} J^{\alpha\beta} + \Gamma_{\alpha\beta} \Gamma^{\alpha\beta}) = 10 \, \delta^{[\rho\sigma]}_{[\mu\nu]} \tag{6}$$

which for N=4 has the value

$$I^{\nu}_{\mu} = \frac{1}{2} (M_{\alpha\beta} M^{\alpha\beta} + \Lambda_{\alpha\beta} \Lambda^{\alpha\beta}) = \frac{15}{2} \delta^{\nu}_{\mu} . \tag{7}$$

All the representations of U(3,1) can be built out of

the regular representation (N=4). Thus for N=10 the generators are given by

$$(J_{\mu\nu})_{[\{\rho\sigma\},\{\alpha\beta\}]}=\frac{1}{2}[g_{\alpha\rho}(M_{\mu\nu})_{\sigma\beta}+g_{\beta\sigma}(M_{\mu\nu})_{\rho\alpha}+g_{\alpha\sigma}(M_{\mu\nu})_{\rho\beta}+g_{\rho\beta}(M_{\mu\nu})_{\sigma\alpha}], \tag{8}$$

$$(\Gamma_{\mu\nu})_{\{\{\rho\sigma\},\{\alpha\beta\}\}}=\frac{1}{2}[g_{\alpha\sigma}(\Lambda_{\mu\nu})_{\rho\beta}+g_{\beta\rho}(\Lambda_{\mu\nu})_{\alpha\sigma}+g_{\alpha\rho}(\Lambda_{\mu\nu})_{\sigma\beta}+g_{\sigma\beta}(\Lambda_{\mu\nu})_{\rho\alpha}]. \tag{9}$$

The Casimir operator is

$$I^{\{\rho\sigma\}}_{\{\mu\nu\}} = 18 \quad \delta^{\{\rho\sigma\}}_{\{\mu\nu\}} ,$$

where

$$\delta^{\{\rho\sigma\}}_{\{\mu\nu\}} = \frac{1}{2}(\delta^{\rho}_{\mu}\delta^{\sigma}_{\nu} + \delta^{\sigma}_{\mu}\delta^{\rho}_{\nu})$$

is the metric of the 10-dimensional vector space spanned by the vectors $\Psi_{\{\mu\nu\}}$.

The N=15 can be expressed in terms of the six-vector notation as

$$(J_{\mu\nu})_{[[ab],[cd]]}=\frac{1}{2}[\delta_{ac}(J_{\mu\nu})_{bd}+\delta_{bd}(J_{\mu\nu})_{ac}-\delta_{bc}(J_{\mu\nu})_{ad}-\delta_{ad}(J_{\mu\nu})_{bc}] \tag{10}$$

$$(\Gamma_{\mu\nu})_{\{[ab],[cd]\}}=\frac{1}{2}[\delta_{ac}(\Gamma_{\mu\nu})_{bd}+\delta_{bd}(\Gamma_{\mu\nu})_{ac}-\delta_{bc}(\Gamma_{\mu\nu})_{ad}-\delta_{ad}(\Gamma_{\mu\nu})_{bc}] \tag{11}$$

which yield the Casimir invariant

$$I_{\{[ab],[cd]\}} = 16 \quad \delta_{\{[ab],[cd]\}} , \tag{12}$$

where

$$\delta_{\{[ab],[cd]\}} = \frac{1}{2}(\delta_{ac}\,\delta_{bd} - \delta_{ad}\,\delta_{bc}) \tag{13}$$

is the metric of the 15-dimensional space spanned by the

vectors $\Psi_{[ab]}$. If we wished we could rewrite the N=15 representation in terms of $M_{\mu\nu}$ and $\Lambda_{\mu\nu}$ representation of N=4 by the same technique used for the N=6 and N=10 representations.

The other relevant matrix for N=15 is defined by

$$\Gamma_5 = \frac{i}{64} \varepsilon^{\mu\nu\rho\sigma} J_{\mu\nu} J_{\rho\sigma} = \frac{1}{4} (g_{ac}\delta_{bd} - g_{bc}\delta_{ad} + g_{bd}\delta_{ac} - g_{ad}\delta_{bc}), \tag{14}$$

where g_{ab} is the matrix element of the Γ_5-matrix for N=6 and is discussed in I. However the Γ_5 for N=15 is a singular matrix since

$$(\Gamma_5)^2 = \Delta_+ \quad , \quad \Delta_+^2 = \Delta_+$$

where

$$(\Delta_+)_{\{[ab],[cd]\}} = -\frac{1}{8} (J_{\mu\nu})_{ab} (J^{\mu\nu})_{cd}$$

$$= \frac{1}{4} (\delta_{ac}\delta_{bd} - \delta_{ad}\delta_{bc} + g_{ac}\, g_{bd} - g_{ad}\, g_{bc}) \ . \tag{15}$$

For one of the N=20 representations we have

$$(J_{\mu\nu})_{[\{ab\},\{cd\}]} = \frac{1}{2}[\delta_{ac}(J_{\mu\nu})_{bd} + \delta_{bd}(J_{\mu\nu})_{ac} + \delta_{bc}(J_{\mu\nu})_{ad} + \delta_{ad}(J_{\mu\nu})_{bc}], \tag{16}$$

$$(\Gamma_{\mu\nu})_{\{\{ab\},\{cd\}\}} = \frac{1}{2}[\delta_{ac}(\Gamma_{\mu\nu})_{bd} + \delta_{bd}(\Gamma_{\mu\nu})_{ac} + \delta_{bc}(\Gamma_{\mu\nu})_{ad} + \delta_{ad}(\Gamma_{\mu\nu})_{bc}]$$
$$- \frac{1}{3} [\delta_{ab}(\Gamma_{\mu\nu})_{cd} + \delta_{cd}(\Gamma_{\mu\nu})_{ab}] \pm \frac{1}{3} [g_{ab}(\Gamma_{\mu\nu})_{cd} + g_{cd}(\Gamma_{\mu\nu})_{ab}] , \tag{17}$$

which yield the Casimir invariant

$$I_{\{\{ab\},\{cd\}\}} = 24 \quad \delta_{\{\{ab\},\{cd\}\}} \quad ,$$

where

$$\delta_{\{\{ab\},\{cd\}\}} = \frac{1}{2} (\delta_{ac}\delta_{bd} + \delta_{ad}\delta_{bc} - \frac{1}{3} \delta_{ab}\delta_{cd}) , \tag{18}$$

is the metric of the N=20 dimensional space spanned by the vectors of the type $\Psi_{\{ab\}}$, where $\Psi_{\{aa\}} = 0$. In this case the Γ_5-operator is given by

$$\Gamma_5 = \frac{i}{3\times 64} \varepsilon^{\mu\nu\rho\sigma} J_{\mu\nu} J_{\rho\sigma} = \frac{1}{4} [\delta_{ac}g_{bd} + \delta_{bd}g_{ac} + \delta_{ad}g_{bc} + \delta_{bc}g_{ad}$$
$$- \frac{2}{3}(g_{ab}\delta_{cd} + g_{cd}\delta_{ab})], \quad (19)$$

where

$$\Gamma_5^2 = \frac{1}{4}(\delta_{ac}\delta_{bd} + \delta_{ad}\delta_{bc} + g_{ac}g_{bd} + g_{bc}g_{ad}) - \frac{1}{6}(\delta_{ab}\delta_{cd} + g_{ab}g_{cd}), \quad (20)$$

is a projection operator.

The Young patterns for U(3,1) up to N=84 are given by

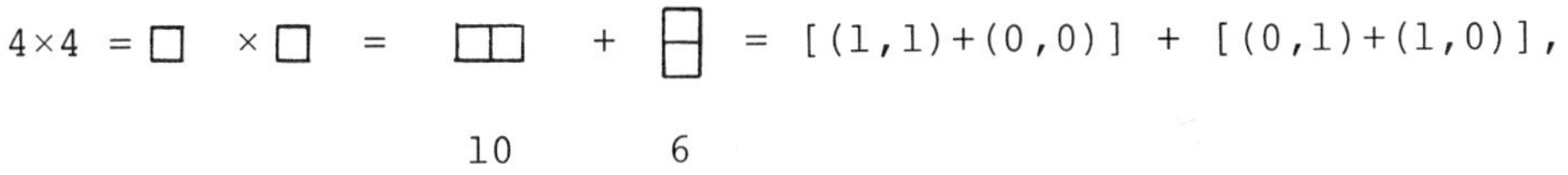

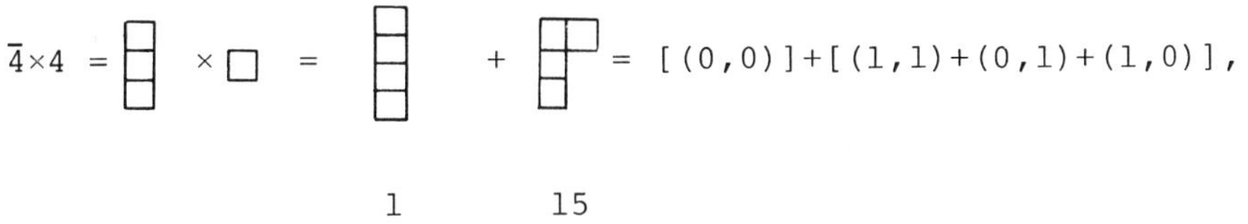

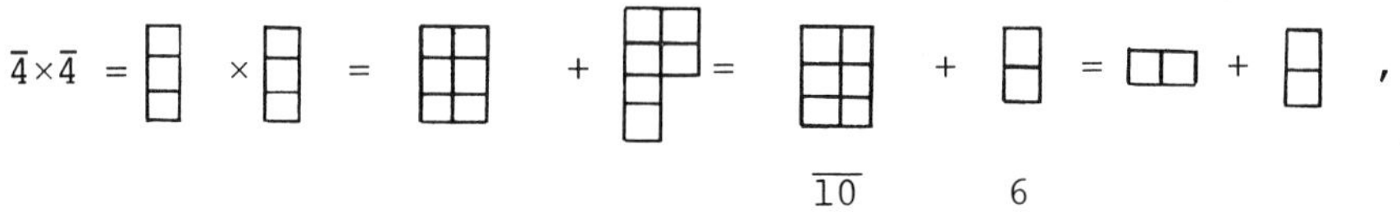

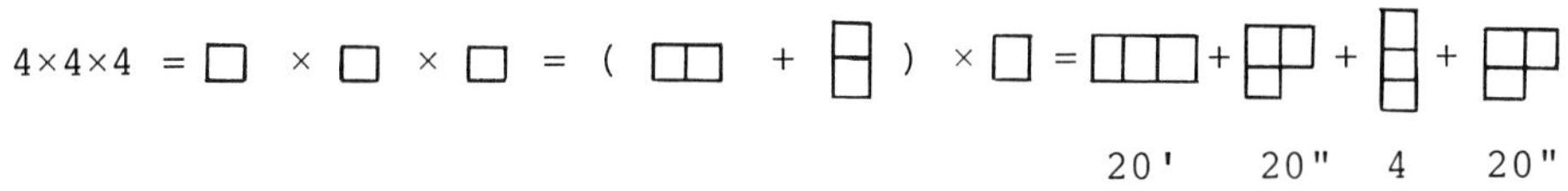

$4 \times 4 = \Box \times \Box = \Box\Box + \begin{array}{c}\Box\\\Box\end{array} = [(1,1)+(0,0)] + [(0,1)+(1,0)]$,

 10 6

$\bar{4} \times 4 = \begin{array}{c}\Box\\\Box\\\Box\end{array} \times \Box = \begin{array}{c}\Box\\\Box\\\Box\\\Box\end{array} + \begin{array}{c}\Box\Box\\\Box\\\Box\end{array} = [(0,0)] + [(1,1)+(0,1)+(1,0)]$,

$\phantom{\bar{4} \times 4 =}$ 1 15

$\bar{4} \times \bar{4} = \begin{array}{c}\Box\\\Box\\\Box\end{array} \times \begin{array}{c}\Box\\\Box\\\Box\end{array} = \begin{array}{c}\Box\Box\\\Box\Box\\\Box\Box\end{array} + \begin{array}{c}\Box\Box\\\Box\\\Box\end{array} = \begin{array}{c}\Box\Box\\\Box\Box\\\Box\Box\end{array} + \begin{array}{c}\Box\\\Box\end{array} = \Box\Box + \begin{array}{c}\Box\\\Box\end{array}$,

$\phantom{\bar{4} \times \bar{4} =}$ $\overline{10}$ 6

$4 \times 4 \times 4 = \Box \times \Box \times \Box = (\Box\Box + \begin{array}{c}\Box\\\Box\end{array}) \times \Box = \Box\Box\Box + \begin{array}{c}\Box\Box\\\Box\end{array} + \begin{array}{c}\Box\\\Box\\\Box\end{array} + \begin{array}{c}\Box\Box\\\Box\end{array}$

 20' 20" 4 20"

$= [(\frac{3}{2},\frac{3}{2})+(\frac{1}{2},\frac{1}{2})] + 2[(\frac{1}{2},\frac{3}{2})+(\frac{3}{2},\frac{1}{2})+(\frac{1}{2},\frac{1}{2})] + [(\frac{1}{2},\frac{1}{2})]$,

$4\times\bar{4}\times 4 =$ 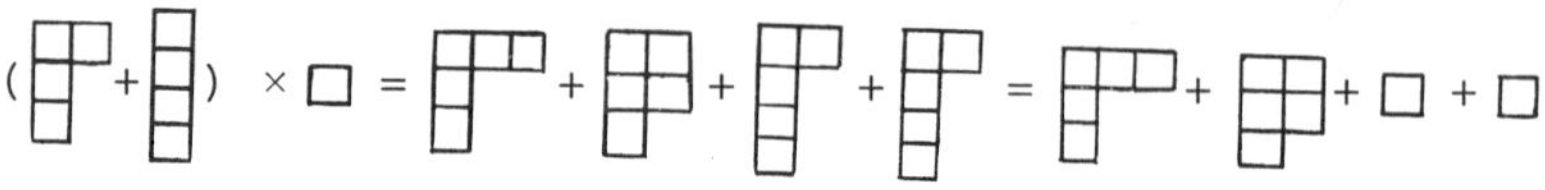

36 $\quad$ 20''' $\quad$ 4 $\quad$ 4

$$= [(\tfrac{1}{2},\tfrac{1}{2})+(\tfrac{3}{2},\tfrac{3}{2})+(\tfrac{1}{2},\tfrac{3}{2})+(\tfrac{3}{2},\tfrac{1}{2})]+[(\tfrac{1}{2},\tfrac{1}{2})+(\tfrac{1}{2},\tfrac{3}{2})+(\tfrac{3}{2},\tfrac{1}{2})]+2[(\tfrac{1}{2},\tfrac{1}{2})]$$

$\bar{4}\times\bar{4}\times 4 =$ 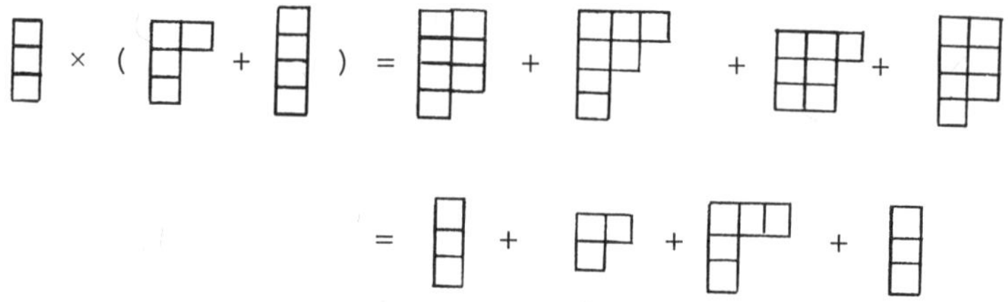

$\bar{4}$ $\quad$ 20'' $\quad$ 36 $\quad$ $\bar{4}$

$\bar{4}\times\bar{4}\times\bar{4} =$ 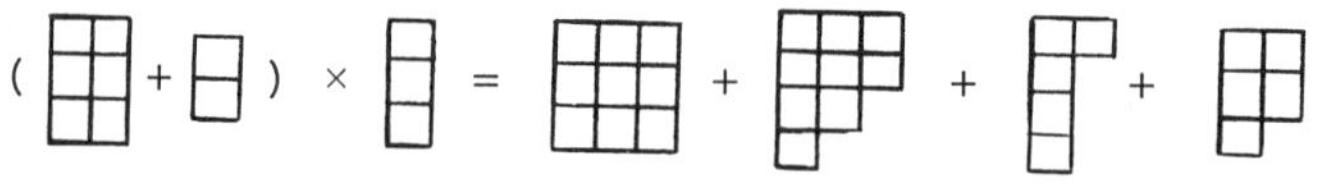

20' $\quad$ 20''' $\quad$ 4 $\quad$ 20'''

$4\times 4\times 4\times 4 =$ 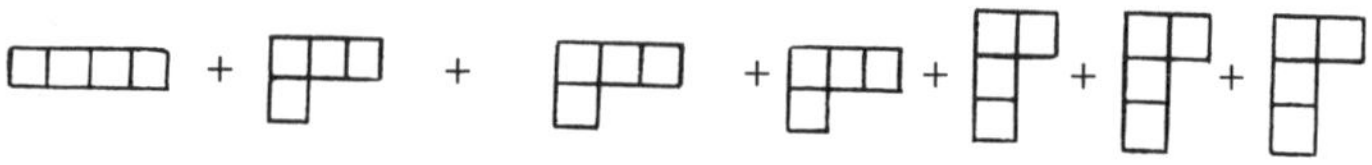

35 $\quad$ 45 $\quad$ 45 $\quad$ 45 $\quad$ 15 $\quad$ 15 $\quad$ 15

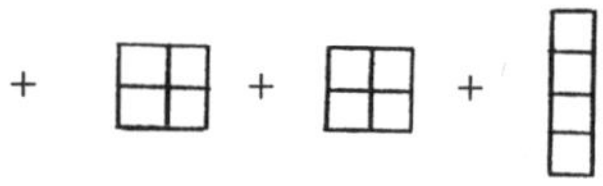

20 $\quad$ 20 $\quad$ 1

$$=[(2,2)+(1,1)+(0,0)+3[(1,2)+(2,1)+(0,1)+(1,0)+(1,1)] +$$

$$3[(1,1)+(0,1)+(1,0)]+2[(1,1)+(0,2)+(2,0)+(0,0)]+[(0,0)] ,$$

where the tableau  for N=20 is represented by $\Psi_{\{ab\}}$,

$\Psi_{\{aa\}} = 0$ and the corresponding generators are given by (16) and (17).

4×4×4×4×4 = 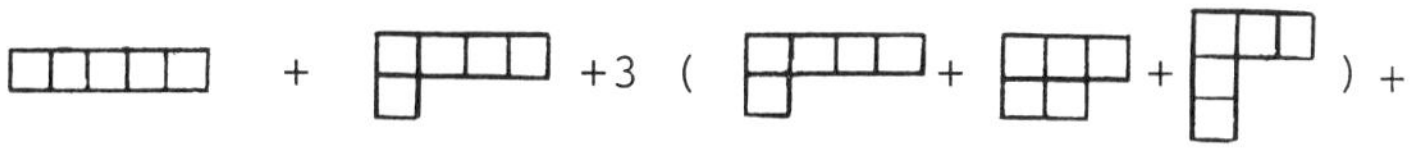

56 84' 84 60 36

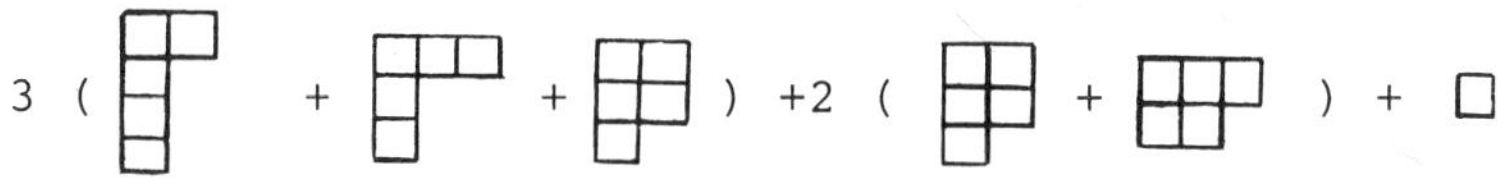

4 36 20''' 20''' 60 4

$$= [(\tfrac{5}{2},\tfrac{5}{2})+(\tfrac{3}{2},\tfrac{3}{2})+(\tfrac{1}{2},\tfrac{1}{2})]+4[(\tfrac{5}{2},\tfrac{3}{2})+(\tfrac{3}{2},\tfrac{5}{2})+(\tfrac{3}{2},\tfrac{3}{2}) +$$

$$(\tfrac{3}{2},\tfrac{1}{2})+(\tfrac{1}{2},\tfrac{3}{2})+(\tfrac{1}{2},\tfrac{1}{2})]+5[(\tfrac{5}{2},\tfrac{1}{2})+(\tfrac{1}{2},\tfrac{5}{2})+(\tfrac{3}{2},\tfrac{3}{2})+(\tfrac{3}{2},\tfrac{1}{2})+(\tfrac{1}{2},\tfrac{3}{2})+(\tfrac{1}{2},\tfrac{1}{2})]$$

$$+6[(\tfrac{3}{2},\tfrac{3}{2})+(\tfrac{3}{2},\tfrac{1}{2})+(\tfrac{1}{2},\tfrac{3}{2})+(\tfrac{1}{2},\tfrac{1}{2})]+5(\tfrac{3}{2},\tfrac{1}{2})+(\tfrac{1}{2},\tfrac{3}{2})+(\tfrac{1}{2},\tfrac{1}{2})]+4[(\tfrac{1}{2},\tfrac{1}{2})].$$

* * * * * *

It was by introducing the concept of the positron that P.A.M. Dirac first exposed us to Alice's looking glass world of antiparticles. It did not take long before the mirror nature of particle-antiparticle conjugations "C" became even more enhanced through the elaboration of the CP and CPT reflections; and since the 1964 Fitch-Cronin experiment we are faced with the possibility that these reflections at the microscopic level may even contain some of the future answers to the century-old quest for the physical origin of irreversibility.

It is with the greatest pleasure and admiration that we thereby dedicate this study to Professor P.A.M. Dirac on his 70th birthday.

* * * * * *

THERMODYNAMICS AND STATISTICAL MECHANICS OF THE CP VIOLATION

Y. Ne'eman and A. Aharony*
Tel-Aviv University
Tel-Aviv, Israel

(Presented by Y. Ne'eman)

1. Introduction

Two of the main results in the theory of statistical mechanics of irreversible processes are the H-theorem, or the Second law of Thermodynamics, and the Onsager relations between such processes. In both cases, the proofs of these results are usually based on the assumptions of microscopic reversibility, namely that the Hamiltonian of the system is invariant under time reversal, T.

The discovery of CP violation in the decay of the neutral kaon[1], with the assumption of CPT conservation, implied a violation of time reversal symmetry in this decay. Moreover, it has recently been established, that the experimental results are indeed consistent with a violation of T, and that they cannot be explained assuming T conservation[2]. There is thus evidence for a microscopic "arrow of time", defined by the behaviour of the neutral kaon system[3]. Although no other experimental proof of time reversal violation has yet been found[4], it is still of interest to investigate the necessity of the assumption of time reversal symmetry in the proofs of both the H-theorem and the Onsager relations. Since the system of the neutral kaons is the only known time reversal noninvariant system, we shall discuss these questions in relation with it.

The basic idea of the present discussion is to describe

*Also: Centre for Particle Theory, University of Texas - Austin, Texas, USA.

the K^0-$\bar{K}^0$ system as interacting with one or more thermal baths, which contain the kaon's decay products. The derivation of a Master Equation for the reduced density matrix of the system is reviewed in Sec. 2. In Sec. 3, this reduced density matrix is used to define an H-function, for the description of the irreversible approach of the system to thermal equilibrium; the validity of the H-theorem is then discussed. Sec. 4 includes a generalization to the case of several thermal baths. The Onsager coefficients for the energy currents between the system and the baths in a stationary state are defined, and the Onsager relations are checked. In Sec. 5, our results are compared with those for a time inverted coordinate scheme, and the relations between the microscopic, thermodynamical and cosmological arrows of time are discussed.

2. The Master Equation Description of the Neutral Kaon System

The derivation of the Master Equation for the reduced density matrix of the neutral kaons' system has been described in detail elsewhere[5], so that we shall give here only a short review.

We describe the kaon's decay as being due to the interaction between the kaonic system, which has the three basis states $|0\rangle$, $a_1^\dagger|0\rangle$ and $a_2^\dagger|0\rangle$ ($|0\rangle$ is the vacuum, $a_i^\dagger$ creates K_i^0, K_1^0 and K_2^0 being the CP eigenstates with eigenvalues +1 and -1), with a thermal bath which contains all the possible final decay states, e.g. 2π, $\pi\ell\nu$, etc., in thermal equilibrium.

The Hamiltonians of the kaonic system, the bath and the interaction are respectively

$$H_o^a = \hbar\omega_K \,(a_1^\dagger a_1 + a_2^\dagger a_2) \tag{2.1}$$

$$H_o^B = \sum_{rk} \hbar\omega_{rk}\, b_{rk}^\dagger\, b_{rk} \tag{2.2}$$

and

$$H_I = -\, i\hbar \sum_i \sum_{rk} g_{rk}^i\, a_i^\dagger\, b_{rk} + \text{h.c.} \tag{2.3}$$

($b_{rk}^\dagger$ creates an eigenstate of the strong Hamiltonian in the decay channel r with quantum numbers k, $-i\hbar g_{rk}^{i}$ is the appropriate matrix element of the weak Hamiltonian).

The reduced density matrix of the kaon system is now defined as

$$\rho_a = \mathrm{Tr}_B \, \rho \tag{2.4}$$

where ρ is the total density matrix, and the trace is taken over the states of the bath. The time evolution of ρ_a is given by the Wangness-Bloch master equation[6]. If we start at t=0 with

$$\rho_a(0) = \begin{pmatrix} 0 & 0 & 0 \\ 0 & \bar{\rho}(0) & \\ 0 & & \end{pmatrix} \tag{2.5}$$

(the first row and column stand for $|0\rangle$), then at time t we have

$$\rho_a(t) = \begin{pmatrix} 1-\mathrm{Tr}\,(t) & 0 & 0 \\ 0 & \bar{\rho}\;(t) & \\ 0 & & \end{pmatrix} \tag{2.6}$$

with

$$i\,\frac{\partial\bar{\rho}}{\partial t} = \Lambda\bar{\rho} - \bar{\rho}\Lambda^{\dagger} + i\Omega(1-\mathrm{Tr}\bar{\rho}) \,, \tag{2.7}$$

$$\Lambda_{ij} = \omega_K\,\delta_{ij} - \int \frac{d\omega}{\omega - \omega_K + i\varepsilon} \sum_r N_r(\omega)\left(\overline{g_{rk}^{i}\;g_{rk}^{j*}}\right)_{\omega=\omega_k} (e^{-\beta\hbar\omega}+1) \tag{2.8}$$

$$\Omega_{ij} = 2\pi \sum_r N_r(\omega_K)\; e^{-\beta\hbar\omega_K}\; \mathrm{Re}\left(\overline{g_{rk}^{i}\;g_{rk}^{i*}}\right)_{\omega_K=\omega_k} + \tag{2.9}$$

$$+ 2P \int \frac{d\omega}{\omega-\omega_K} \sum_r N_r(\omega) e^{-\beta\hbar\omega}\; \mathrm{Im}\left(\overline{g_{rk}^{i}\;g_{rk}^{i*}}\right)_{\omega=\omega_k}$$

($N_r(\omega)$ is the density of states in the channel r, $\beta = 1/k_BT$).

Clearly, for zero temperature Eq. (2.7) reduces to the usual Wigner-Weisskopf result[7]

$$i \frac{\partial \bar{\rho}}{\partial t} = \Lambda^{o}\bar{\rho} - \rho\Lambda^{o\dagger} \quad . \tag{2.10}$$

Since $h\omega_K \simeq 500$ MeV, $e^{-\beta\hbar\omega_K}$ will be very small compared to unity, unless the temperature approaches 10^{12} degrees. In this case, we can express Λ_{12} and Λ_{21} by the usual[8] T- and CPT- violation parameters ε and δ,

$$\Lambda_{12} = [\Delta m - \frac{i}{2}(\gamma_L - \gamma_S)](\varepsilon-\delta) \tag{2.11}$$

$$\Lambda_{21} = [\Delta m - \frac{i}{2}(\gamma_L - \gamma_S)](\varepsilon+\delta) \tag{2.12}$$

where γ_S, γ_L are the inverse lifetimes of K^o_S, K^o_L, $\Delta m = m_L - m_S$ their mass difference (K^o_L and K^o_S are the eigenstates of Λ). Since ε and δ are of the order 10^{-3}, ε, δ and Ω_{ij} may be regarded as small perturbations, and Eq. (2.7) may be solved, to first order in them.

For any initial condition the solution approaches the same equilibrium,

$$\rho_a^{eq} = \begin{pmatrix} 1-2e^{-\beta\hbar\omega_K} & 0 & 0 \\ 0 & e^{-\beta\hbar\omega_K} & 0 \\ 0 & 0 & e^{-\beta\hbar\omega_K} \end{pmatrix} \quad . \tag{2.13}$$

Explicit expressions for the time dependence of ρ_{ij} may be found in (5).

3. The H-Theorem

We define the H-function of the total system as

$$H_a(t) = \mathrm{Tr}[\rho_a(\ln \rho_a - \ln \rho_a^{eq})] \quad . \tag{3.1}$$

Inserting the solutions of (2.7) for a pure K^o at t=0 we find

$$H_a(t) = K^o_a(t)[1 + \mathrm{Re}\varepsilon . K^1_a(t) + \mathrm{Re}\delta . K^2_a(t) + \mathrm{Im}\delta . K^3_a(t)] \quad . \tag{3.2}$$

$K^o_a(t)$ and $-\mathrm{Re}\varepsilon . K_a(t)$ are given in Figs. 1 and 2 for $T=300^oK$ and for the experimental values Re $=1.42.10^{3}$ [9], $\delta=0$

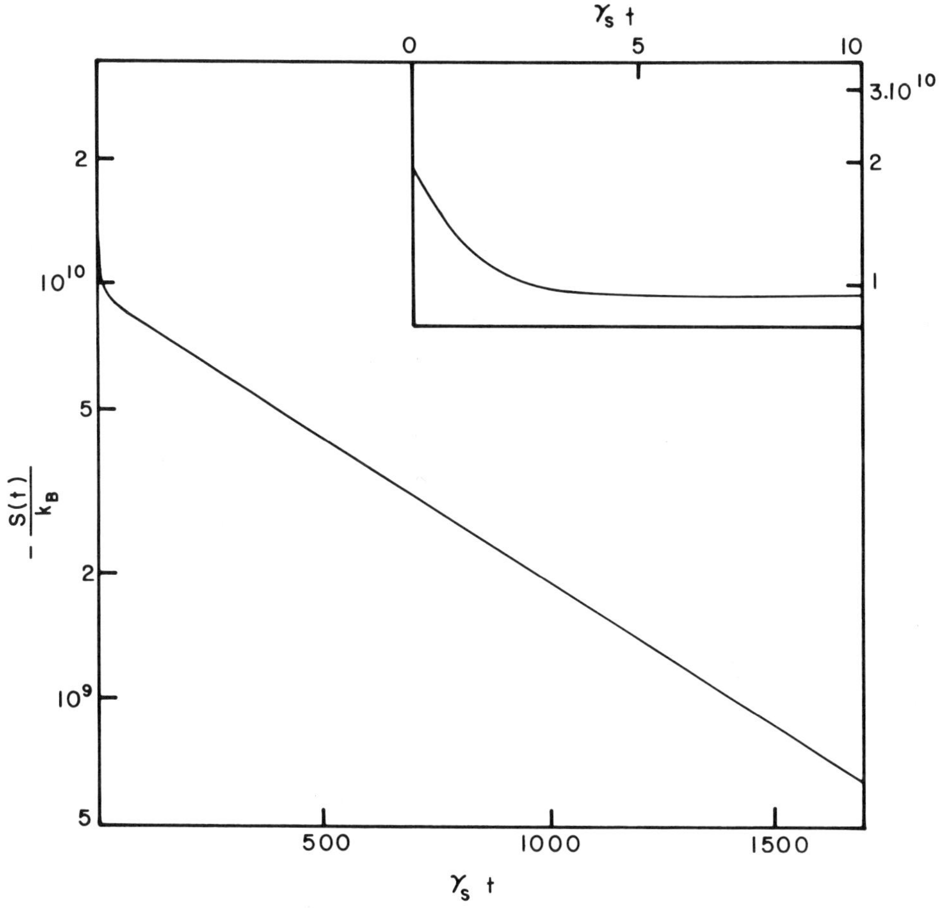

Fig. 1. The Inverse entropy $H_a = -\frac{S}{k_B}$ for K^o-decay with T and CPT symmetry.

(detailed expressions may be found in Ref. 5). As seen from the figures, the violation of time reversal symmetry, $\varepsilon \neq 0$, results in an oscillation in the entropy, superimposed on the otherwise regularly monotonic increase. Obviously, $H_a(t)$ described in Figs. 1 and 2 obeys the H-theorem, namely

$$\frac{dH_a}{dt} \leqslant 0 \qquad . \tag{3.3}$$

In order to check the possibility of a violation of the H-theorem for higher values of $\mathrm{Re}\varepsilon$, Fig. 3 gives $\frac{dH_a}{dt}$ for various values of $\mathrm{Re}\varepsilon$. It is seen that $\frac{dH_a}{dt}$ becomes negative only for $\mathrm{Re}\varepsilon$ higher than $3.4.10^{-2}$. This value is higher than the limit given by the Unitarity Sum Rule[9]:

$$\gamma_S = \sum_f | \langle f|T|K_S\rangle |^2 \tag{3.4}$$

$$\gamma_L = \sum_f | \langle f|T|K_L\rangle |^2 \tag{3.5}$$

$$[(\gamma_S-\gamma_L) - i2\Delta m \,|(\mathrm{Re}\varepsilon - i\mathrm{Im}\delta)| = \sum_f \langle f|T|K_L\rangle^* \langle f|T|K_S\rangle \ . \tag{3.6}$$

The Schwartz inequality gives - for $\delta=0$ -

$$|\mathrm{Re}\varepsilon| < \frac{(\gamma_L\gamma_S)^{\frac{1}{2}}}{|(\gamma_S-\gamma_L) - 2i\Delta m|} \simeq 0.03 \tag{3.7}$$

where the experimental values $\gamma_S=(1.17\pm0.1).10^{10}$ sec^{-1}, $\gamma_L=(1.89\pm.05).10^7$ sec^{-1} and $\Delta m/\gamma_S=.46\pm.02$ [10] have been used.

Thus, unitarity is sufficient to ensure the validity of the H-theorem for this case[12]. In fact, it may be shown that the proof of the H-theorem may be based on the unitarity of the S-matrix, and is thus independent of time reversal symmetry.[13]

4. The Stationary State and the Onsager Relations

In order to discuss the Onsager relations, we now extend the above formalism to the case where the bosons may form several heat baths, with unequal temperatures.

We assume that the particles created in the channel r enter a bath with temperature T_r. All the expressions in Eqs. (2.7)-(2.9) remain unchanged, except that β is replaced

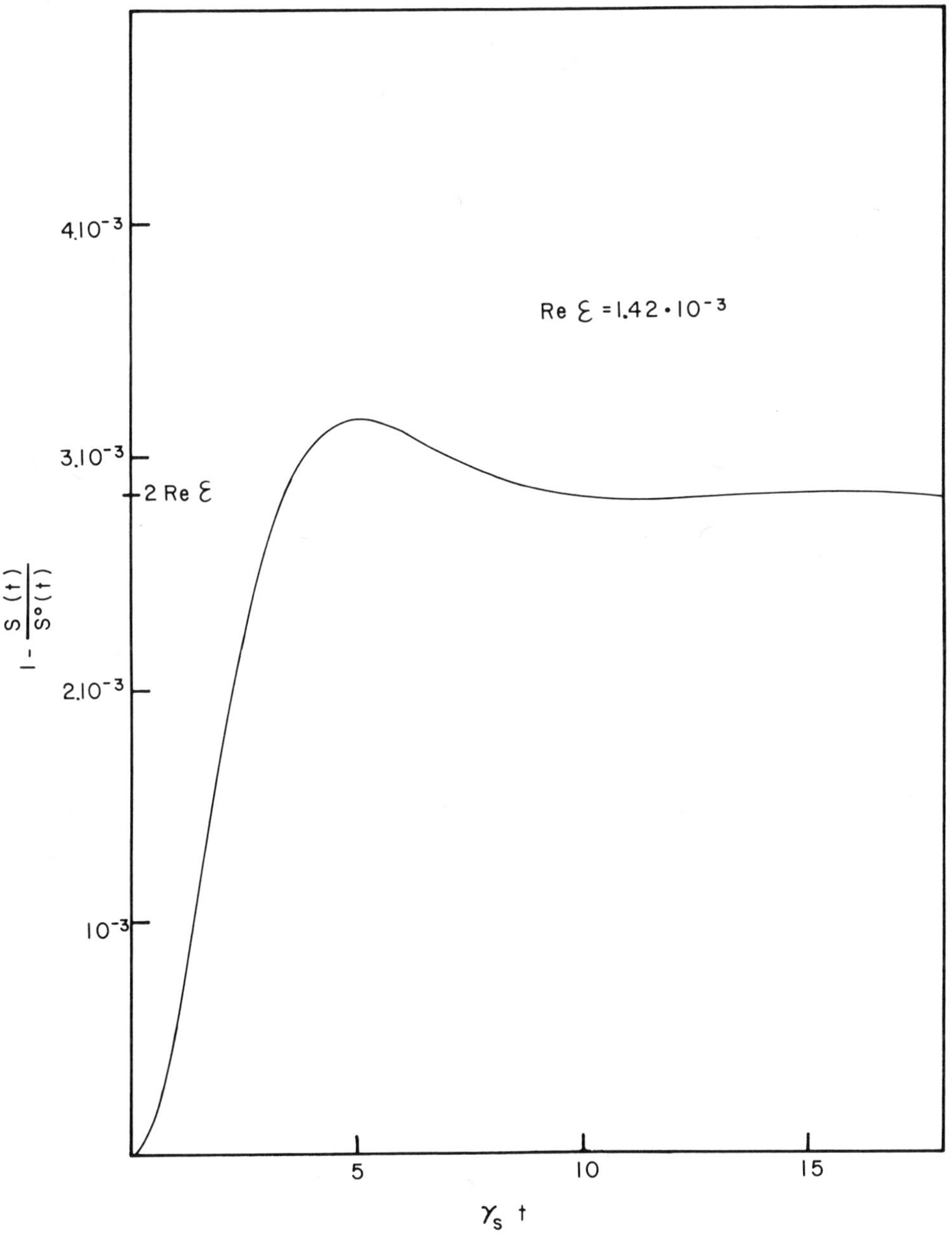

Fig. 2. The oscillatory part in the entropy function due to T-violation.

everywhere by $\beta_r = 1/k_B T_r$. The stationary state, given above by Eq. (2.13), will now change, and we shall have

$$\bar{\rho}^{st}_{ii} = \sum_r a^r_i \, e^{-\beta_r \hbar\omega_K} \quad , \tag{4.1}$$

with

$$a^r_i = \frac{\pi N_r(\omega_K) \, \overline{|g^i_{rk}|^2}_{\omega_k=\omega_K}}{\sum_{r'} \pi N_{r'}(\omega_K) \, \overline{|g^i_{r'k}|^2}_{\omega_k=\omega_K}} = \frac{A^r_i}{\sum_{r'} A^{r'}_i} \quad . \tag{4.2}$$

Clearly, if all β_r's are equal this reduces to (2.13).

As the thermodynamical currents we now define the energy flows from the system to the r'th bath,

$$J_r = \hbar\omega_K \left(\frac{\partial}{\partial t} \operatorname{Tr} \bar{\rho}\right)_r = \hbar\omega_K \sum_i \left(A^r_i \, e^{-\beta_r \hbar\omega_K} - A^r_i \, \bar{\rho}^{st}_{ii}\right) \tag{4.3}$$

(we have neglected here terms of the order of $e^{-2\beta_r \hbar\omega_K}$).

Since the sum of all the J_r's is zero, we can express one particular current J_o in terms of the others. By the first law of Thermodynamics, the entropy production in the system due to the energy flow is

$$\dot{S} = -\sum_r k_B \, \beta_r \, J_r = \sum_{r\neq 0} k_B(\beta_o - \beta_r) \, J_r \quad . \tag{4.4}$$

Thus, the J_r's and the $k_b(\beta_o - \beta_r)$'s are the currents and the forces respectively, which are used in the Onsager theory[14]. The Onsager coefficients are now defined, by

$$L_{rr'} = -\left(\frac{1}{k_B} \, \frac{\partial J_r}{\partial \beta_{r'}}\right)_{\beta_{r'}=\beta_o} \quad . \tag{4.5}$$

By Eqs. (4.2), (4.3) we now find

$$L_{rr'} = -\frac{(\hbar\beta_K)^2}{k_B} \sum_i \frac{A^r_i \, A^{r'}_i}{(\sum_{r''} A^{r''}_i)} \, e^{-\beta_o \hbar\omega_K} \tag{4.6}$$

and clearly, to the first order in ε, δ and $e^{-\beta_o \hbar\omega_K}$ the Onsager relations hold,

$$L_{rr'} = L_{r'r} \quad . \tag{4.7}$$

To obtain a higher approximation, we retain terms to order $\varepsilon e^{-\beta_r \hbar\omega_K}$ and $\delta e^{-\beta_r \hbar\omega_K}$, and find that we must keep $\bar{\rho}^{st}_{12}$,

$$\bar{\rho}^{st}_{12} = [\Omega_{12} + i\,(\Lambda^*_{21}\,\bar{\rho}^{st}_{11} - \Lambda_{12}\,\bar{\rho}^{st}_{22})]/[i\Delta m - \tfrac{1}{2}(\gamma_L + \gamma_S)] \ . \tag{4.8}$$

A simple manipulation of the additonal part in J_r shows that

$$L_{rr'} - L_{r'r} = \frac{4(\hbar\omega_K)^2}{k_B}\, e^{-\beta_o \hbar\omega_K}\, \mathrm{Im}\left\{\frac{B^{12}_r(\Lambda_{12}a^{r'}_2 - \Lambda^*_{21}a^{r'}_1) - B^{12}_{r'}(\Lambda_{12}a^r_2 - \Lambda^*_{21}a^r_1)}{i\Delta m - \frac{1}{2}(\gamma_L + \gamma_S)}\right\} \tag{4.9}$$

with

$$B^{12}_r = \pi\, N_r(\omega_K)\, \overline{(g^1_{rk}\, g^{2*}_{rk})}_{\omega_k = \omega_K} \quad . \tag{4.10}$$

Clearly, if both T and CPT are conserved, this difference will vanish, since in this case $\Lambda_{12} = \Lambda_{21} = 0$ (Eqs. (2.11)-(2.12)). In all other cases, the violation of the Onsager relations is of the order $e^{-\beta_o \hbar\omega_K}\varepsilon$. (Note that only Λ_{12}, which is related with the sum $\sum_r B_r^{12}$, is of order ε or δ, while B_r^{12} may be quite large[5]).

5. The Arrows of Time

From the definition of ε and δ it may be shown[5], that under the transformation of time reversal T,

$$\begin{aligned} \varepsilon &\overset{T}{\rightarrow} \varepsilon_T = -\,\varepsilon \\ \delta &\overset{T}{\rightarrow} \delta_T = \delta \quad . \end{aligned} \tag{5.1}$$

Therefore, if we consider a time-inverted coordinate scheme, many of the experiments on the K^0-$\bar{K}^0$ system will give different results. In particular, the oscillation in the

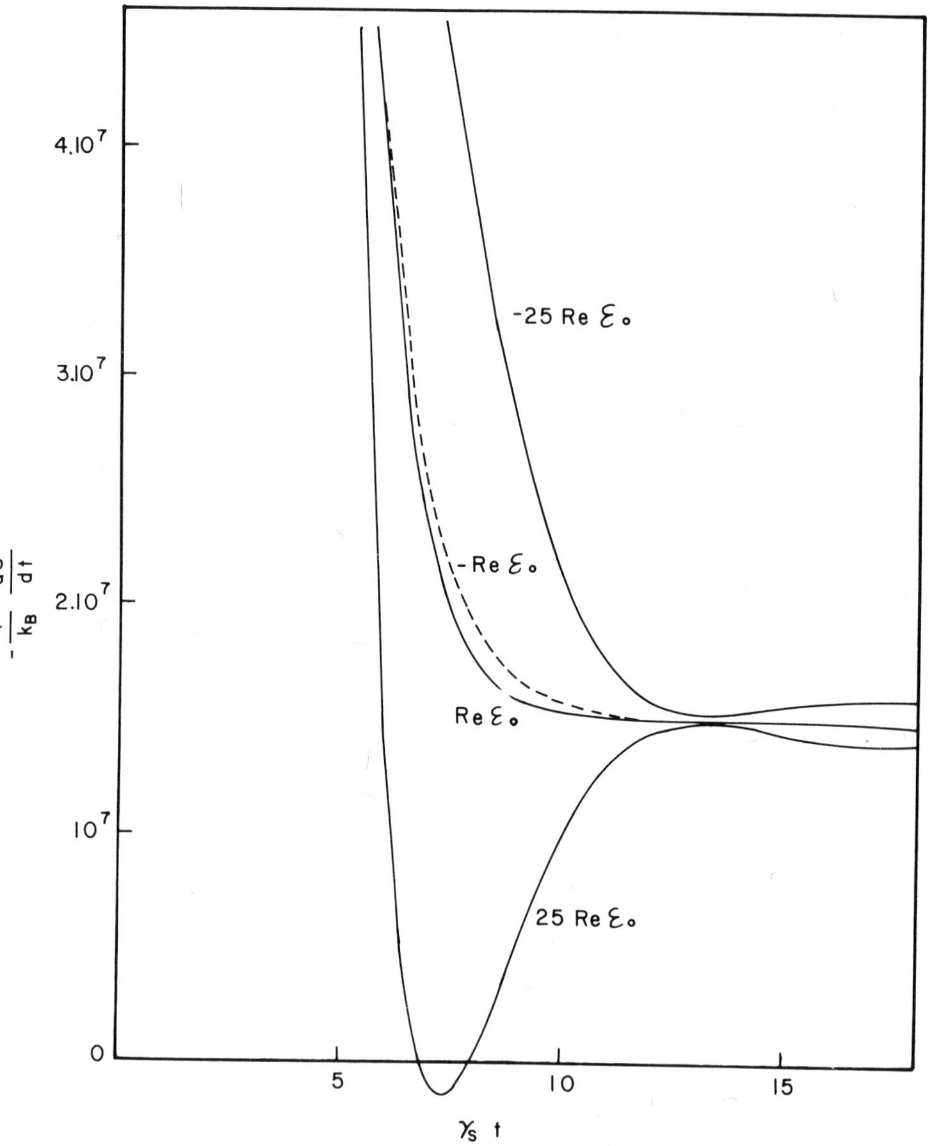

Fig. 3. The time derivative of the entropy function for K^0-decay, with several values for the time reversal violation parameter Reε. The numbers indicate the values of this parameter, with $\text{Re}\varepsilon_0 = 1.42.10^{-3}$.

evolution of the H-function will have a different sign, (see Fig. 3), the fraction of $\bar{K}^0$'s in a beam of K^0's etc., will have a different value[15] etc.

Such an inverted coordinate scheme is suggested in the theories of an oscillating universe[16]: In such theories, if τ is the oscillation period, there must exist a complete identity of physical behaviour at time $-t$ and at time $(\tau-t)$. Therefore, in a contracting phase one should observe different physical laws than in the expanding phase, due to the asymmetry in the microscopic behaviour discussed above.

In all the expressions we discussed the change of the initial state from K^0 to $\bar{K}^0$ results in a change in the signs of both ε and δ. Therefore, if CPT is conserved ($\delta=0$), the behaviour of matter in a direct scheme will be the same as that of antimatter in an inverted scheme. In this case, the definition of the microscopic direction of time is not absolutely defined. But if $\delta\neq 0$, it is clear that no such ambiguity exists; the behaviour of a K^0 decaying in a direct time scheme is clearly different (e.g. in the features of the entropy time dependence) from that of a $\bar{K}^0$ decaying in an inverted scheme.

Thus we arrive at a clear connection between the microscopic, thermodynamical and cosmological arrows of time.

REFERENCES

1. J.H. Christenson, J.W. Cronin, V.L. Fitch and R. Turlay, Phys. Rev. Lett. 13 (1964), 138.
2. For a review, see G.V. Dass, preprint TH-1373-CERN (1971).
3. Y. Ne'eman, paper presented at the March 1968 Session of the Israel Academy of Sciences; G. Zweig, paper presented at the Conf. on Decays of K-mesons, Princeton-Pennsylvania Accelerator, Nov. 1967; A. Aharony and Y. Ne'eman, Lett. Nuovo Cimento 4 (1970), 862.
4. For a review, see F. Cannata, R. Del Fabro and O. Signore, preprint INFN/AE-70/6 (1970).
5. A. Aharony, Ann. Phys. 67 (1971), 1; 68 (1971), 163.
6. K. Wangness and F. Bloch, Phys. Rev. 89 (1953), 728; A. Aharony, Ann. Phys. 62 (1971), 343.
7. T.D. Lee, R. Oehme and C.N. Yang, Phys. Rev. 106 (1967), 340;
   L. Michel, in "Proceedings of the VIIIth Nobel Symposium, 1968" (N. Svartholm, Ed.), p. 345, John Wiley, New York (1968).
8. T.D. Lee and C.S. Wu, Ann. Rev. Nucl. Sci. 17 (1967), 513.
9. J. Steinberger, in Proc. of the Topical Conf. on Weak Interactions (Geneva, 1969), CERN 69-7.
10. J.S. Bell and J. Steinberger, Oxford International Conf. on Elementary Particles, Sept. 1965;
    J. Steinberger, CERN 70-1.
11. Particle Data Group, Rev. Mod. Phys. 42 (1970), 87.
12. See also E. Teller, in "Symmetry Principles at High Energy", Coral Gables Conf. 1968, (A. Perlmutter et al., Eds.), p. 1, W.H. Freeman and Co. (1968).
13. A. Aharony, to be published.
14. L. Onsager, Phys. Rev. 37 (1931), 405; ibid. 38 (1931), 2265.
15. A. Aharony, Lett. Nuovo Cimento 3 (1970), 791.
16. A. Aharony and Y. Ne'eman, Int. J. Theor. Phys. 3 (1970), 457.

DISCUSSION

TELLER: I think I understand most of the things that have been said except your very last words. Assume that there were a time asymmetry whose strength is a millionth of the time symmetric Hamiltonian. Can this explain the observed asymmetry in our experience?

NE'EMAN: I agree that the quantitative aspect baffles me. In the first diagram you saw that only when in my example beta, the time asymmetry parameter's value was point one, only then did we really get a smooth continuation of the curves in the second half of the diagram and irreversibility looked natural. On the other hand many of the cosmologists and some of the people working in statistical mechanics point for example to a publication by Morrison a few years ago in the Weisskopf "Festschrift" and to many of the astronomers' comments, such as Tommy Gold's in the Infeld volume claim that they would be satisfied with an effect which is not really larger at the local level, except that it comes from the cosmology. Personally I still feel I would prefer information theory to give the complete result but then again why does it require from nature to add our CP effect? Why does it need this help?

CARRUTHERS: I wonder if you would explain in more detail how the K meson system can retain its phase coherence in the presence of the interaction with the bath. In order to derive the master equation you make an initial random phase assumption which destroys the phase memory of the system. It seems that the results on coherent regeneration would put a strong constraint on the interaction with the bath.

NE'EMAN: Unlike many approaches to irreversible equations of motion in Quantum Statistical Mechanics, our formalism does not assume random phases. Following Wangness and Bloch[6], we obtain irreversibility by averaging over the states of the thermal bath (e.g. Eq. (2.4)). We do not have any restrictions on the phases of the reduced density matrix elements $\bar{\rho}_{ij}$, in which all the information related with the states of K^0 and $\bar{K}^0$ is contained[15].

Moreover, in our formalism we can deal both with coherent and with non-coherent mixtures of K^o and $\bar{K}^o$. The latter is obviously the equivalent of the random phased case (zero off-diagonal elements). As shown in ref. 5, in this case no effects of order ε or δ are observed. Thus, a formalism based on random phases is not appropriate for studying the effects we are looking for.

(Note that the phase-shifts due to the strong interaction between the pions are also included in the formalism, through the definition of the bath's eigenstates $|rk\rangle$[5]).

SCHNITZER: Is it not true that all the T violating effects of the statistical mechanics of the K_1^o, K_2^o, 2π system are "washed out" in times characteristic of macroscopic systems? After all if you wait long enough all you have are decay products, such as 2π, etc. The effects you speak of seem relevant only on level of microscopic time scales.

NE'EMAN: This is why I was mentioning the need for an "exotic" medium. In an actual laboratory experiment you are certainly right. We have to invent or discover situations which might produce stable baths. For instance one could take decay products which contain fermions, as in K going into π plus electron and neutrino. It would then be conceivable to have a situation in which in a star or in the early universe we would have a degeneracy, a Pauli principle helping us to conserve the bath. It is true that the main modes are two pi and three pi, which seem more difficult to preserve. However, the violation exists in all the other modes as well.

One additional comment: the formalism we have developed and Aharony's calculations are also good for a more practical situation, that of a damped-degenerate level atom which could be excited by a laser. Moreover, if CP violation were due to electromagnetism our irreversibility experiments might become more practical.

GAUGE INVARIANT SPINOR THEORIES AND THE GRAVITATIONAL FIELD

H. P. Dürr
Max-Planck-Institut für Physik und Astrophysik
München, Germany

Mass zero solutions present a rather peculiar problem in a fundamental dynamical theory because one cannot really hope to obtain zero energy except by dynamical accident. On the other hand the mass zero boson solutions are of great importance in nature because they are connected with electrodynamics and gravitation. Hence one would like to have a general condition on the dynamics which enforces such solutions. In the past we have emphasized the connection between spontaneously broken symmetries (asymmetrical groundstate) and mass zero modes[1]. However, this mechanism produces a mass zero condition only for nontransverse modes which correspond to spin zero particles, at least if no long range interaction is present from the beginning (gauge field)[2]. Another possibility to infer mass zero solutions is to require gauge symmetries for the theory. This will be discussed in the present talk. The gauge approach has the advantage that it does not only provide a handle to enforce mass zero particles with nonzero spin but introduces these particles with the correct coupling (minimum coupling) which automatically contains the non-particle contributions as Coulomb interaction and Newtonian gravitational interaction, respectively.

Gauge symmetries as used here are defined as continuous symmetry transformation in which the Lie parameters are arbitrary functions of space and time, i.e.

$$\psi(x) \rightarrow e^{\varepsilon^a(x) T_a} \psi(x) \qquad (1)$$

with T_a the Lie generators. Yang and Mills[3] and Utiyama[4] have demonstrated that a n-parameter (function) gauge symmetry group, in general, can only be established if n vector fields $A_\mu^a(x)$ are introduced as additional dynamical variables which transform inhomogeneously under these transformations

$$A_\mu^a(x) \rightarrow V_b^a(\varepsilon^c(x))A_\mu^b(x) + \varepsilon^a_{,\mu}(x) \quad . \tag{2}$$

These gauge fields are coupled to the source field at least in the "minimal" coupling form and - under additional assumptions regarding their dynamical behavior - can be connected to mass zero modes.

On the other hand in a former paper[5] concerned with a nonlinear spinor theory we have suggested that electrodynamics may already be contained in such a theory as a special solution. In the nonlinear spinor theory based on a field equation of the form

$$i\gamma^\mu\partial_\mu\psi(x) + O_i^{(1)}:\psi(\bar{\psi}O_i^{(2)}\psi):(x) = 0 \tag{3}$$

the electromagnetic vector potential seems to occur as a "bound state" of the noncanonical spinor fields

$$A_\mu(x) \sim :\bar{\psi}\gamma_\mu\psi:(x) \tag{4}$$

(suppressing isospin properties).

Hence the following question arises: Is it possible to establish gauge symmetries in nonlinear spinor theories without introducing additional vector fields?

This, in fact, seems to be possible if the following conditions are fulfilled[6]:

1. The spinor field $\psi(x)$ has to be a subcanonical field with mass dimension 1/2 and not a canonical spinor field which has dimension 3/2. As a consequence the $:\psi\gamma_\mu\psi:(x)$ are "fields" and not currents and the nonlinear spinor theory becomes locally scale invariant and formally renormalizable.
2. The nonlinear interaction term has to be defined as a "finite part" or generalized Wick product according to the prescription of Zimmermann[7] and Wilson[8].

Under these conditions the gauge invariance of the field equation can be established if the dimensionless vector coupling constants of the theory are numerically fixed in a particular way[6].

In a nonlinear spinor theory of the Heisenberg type[9]

$$i\sigma^{\mu}\partial_{\mu}\psi(x) + g\sigma^{\mu}:\psi(\psi^{*}\sigma_{\mu}\psi):(x) = 0 \tag{5}$$

or its parity symmetrical analogue given by Dürr[10]

$$i\gamma^{\mu}\partial_{\mu}\psi(x) + \tfrac{1}{2}g:[\gamma^{\mu}\psi(\bar{\psi}\gamma_{\mu}\psi) - i\gamma_{5}\gamma^{\mu}\psi(\bar{\psi}i\gamma_{5}\gamma_{\mu}\psi)]:(x) = 0 \tag{6}$$

which is chiral symmetrical, only vector (or axial vector) coupling terms occur. In these cases all coupling constants can be fixed by the gauge symmetry requirement.

The gauge invariance of the field equation results from the fact that the "formal currents" $\sim :\bar{\psi}\gamma_{\mu}T_{a}\psi:(x)$ or $\sim :\bar{\psi}\gamma_{5}\gamma_{\mu}T_{a}\psi:(x)$ transform inhomogeneously under the corresponding gauge transformations like ordinary gauge fields because of the slightly nonlocal structure of the finite part. The finite Weyl vector field

$$V_{\mu}(x) \equiv -:\psi^{*}\sigma_{\mu}\psi:(x) \tag{7}$$

for example, implicitly appearing in (5) and defined as

$$V_{\mu}(x) \equiv -\overline{\lim_{\zeta\to o}}\{E_{o}(\zeta)\psi^{*}(x-\zeta/2)\sigma_{\mu}\psi(x+\zeta/2) - \frac{1}{4\pi^{2}i}\frac{\zeta_{\mu}}{\zeta^{2}}I\} \tag{8}$$

($E_{o}(\zeta)$ at most a logarithmic type function) varies as

$$V_{\mu}(x) \rightarrow V_{\mu}(x) + \frac{1}{16\pi^{2}}\alpha,_{\mu}(x) \tag{9}$$

under the phase gauge transformation

$$\psi(x) \rightarrow e^{-i\alpha(x)}\psi(x) \quad . \tag{10}$$

In order to establish phase gauge invariance of the nonlinear Weyl equation (5) the vector coupling constant has to be fixed to

$$g = 16\pi^2 \tag{11}$$

In a "geometrical" interpretation the gauge fields are closely related to the spinor affinities connected with the parallel displacement of the "interior coordinate frame" and hence are always coupled in the "minimal" form. For the Weyl case the affinity is

$$\bar{A}^{\alpha}{}_{\beta\mu}(x) \equiv i\delta^{\alpha}{}_{\beta}\, 16\pi^2\, V_{\mu}(x) \tag{12}$$

and the nonlinear equation (5) can be written with a corresponding covariant derivative

$$i\sigma^{\mu}\psi_{;\mu}(x) \equiv i\sigma^{\mu}[\psi_{,\mu}(x) + :\bar{A}_{\mu}\psi:(x)] = 0 \quad . \tag{13}$$

Since in our description the "formal currents" constructed from the spinor fields $\sim :\bar{\psi}\,\gamma_{\mu}\,T_a\,\psi:(x)$ are the gauge fields rather than the currents because of the subcanonical dimension of the spinor field the important question now arises how the currents should be constructed. To simplify this discussion we will only consider the construction of the current in the 2-component Weyl theory[11]. The generalization to other cases will be straightforward. The current $j_{\mu}(x)$ is required to obey the following conditions:

1.) Finite operator (finite matrix elements)
2.) Local generator of the unitary representation of the symmetry group (here constant phase transformation), i.e.

$$\lim_{x^0\to 0} [j^0(x), \psi(0)] = -\lim_{x^0\to 0} \frac{\partial}{\partial x^0} \frac{1}{2\pi}\, \mathrm{sgn}x^0\, \delta(x^2)\, \psi(0) + \ldots$$
$$= -\,\delta(\vec{x})\,\psi(0) + \ldots \tag{14}$$

with + ... referring to less singular terms which vanish upon space integration.

3.) Gauge invariance
4.) Conservation $\partial_{\mu} j^{\mu}(x) = 0$ if possible.

A lengthy investigation[11] reveals that the current should

be identified with the gauge invariant operator

$$j^{\mu}(x) = - V^{\mu\kappa}{}_{\kappa}(x) \equiv (\frac{i}{2})^2 :\psi^* \sigma^{\mu} \overleftrightarrow{\partial}^{\kappa} \overleftrightarrow{\partial}_{\kappa}\psi:(x) \quad . \tag{15}$$

On the basis of the field equation one deduces

$$\partial_{\mu} j^{\mu}(x) = - 16\pi^2 \varepsilon^{\mu\nu\kappa\lambda} : V_{\mu,\nu} V_{\kappa,\lambda}:(x) \quad , \tag{16}$$

i.e. the gauge invariant current is not conserved in this case due to Adler type terms[12]. The current (15) is closely related to the gauge invariant current

$$j'^{\mu}(x) \equiv : \psi^*_c \bar{\sigma}^{\mu} \psi_c:(x) \tag{17}$$

constructed from the canonical fields

$$\psi_c(x) \equiv i\sigma^{\mu}\partial_{\mu}\psi(x) \quad . \tag{18}$$

For the energy-momentum tensor one finds in a similar way[11] the gauge invariant operator

$$T^{\mu\nu}(x) = \Box V^{\mu\nu}(x) \tag{19}$$

with the gauge invariant operator

$$V^{\mu\nu}(x) \equiv - \frac{i}{2}: \psi^* \sigma^{\mu} \overleftrightarrow{\partial}^{\nu} \psi:(x) \quad . \tag{20}$$

The energy momentum tensor is conserved

$$\partial_{\mu} T^{\mu\nu}(x) = 0 \quad . \tag{21}$$

These considerations can be generalized to Dirac type spinor theories and theories involving higher interior symmetries.

Let us return now to our original question regarding mass zero solutions. In the conventional approach the mass zero aspect of the gauge fields enters only at a second stage namely by postulating the gauge field to be a genuine vector field $A^{\mu}(x) \neq \partial^{\mu}\Lambda(x)$ which also obeys a gauge invariant field equation. In our case no additional assumptions are possible because our gauge field is not an independent dynamical field. Therefore it is not obvious why mass zero solutions should be connected with this gauge field. However, H. Saller[13] has shown by a detailed analysis of the boson solutions resulting from the spinor equation that under certain conditions mass zero solutions, in fact, occur as a direct consequence of gauge invariance.

To demonstrate this it is, at first, necessary to start with a theory which contains some mass scale at finite distances because otherwise the statement "mass zero" is rather trivial. In introducing a mass scale we have to distinguish between theories of the Dirac type in which mass scales enter linearly and hence may be incorporated in the definition of the nonlinear term of (3), and theories of the Weyl type (including chiral invariant Dirac type theories) in which the mass enters only quadratically. In the latter case mass terms cannot be contained in the interaction term but enter for the first time in operator products, namely $:\psi^*\psi\psi^*\psi:(x)$ or $:\psi^*\partial\psi:(x)$. In both cases it can be shown that mass zero solutions do automatically occur as a consequence of gauge invariance provided the fermion propagator does not contain any strong infrared singularity.

We may, perhaps, phrase this interesting result in more physical terms as stating that massless fermions cannot couple to massless vector bosons. It may very well turn out that this is only a special case of a more general rule that "charged" particles (in the sense of being minimally coupled to a massless vector field) can never be massless. If this is true there would be some important consequences for the infrared behavior for the massless Yang-Mills theory. This also means that the massless gauge fields in this approach appear as S-wave, J = 1 "bound states" of the abstract spinor-antispinor field which itself is not massless.

On the other hand it seems to imply that if we force the fermion field coupled to the gauge field to be massless the gauge field cannot contain massless modes. In this context there appears to be a close analogy to the situation described earlier by Higgs and Kibble[2)]. Unfortunately I cannot go into details here.

Let me instead turn to another question which immediately comes up in this connection, namely the question whether our gauge approach can be generalized to space-time symmetries like the Poincaré group. It is well-known[4,14] that in formulating Poincaré gauge invariant theories ten additional gauge fields have to be introduced which are closely related to the gravitational field in Einstein's general theory of relativity. Hence a possibility to express these gauge fields directly in terms of the spinor fields would imply that also the gravitational field should be considered in a certain sense as a "bound state" of the spinor-antispinor fields. This possibility, indeed, seems to exist. To show this[15] we have to make several steps.

The first step is to show similar as Kibble[14] that Poincaré gauge invariance requires the introduction of the four translation gauge fields $g_{\mu m}(x)$ and the six Lorentz gauge fields $A_{[\kappa\lambda]m}(x)$ which are "minimally" coupled to the spinor field. E.g. the Poincaré and phase gauge invariant Weyl equation has the particular form

$$i\sigma^{\mu}\psi_{;\mu}(x) = 0 \tag{22}$$

with the Poincaré gauge covariant derivative

$$\psi_{;\mu}(x) \equiv :g_{\mu}{}^{m}(\partial_{m} + \frac{1}{4}A_{\kappa\lambda m}\sigma^{\kappa\lambda} + iB_{m})\psi:(x) \quad . \tag{23}$$

The $B_m(x)$ is the gauge field connected to phase gauge transformations. Latin indices refer to the transformation law of the coordinate differentials dx^m, Greek indices to the transformation law of vectors constructed from the spinors, e.g. $:\psi^*\sigma_{\mu}\psi:$ In a geometrical interpretation Latin indices refer to "world coordinates", Greek indices to local Minkowskian coordinates. Hence $g_{\mu m}(x)$ has the meaning of Vierbein components[16] and $A_{[\kappa\lambda]m}$ of the vectorial spinor affinities. As usual the metric tensor can be expressed as bilinear forms of the Vierbein components. The world affinities $\Gamma_{n\ell m}(x)$ are related to $g_{\mu m}$ and $A_{[\kappa\lambda]m}$ by means of metric conditions. We obtain, in general, a Riemann-Cartan space with torsion.

The second step is to show that $A_{[\kappa\lambda]m}(x)$ and also $B_m(x)$ can be expressed in terms of $g_{\mu m}(x)$ and the spinor field $\psi(x)$. It can be demonstrated that due to phase gauge invariance of the field equation no torsion term is possible, i.e. that the world affinities simply reduce to the Christoffel affinities constructed from the metric tensor. The $A_{[\kappa\lambda]m}(x)$ are then simply the Ricci rotation coefficients

$$A_{[\kappa\lambda]m}(x) = {}^{o}A_{[\kappa\lambda]m}(x)$$

$$\equiv g^{k}_{\kappa}g^{\ell}_{\lambda}g^{\mu}_{m}g_{\mu[k,\ell]} - g^{\ell}_{\lambda}g_{\kappa[\ell,m]} + g^{k}_{\kappa}g_{\lambda[k,m]} \quad . \tag{24}$$

The phase gauge field $B_m(x)$ is

$$B_m(x) = {}^{o}B_m(x) = 16\pi^2 :g^{\mu}{}_{m}V_{\mu}:(x) \tag{25}$$

with $V_\mu(x)$ as given before by (7).

The third and final step is to suggest a direct connection between the translation gauge field $g_{\mu m}(x)$ and the spinor field which at least in principle, allows to express the $g_{\mu m}(x)$ in terms of the spinor fields. For the Weyl theory, e.g., we postulate a condition essentially of the form

$$M^2 g_{\mu m}(x) \equiv -\frac{i}{2}:(\psi^*\sigma_\mu\psi_{;m} - \psi^*_{;m}\sigma_\mu\psi):(x) \quad . \tag{26}$$

This relation has to involve a mass scale M.

In physical terms this relation means that in a linear approximation the "gravitons" appear as a P-wave, J = 2 "bound state" of our abstract spinor-antispinor system. The relation becomes more transparent if we perform a double differentiation on both sides of the form $\sim g^{\mu}{}_{n}d_{[k}d_{\ell]}$ with d_k a covariant derivative in the usual sense (translation gauge covariant derivative). On the l.h.s. we produce the Riemann curvature tensor

$$g^{\mu}{}_{n}d_{[k}d_{\ell]}g_{\mu}{}^{m} = -\frac{1}{2}R^{m}{}_{nk\ell} \tag{27}$$

which can be shown to be also a Poincaré gauge invariant

tensor. On the r.h.s. it leads to an expression which after proper contraction has all the properties of the energy-momentum tensor. In particular, it agrees in the linear approximation with the momentum-energy tensor (19) of the Minkowskian space as indicated earlier. This means that condition (26) - in a classical interpretation - agrees with Einstein's equation of gravitation without torsion and without cosmological term if the arbitrary mass parameter M is identified essentially with the inverse Planck length.

Let me summarize our result: It appears that by requiring gauge symmetries and Poincaré gauge symmetry we can establish maximally symmetrical nonlinear spinor equations which contain a minimum number of dynamical fields and no arbitrary dimensionless constants. These equations are suited to produce massless gauge fields of spin one (photons) and spin two (gravitons) as particular solutions. If gravitation is suppressed they are scale invariant at small distances and probably renormalizable.

In closing I wish to emphasize that most of our investigations so far have been carried out on a rather formal level and hence a much more detailed study of our conjecture is necessary before any conclusion to its value can be drawn. The first attempts in this direction, however, seem to be very encouraging.

REFERENCES

1. H.P. Dürr, W. Heisenberg, H. Mitter, S. Schlieder, K. Yamazaki: Z. Naturforschung 14a, 441 (1959).
J. Goldstone: Nuovo Cimento 19, 154 (1961).
J. Goldstone, A. Salam, S. Weinberg: Phys. Rev. 127, 965 (1962).
2. P.W. Anderson: Phys. Rev. 130, 439 (1963).
P.W. Higgs: Phys. Rev. 145, 1156 (1966).
T.W.B. Kibble: Phys. Rev. 155, 1554 (1967).
3. C.N. Yang, R.L. Mills: Phys. Rev. 96, 191 (1954).
4. R. Utiyama: Phys. Rev. 101, 1597 (1956).
5. H.P. Dürr, W. Heisenberg, H. Yamamoto, K. Yamazaki: Nuovo Cimento, 38, 1220 (1965).
6. H.P. Dürr, N.J. Winter: Nuovo Cimento 70A, 467 (1970).
7. W. Zimmermann: Nuovo Cimento 10, 597 (1958); Comm. Math. Phys. 6, 161 (1967); 8, 66 (1968).
8. K.G. Wilson: Cornell Report 1964; Phys. Rev. 179, 1499 (1969).
9. For a review see e.g. W. Heisenberg: Introduction to the Unified Field Theory of Elementary Particles, J. Wiley (London, 1966); German edition: Einführung in die einheitliche Feldtheorie der Elementarteilchen, Hirzel Verlag (Stuttgart, 1967).
H.P. Dürr: Acta Phys. Austr. Suppl. III, 1 (1966).
10. H.P. Dürr: Z. Naturforschung 16a, 327 (1961).
11. H.P. Dürr, N.J. Winter: Construction of currents in a gauge invariant spinor theory. Preprint July 1971, München. To be published in Nuovo Cimento.
12. J. Schwinger: Phys. Rev. 82, 664 (1951).
S.L. Adler: Phys. Rev. 177, 2426 (1969).
13. H. Saller: Nuovo Cimento 4A, 404 (1971).
Gauge Invariance and Mass Scale, Preprint MPI für Physik und Astrophysik, September 1971, München.
14. T.W.B. Kibble: Journ. Math. Phys. 2, 212 (1961).
15. H.P. Dürr: For details see: Poincaré gauge invariant spinor theory and the gravitational field, preprint MPI für Physik und Astrophysik, October 1971. To be

published in JGRG.
16. H. Weyl: Zeits. Physik 56, 330 (1929).
V. Fock, D. Ivanenko: Zs. f. Physik 54, 798 (1929).
V. Fock: Zs. f. Physik 57, 261 (1929).

PROGRESS IN LIGHT CONE PHYSICS*

Giuliano Preparata
Universitá di Roma
Roma, Italy

and

New York University
New York, New York

This will be a very brief (probably biased) review of the progress which has been made in the physics of the light cone this year. The interested reader is advised to turn for a more detailed and up to date exposition to the Hamburg Lectures of R. Brandt and myself.[1]

As has been emphasized so many times, the light cone techniques have been developed in order to deal with the interaction of weak probes (e,μ,ν) with hadrons by means of currents carrying a very large "mass" and at very high energy; such kinematical configuration is called the Bjorken limit or the A-limit.

In the Bjorken limit the dominance of the light cone region can be simply proved by use of a sort of Riemann-Lebesgue theorem for the relevant four-dimensional Fourier transform. It is easy to show, however, that this argument is incomplete, and that one can construct simple counter examples; so that for light cone dominance to hold one needs some extra assumptions. The hole in the simple argument can however be filled[2] by the reasonable physical requirement that Regge behavior should dominate the high energy fixed mass limit for amplitudes involving currents, which seems to be true experimentally. Thus we see a further remarkable connection between Regge and light cone dominance, besides the one, al-

*Supported in part by the National Science Foundation Grant No. GU-3186 and the Air Force Office of Scientific Research Contract No. F 61/052 67 C0084.

ready reported at the 1971 Coral Gables Conference,[3] between usual Regge behavior and the small $\omega = \frac{q^2}{2\nu}$ limit of scaling functions.

Although very vigorous attacks have been made on the problem of getting detailed results about light cone expansion with Callan-Symanzik renormalization group techniques, not much has yet been accomplished. Let me here mention, however, the interesting work of Gatto and collaborators[4] which used the information contained in the conformal group to constrain the general form of light cone expansions, thus yielding very elegant and possibly useful operator expressions, which may have some bearing on the previous problem.

On the analysis of light cone expansion in soluble models, like the Thirring model, one should notice the paper by Dell' Antonio, Frishman and Zwanziger,[5] where they show the explicit form of field-correlation functions near the light cone.

Some work has been also devoted to extending SLAC scaling results to more complicated situations like the description of the various neutrino deep inelastic scattering scaling functions[1] as well as of the spin dependent effects in inelastic electroproduction.[6] In particular the problem of implementing gauge invariance and the consequences of precocious scaling near threshold have received some attention.[1] For these inelastic processes we are now in a position where we have available a large number of predictions, some of which are obeyed experimentally, while some others are awaiting confirmation. It seems that the canonical light cone expansions do represent a very effective tool for predicting the basic features of highly inelastic semileptonic processes. This is something that only three years ago seemed quite far fetched.

Some progress[7] has also made in understanding better the intricate physical ideas involved in the light cone description of the Columbia-BNL μ-pair experiment, i.e.,

$$p + p \to \mu^+\mu^- + \text{anything} \quad .$$

This process, which provided for the original motivation for writing an operator light cone expansion, also turned out to stir up ideas which have been extremely useful to purely hadronic physics.[8] In fact, besides its light cone content this process has also a most interesting Regge structure as an inclusive process involving "one-particle" distribution of a hard photon in pp collisions, which is no different from usual hadronic one-particle inclusive distributions. The new[7] refined treatment of Reggeism and the better handling of the light cone spectral properties show that this process is likely to be very inefficient (i.e., there is a rapid fall-off with mass) in producing high mass currents and all the more or less exotic beasts which may couple to them (W-bosons, heavy leptons, monopoles, etc.). Within the present data[9] a very satisfactory description of the various experimental distributions is obtained. It is very interesting that an experiment of this sort is now under way at CERN at extremely high energies ($s \simeq 1600$ GeV^2), so that very soon a verification of these ideas will be possible.

The canonical approach[1] to light cone expansions has made it possible to undertake the more speculative program of writing down, within the quark-gluon model, expansions involving operators different from e.m. or weak currents. This has proved quite useful for estimating the high mass behavior of certain amplitudes involving the divergence of axial vector currents $\partial_\mu A^\mu$, e.m. currents, etc.

The control over the high mass limits of these amplitudes together with the ideas of precocious scaling represents quite a strong tool to handle quantitatively, in the framework of mass dispersion relations, off-shell corrections. Among the interesting results one finds

a. A promising approach to the breaking of chiral $SU_3 \times SU_3$.[10] The idea is to derive low energy theorems for amplitude involving divergences of $SU_3 \times SU_3$ non-conserved currents. One then relates the values at zero energy to the on shell amplitudes through a dispersion relation, whose high mass behavior is strongly controlled by the light cone expressions and their simple dependence on the symmetry breaking lagrangian.

b. An understanding of the basic pattern of violation of the Vector Meson Dominance model (VMD),[1] according to the mass dispersion relation formalism which connects real photon

($q^2 = 0$) - with $\rho(q^2 = m_\rho^2)$ - amplitudes through the light cone control of high mass behavior, much in the same way as in a.

c. Description of electroproduction in the low q^2 region where strict light cone dominance is obviously inapplicable.[1,11]

d. Neat connection between Regge behavior and the small $\omega = \frac{q^2}{2\nu}$ limit of scaling functions.[1]

Another interesting development has been the application of Fritsch and Gell-Mann bilocal algebra[12] to various interesting processes like electroproduction of high mass μ-pairs.[13] Although the feasibility of these experiments looks quite remote at present, one may hope that these predictions will act as an incentive to experimentalists to do their best to make such experiments possible.

Lastly I would like to spend a few words on a new direction of research in the physics of the light cone, i.e., the assessment and the exploitation of analyticity properties in both mass and energy of light cone amplitudes. A knowledge of such properties will make it possible to enlarge considerably the field of applications of the light cone expansions to processes which are not directly described in terms of matrix elements of operator products, i.e.,

$$e^+e^- \longrightarrow \text{hadron + anything} \quad (*)$$

It seems at present possible[14] to formulate a consistent and particularly simple set of analyticity properties for light cone amplitudes, so that with little effort one can relate light cone dominated amplitudes to processes like (*). This will provide a far reaching connection between electroproduction (q spacelike) and the various aspects of the physics of colliding beams (q-timelike) which I believe are going to play a major role in the physics of the Seventies.

REFERENCES

1. R.A. Brandt and G. Preparata, Proceedings of the 1971 Hamburg Summer School on Electromagnetic Interactions, (to be published).
2. G. Preparata, Proceedings of the International School of Subnuclear Physics, Erice (1971), (to be published). See also C. Orzalesi, N.Y.U. Preprints (1972).
3. R.A. Brandt and G. Preparata, Proceedings of the 1971 Coral Gables Conference, (Gordon and Breach, New York 1971).
4. S. Ferrara, R. Gatto and A. Grillo, to be published in Springer Tracts.
5. G.F. Dell'Antonio, Y. Frishman, and D. Zwanziger, to be published.
6. J. Hey and J. Mandula, CALTECH Preprint.
7. R.A. Brandt and G. Preparata, (to be published). This paper slightly modifies the original analysis of G. Altarelli, R.A. Brandt and G. Preparata, Phys. Rev. Letters 26, 42 (1971).
8. A.H. Mueller, Phys. Rev. D2, 2963 (1970).
9. J.H. Christenson et al, Phys. Rev. Letters 25, 1523 (1970). See also B.G. Pope's Thesis (unpublished).
10. R.A. Brandt and G. Preparata, Phys. Rev. Letters 26, 1605 (1971).
11. G. Preparata, Phys. Letters 36B, 53 (1971) and G. Altarelli and G. Preparata, Phys. Letters (to be published).
12. H. Fritzsch and M. Gell-Mann, Proceedings of the 1971 Coral Gables Conference, (Gordon and Breach, New York 1971).
13. D. Gross and S. Treiman, Phys. Rev. (to be published).
14. G. Preparata, to appear.

DISCUSSION

KASTRUP: I should like to make a comment and an appeal to the more mathematically informed people in the audience. My impression is that one of the major problems in connection with light cone physics is the following: In this field one has the situation that one is interested in some asymptotic limit in momentum space. Then the question is, which region in coordinate space is actually dominating the limit considered in momentum space? As far as I know this is a nontrivial mathematical problem and the answer is difficult to extract from the mathematical literature. As the light cone physics is so important, it would be very helpful to have sound mathematical assertions and results in this field, not just intuitive or handwaving ones. I think this is one of the rare occasions where experimentally oriented and mathematically oriented theoreticians can join forces.

SIMON: There is a fairly new technique in the theory of partial differential equations that has the "feel" of being relevant to a rigorous formulation of light cone singularities. This technique of "wave front sets of distributions" involves a mixed x-space, p-space analysis of singularities. The interested person can get a start on the mathematical literature by consulting a recent paper of L. Hörmander in Acta Mathematica ("Fourier Integral Operators, I").

BREIT: I should like to ask just what was meant by your remarks regarding vector-meson dominance. One sees in the literature so many indications of vector dominance not being in good agreement with experiment, for instance recent work at DESY regarding the proton Compton effect. Just what did you mean?

PREPARATA: This is precisely the point. The formalism of light-cone controlled mass dispersion relations gives one the possibility of studying the effects of off-shell extrapolation in amplitudes involving emission or absorption of virtual photons. One can therefore estimate quantitatively the relation between "on-shell" ρ-amplitude with the corresponding photon amplitude. In this way, in certain circumstances, one obtains significant deviations from simple VMD; and fortunately this just happens when the experiments seem to show the failure of VMD.

TESTS OF CHARGE INDEPENDENCE BY NUCLEON-NUCLEON SCATTERING*

G. Breit
State University of New York at Buffalo
Buffalo, New York 14214

Dedicated to Professor P.A.M. Dirac on the occasion of his 70th birthday anniversary with profound admiration of his ability to separate the wheat from the chaff in theoretical physics, of formulating the transformation theory giving the connection between quantum mechanics and classical Hamiltonian dynamics, of the discovery of the Dirac equation for the electron, the associated introduction of negative energy states, of his mathematical derivation of Einstein's spontaneous emission probability thus providing the essential step for the development of quantum electrodynamics, and continued leadership in providing a logical foundation for the structure of physical theory.

* * * * * * *

Improvements in search procedures for nucleon (N-N) phase shifts and in the determination of the pion-nucleon coupling constant g^2 have been recently reported.[1] The present paper is partly concerned with: (a) the related improvement[1] of the tests of the charge independence (CI) of the long-range (CLR) nucleon-nucleon (N-N) interaction and partly, (b) with the possible existence of coupling of mesonic origin[2] between states with the same orbital quantum number L but different values of the isospin and ordinary spin quantum numbers I and S respectively. As previously pointed out[2] such coupling could conceivably affect the phase shift analysis of n-p scattering data and hence also conclusions regarding long-range CI (LRCI). The work reported has been done in the earlier stages in collaboration with Messrs. M. Tischler, M. Mukherjee and J. Lucas; in the later stages with M. Tischler, R. Nisley and G. Pappas.

*Work supported by the U.S.A. Atomic Energy Commission (COO-3475-1).

The latest phase-shift analysis (Y-IV) of the Yale group[3,4] has been used for (a). The pion-nucleon coupling constant $g^2 = \hbar e g_o^2$ has been determined[1] as usual by assigning phase shifts and coupling constants, referred to collectively as phase parameters, at sufficiently low collision energies E and high L to the one-pion exchange (OPE) group of "phases" (a term used as an abbreviation for phase parameters).[5] The energy above which a "phase" δ is subjected to a phenomenological search and below which it is taken to be in the OPE group is referred to as a transition energy and is written $E_{tr}(\delta)$. One of the improvements of Ref. 1 makes it possible to determine g^2 treating it formally as a phase parameter, thus simplifying and accelerating the work. Another is the more careful enforcement of continuity of any δ, say δ_p, at its transition energy $E_{tr}(\delta_p)$, not only for the statistically best g^2, $(g^2)_{best}$, but also in the g^2 variations used to find $(g^2)_{best}$. This continuity was not sufficiently well satisfied in the applications of the parabola method[6,7] made in the work for Ref. 3. This permits us to take better advantage of the greater influence of experimental data on g^2 made possible by the graded introduction of the $E_{tr}(\delta)$ first introduced in Refs. 3 and 4. The effect of corrections to $(g_o^2)_{p-p} - (g_o^2)_{n-p}$ for two-pion exchange was estimated as ≈ 1 in Ref. 1. Even though the methods were appreciably improved, their application was not completed. In the work reported here, some omitted or too briefly considered effects were more carefully evaluated. Employing ps coupling in the comparison, the agreement between the p-p and n-p values of g^2 is, perhaps fortuitously, no worse than previously.

In the tables below, the values of g_o^2 derived from p-p scattering are compared with those from n-p. The procedures in the two cases were kept the same for the two cases as much as possible therefore. For example, the same E_{tr} were used for the I = 1 phases and the same formulae were used to secure continuity of the "phases" at their E_{tr}. It is believed that one may expect therefore a somewhat smaller absolute uncertainty in the values of $(g^2)_{p-p} - (g^2)_{n-p}$ obtained in this work than in the values of $(g^2)_{p-p}$ and $(g^2)_{n-p}$ individually.

For forming a judgment regarding freedom from spurious effects such as those caused by two-pion exchange (TPE) and of exchanges of vector mesons it may be well to be reminded that the effects of physically observed vector mesons on account of their large masses should decrease faster with the impact parameter of classical mechanics than those of TPE. Since the effect of the latter is small, as will be seen below, the choice of transition energies is sufficiently conservative. It should be pointed out, however, as has been done previously, that the estimates of statistical uncertainties are difficult to make reliably especially because of the possible presence of unknown systematic errors not only within individual angular distributions of the same experimental observable but also for groups of angular distributions of the same observable or of related observables at neighboring energies. The effect of such errors on $(g^2)_{p-p}$ may well be different from that on $(g^2)_{n-p}$ and $(g^2)_{p-p} - (g^2)_{n-p}$ may, therefore, be affected.

Table I shows a comparison of $(g_o^2)_{p-p}$ with $(g_o^2)_{n-p}$ and lists the difference between them, with and without the TPE corrections. The absolute value of the difference is seen to have decreased as a result of the introduction of TPE. The latter was calculated employing the ps coupling calculations of Gupta[8] and of Gupta, Haracz, et. al.[9]

TABLE I

Values of g_o^2 from $(Y\text{-}IV)_{p-p}$ and the $(Y\text{-}IV)_{n-p}$ M data, with and without TPE corrections.

	$(g_o^2)_{p-p}$	$(g_o^2)_{n-p}$	$(g_o^2)_{n-p} - (g_o^2)_{p-p}$
With TPE	15.11±0.62	14.67±0.51	-0.44±0.80
Without TPE	13.74±0.53	14.70±0.52	0.96±0.74

In these calculations the corrections for magnetic moments were made by the procedure[10] used by the Yale group. Although the calculation of the effect of wave function distortion on magnetic moment corrections (MAG, for short) has been formally circumvented in that procedure it has some weak points. Thus, for example, the procedure supposes that "phases" in the "searched" group as obtained by phenomenological adjustment to scattering data fully include the influence of nucleon-magnetic moment effects. The correction functions used in practice (the f_{pq}) are not sufficiently flexible, however, to make this more than partially true, the experimental data being too inaccurate to justify the use of more realistic f_{pq}. Furthermore, even if such an adjustment could be made, the enforcement of continuity at the E_{tr} could not be made quite properly without additional calculations because in g^2 variations a change in g^2 should not affect the magnetic-moment part of the searched phase, the mesonic part of the interaction being actually affected by the change in g. Furthermore, the I = 1 "searched phases" for n-p scattering are usually taken from the analysis of p-p scattering after applying Coulomb corrections to these "phases". A systematic error is thus introduced because the MAG effects are different for n-p and p-p scattering. The error is not clearly negligible because MAG effects are even more "long-range" than those of OPE. In point of fact one of the main reasons for introducing MAG into the analysis was[10] its possible effect on g^2 determinations, in view of the long-range character of the MAG interaction. A "new MAG" was therefore introduced at SUNY/AB. Potential models are used for it. These are admittedly not justifiable on purely theoretical grounds. But this weakness is presumably not of primary importance for intermediate ranges, say around 1.5 - 2.0 F. Present experience indicates only a mild sensitivity to the choice of the potential. The values below have been obtained employing Y_{HJ}, the Yale modification of the "Hamada-Johnston" potential, a modification which has been obtained by fitting the Y-IV "phases".

TABLE II

Values of g_o^2 from $(Y\text{-}IV)_{p-p}$ and $(Y\text{-}IV)_{n-p}$ M data, employing "old MAG" and "new MAG".

Procedure	$(g_o^2)_{p-p}$	$(g_o^2)_{n-p}$	$(g_o^2)_{n-p} - (g_o^2)_{p-p}$
"new MAG"	14.59 0.62	14.69 0.51	0.10 0.80
"old MAG"	15.11 0.62	14.67 0.51	-0.44 0.80

Table II shows the effect of changing from the "old" to the "new" MAG. The second line of numbers in Table II is the first such line in Table I. Comparison of that line in Table II with the second line of numbers in Table I gives the overall effect of including TPE and the improvement in MAG. The statistical uncertainties of the difference in the values of g^2 shown in the last columns of the two tables have been obtained by adding in quadrature the uncertainties of the two preceding columns.

It has been stated in Ref. 1 that $(g_o^2)_{n-p}$ is largely determined by the π^c - N coupling and that n-p scattering contains, therefore, primarily information regarding the coupling of the charged rather than that of the neutral pion to the nucleon. This situation is illustrated in Table III. In that table the values o and c of the subscript of g indicate whether the coupling of the nucleon considered is with the neutral or charged pion. The first column describes conditions under which the phase-parameter search was made regarding enforcement of continuity of the "phases" at the transition energies E_{tr} and presence or absence of corrections for TPE.

TABLE III

Separation of π^o from π^c effects in n-p scattering. Values of g_o^2 for π^o and π^c denoted by g_{oo}^2 respectively are listed for several searches on $(Y\text{-}IV)_{n-p}$ M data together with those of $(g_o^2)_{n-p}$ obtained from the same data assuming that $g_{oo}=g_{oc}=(g_o)_{n-p}$. Last line obtained by setting $g_{oo}^2 = (g_o^2) = 13.74$ (cf. Table I). With last entry in last column it reproduces last line of Table 1.

Type of search	g_{oo}^2	g_{oc}^2	$(g_o^2)_{n-p}$
No continuity imposed at E_{tr} No TPE corrections	12.86±4.7	13.47±0.74	13.49±0.74
With continuity at E_{tr} With TPE corrections	12.59±3.5	14.62±0.52	14.67±0.51

(Table III continued)

TABLE III (CONTINUED)

With continuity at E	15.14±3.3	14.69±0.52	14.70±0.52
No TPE corrections			
With continuity at E	13.74	14.64±0.51	
No TPE corrections; g_{oo}^2 frozen			

The second and third columns give the values of g_o^2 derived for π^o and π^c respectively. The last column lists the values of g_o^2 derived for the same type of phase-parameter searches as those for g_{oo}^2 and g_{oc}^2 in the same row but on the assumption that $g_{oo} = g_{oc}$. The uncertainty in the value of g_{oo}^2 is seen to be much greater than that in the other two cases. This indicates that much less information regarding π^o - N coupling is contained in n-p scattering data than regarding π^c - N coupling. The near equality of g_{oc}^2 and of $(g_o^2)_{n-p}$ apparent in the inter-comparison of the last two columns shows that the usual $(g_o^2)_{n-p}$ is very nearly the value of g_o^2 for π^c - N coupling. If the accuracy of n-p scattering data were high enough a comparison of g_{oo}^2 obtained from it with the usual (g_o^2)p-p could be used for obtaining the difference in the coupling of π^o with n from that with p. The large uncertainties of g_{oo}^2 in Table III show that large improvements in the accuracy of n-p scattering are needed to make the uncertainty in $(g_{oo}^2)_{n-p}$ comparable with the relatively small uncertainty of $(g_{oo}^2)_{p-p} = (g_o^2)_{p-p}$ obtainable from p-p data such as appears in Table I and II. The present evidence gives no definite indication of a statistically significant difference between these two quantities, i.e., between the coupling constants of π^o to the proton and the neutron; nor is there a statistically significant difference in the values $(g_{oo}^2)_{n-p}$ and $(g_{oo}^2)_{p-p}$, i.e., the value of the coupling constant for the interaction of the nucleon with the charged pion π^c as compared with its interaction with the neutral pion π^o, if it is assumed that the coupling constant does not depend on the nucleon's charge. The statistical accuracy of the values of g dealt with in the comparison of π^c with π^o is of the order of 2% (cf. Tables II and III) in the standard deviation convention.

The significance of the conclusions depends on the applicability of the mathematical form of the π-N interaction used in the work. Tests of the correctness of that form have been made previously.[11] The experimental data then available were less complete than those in the Y-IV data collection. The "phase"-search and g^2-determination techniques were not as good as those available now. The motivation for those tests, while having a bearing on the charge-independence tests, was primarily concerned with the general validity of the theory of the OPE interaction. It appeared desirable, therefore, to supplement the tests by an additional simple one, concerned with the detection of a linear dependence of g^2 on energy. The constant value of g_o^2 was replaced in the data fitting by

$$g_o^2(E) = g_o^2(0) + SE \qquad (0 < E < 350 \text{ Mev})$$

where E is the collision energy in the laboratory system and S is energy independent. Enforcement of continuity at the E_{tr}, TPE corrections and the "new MAG" improvements were made use of. The results are as follows:

$$[g_o^2(E)]_{p-p} = 15.17 \pm 0.62 + (-6.9 \pm 7.4)\, E_{GeV}\,,$$

$$[g_o^2(E)]_{n-p} = 14.59 \pm 0.51 + (0.62 \pm 5.3)\, E_{GeV}\,.$$

There is clearly no definite indication in this test of a systematic effect large enough to matter for the conclusion regarding there being no evidence of the presence of violation of LRCI.

There are many ways, however, in which the difficulties of interpretation of the experiments combined with those in their execution could conceivably combine to give a wrong conclusion. A number of these have been mentioned[2,6] in earlier papers. Among them there stands out one of possibly special interest, namely, that in the n-p interaction there exists nonnegligible coupling of non-electromagnetic origin between states with the same values of the L and J quantum numbers, thus equivalent to a partial breakdown of CI, since the coupled states have different isospins. A small effect of electromagnetic origin which couples the same pairs of states exists in much the same way as in atomic spectra. It can cause a change from Russel-Saunders to j-j coupling. If such an effect should be large enough the analysis of n-p data would be in error because it would be wrong to adopt the I = 1 "phases" as obtained from p-p data for the L = J states after simply correcting them for the Coulomb interaction between two protons. A change in the phases of the searched group puts different requirements on the contributions to the scattering amplitudes of the OPE group, thus conceivably affecting the value of g_o^2 .

The probability of a breakdown of CI of this type appears more probable if it is remembered that the vector meson exchanges between nucleons may take place partly through meson-nucleon coupling caused by the mesonic equivalent of the electric charge and partly because of the presence of a similar equivalent of the magnetic moment of a particle. Since the numerical value of the coupling constant of the latter, the hadronic dipole (hdm) moment, is inaccurate[12] only crude estimates are possible. It would be very difficult to claim equality of coupling through that dipole moment of the same vector meson to the proton and to the neutron. In point of fact some one-boson-exchange (OBE) fits to N-N scattering are made without invoking such coupling. It is conceivable, therefore, that such coupling exists but has effectively different strength in n-p and p-p scattering in such a way that LRCI is violated. This would not be unreasonable because the hdm depends for its existence on relatively subtle features of nucleon and vector meson structure.

It is within the range of possibilities that the breakdown of CI arising in this manner causes coupling between states with the same L but different S. As discussed in the preceding paragraph, there would then be an error in all present day analyses of n-p scattering. Such an error could conceivably vitiate conclusions regarding LRCI also for scattering caused by other causes than interactions with the hdm's.

It appeared desirable, therefore, to estimate the effects that may be expected, even though the input to the calculation is admittedly uncertain. Since the earliest and main attempts to determine the magnetic like hdm coupling constant to the nucleus were made by Bryan and Scott the notation used in their papers[13] is used here. The part of the Lagrangean representing the interaction of nucleon and vector-meson fields contains two coupling constants g and f. In the natural units used ($\hbar = 1$, $c = 1$) g plays the role of the electric charge on the nucleon and f that of the magnetic moment. The derived OBE potential contains in the $(\vec{L}\cdot\vec{S})$ term as a factor the quantity $g(3g + 4f)$ which is analogous to $e^2(3 + 4\mu_a)$ of the $(\vec{L}\cdot\vec{S})$ interaction in the electromagnetic case. Here μ_a is the anomalous, i.e., the extra Diracian part of the proton magnetic moment, expressed in nucleon Bohr magneton $e\hbar/2Mc$ units. The $4\mu_a$ part is caused by the interaction of the electric field of one proton at the other with the electric dipole of the latter caused by the motion of the μ_a part of its magnetic moment. The "3" results from an original 4 after correction for the Thomas term. The 4f/g in the hadronic case thus replaces μ_a. In the last Bryan-Scott paper quoted above the value of f/g for the ρ meson arrived at from fitting phenomenological phase shifts is 1.13. The value quoted there from the second paper is 3.82. In both cases g^2 for the ρ meson is low (1.81 and 1.32) but g^2 for the ω meson is high. It appears possible though not certain that if g^2 for ω were made lower then a higher value would be needed for ρ. Table 1 of Ref. 12 shows a large variety of combinations of f/g values obtained by different investigators qualitatively supporting the view that the meson-nucleon coupling constants are known poorly except for that between pions and nucleons.

The important dimensionless quantity for the calculation of the coupling effect under discussion, may be shown to be

$$F = (g_0^V)^2 \; (f_A^V/g_A^V - f_B^V/g_B^V), \quad (g^V)^2 = (g_0^V)^2 \, \hbar c \; .$$

Here superscript V stands for "vector" and the interacting nucleons are A and B. Remembering that for the electromagnetic case $\mu_p^a - \mu_n^a \approx 4$ and attempting to make a reasonable compromise between various requirements it appears fair to use F = 5, assigning somewhat arbitrarily the value ≈ 10 to $(g_0^V)^2$ and 0.5 to the difference of the two f/g ratios. The factor F is meant to be applied to n-p scattering only. This admittedly crude and somewhat arbitrary assignment gives values as in Table IV.

TABLE IV

Polarization $P(\theta)$ and its expected fractional change $\Delta P(\theta)/P(\theta)$ caused by "mesonic MAG" in n-p scattering for effectiveness factor value 5 with scattering angle θ in degrees and E in MeV. (both Lab.).

E (MeV)	θ	5°	25°	45°	65°	85°
350	P	0.1122	0.418	0.331	0.072	0.1288
	ΔP/P	0.10	0.13	0.18	0.40	0.01
110	P	0.045	0.256	0.431	0.471	0.311
	ΔP/P	0.02(8)	0.02(5)	0.02(2)	0.1(7)	0.01(1)

The values listed have been obtained in a first order calculation of the off-diagonal matrix elements of collision matrices for the coupled states, referred to below as ρ_L' and taking these elements into account in first order only. In this respect the calculation is very similar to that in Ref. 10. The factor F = 0 for p-p and n-n scattering. The scattering amplitudes obtained in the Y-IV data fit were used in the Calculation of P and of ΔP, the latter somewhat similarly to the way they enter Eq. (4.6) of Ref. 10. The absolute value of the ρ_L' in the calculations used for the table is largest for E = 350 Mev, L = 2. In this case the value is $\approx 10^{-2}$, which is not high enough to cause alarm, regarding applicability of first order calculations of the ρ_L'. There is perhaps more of a question regarding the use of the first order effects of the ρ_L' in obtaining the scattering amplitudes. No serious error attributed to this cause is expected, however, for the value of F used at the energies of Table IV. Caution regarding use of values of effect of new MAG in Table II and of hadronic moments in Table IV is advisable because potentials used were exclusively of the hard-core type.

The considerations leading to Table IV are decidedly speculative and qualitative only. The reason for making them is the desirability of not overlooking a non-negligible effect on the value of the coupling constant in n-p scattering. The table and the computed values of ρ_L' indicate that one deals primarily with a low L and high E effect. This circumstance decreases its importance for tests of LRCI especially if it is considered that g^2 determinations emphasize the influence of high L and low E.

If the rather close equality of $(g^2)_{n-p}$ and $(g^2)_{p-p}$ as determined from N-N scattering will persist with the use of more modern data and desirable improvements in their treatment the question of reconciling such a finding with the difference of the charged and neutral pion masses will arise even more seriously than at present. This difference indicates a lack of symmetry (isotropy in isospace) for strong interactions and an explanation for its having no effect would then be called for. The main effect of the pion mass on the tunneling of the pion through the space between the pions is of course taken into account in the calculation of the OPE contributions to the scattering amplitudes. But it is not clear that the difference between π^c and π^o masses does not show itself in other ways. It is conceivable that nucleon structure produces a non-negligible complicating effect, the neutron-proton mass difference being incompletely understood. But an explanation in terms of a modernized version of the Fermi-Yang model of the pion is perhaps also a possibility. The low mass of π^o could conceivably be primarily an effect of an exchange energy between p-$\bar{p}$ and n-$\bar{n}$ while the mechanism of combination with p or n produced mainly by vector meson fields is the same as for π^c.

It is desired to repeat the acknowledgment of helpful collaboration with Messrs. Tischler, Lucas, Mukherjee, Nisley and Pappas.

Thanks are due to Professor Danielli, Director of the Center of Theoretical Biology, State University of New York at Buffalo, for his hospitality in providing working space.

REFERENCES

1. G. Breit, M. Tischler, S. Mukherjee, and J. Lucas, Proc. Nat. Acad. U.S.A. 68, 897 (1971).
2. G. Breit, Rev. Mod. Phys. 34, 766 (1962); p. 791 especially.
3. R.E. Seamon, K.A. Friedman, G. Breit, R.D. Haracz, J.M. Holt, and A. Prakash, Phys. Rev. 165, 1579 (1968).
4. G. Breit, Rev. Mod. Phys. 39, 560 (1967).
5. The data collection, the phase-parameter fit and the value of g_o^2 for p-p scattering in References 3 and 4 are designated as $(Y\text{-}IV)_{p-p}$; those for n-p scattering $(Y\text{-}IV)_{n-p}$ or $(Y\text{-}IV)_{n-p}$ M. The latter differ regarding treatment of p+d data, the second being presumably an improvement on the first.
6. G. Breit and R.D. Haracz, in High Energy Physics edited by E.H.S. Burhap (Academic Press Inc., New York, 1967), Vol. I, p. 21.
7. Subsection B of Section VII of Ref. 6, for the theory of the method.
8. W.S.N. Gupta, Phys. Rev. 117, 1146 (1960).
9. S.N. Gupta, R.D. Haracz, and J. Karkas, Phys. Rev. 138, B1500 (1965); B.M. Barker, S.N. Gupta, and R.D. Haracz, Phys. Rev. 161, 1411 (1967). K.D. Haracz and R.D. Sharma, Phys. Rev. 176, 2013 (1968); B.M. Barker and R.D. Haracz, Phys. Rev. 186, 1624 (1969); Phys. Rev., D1, 3187 (1970); cf. W.R. Wartman, Phys. Rev. 176, 1762 (1968) for a dispersion-theoretical approach.
10. G. Breit and H.M. Ruppel, Phys. Rev. 127, 2123 (1967); 131, 2839 (1963); cf. also Ref. 2 and Ref. 6.
11. G. Breit, M.H. Hull, Jr. K. Lassila and H.M. Ruppel, Phys. Rev. Letters 5, 274 (1960); Proc. Nat. Acad. Sci. 46, 1649 (1960).
12. G. Breit, Proceedings International Conference on Properties of Nuclear States, Montreal, Canada, August 1969, Les Presses de L'Université de Montreal, p. 293.
13. Ronald A. Bryan and Bruce L. Scott, Phys. Rev. 135, B 434 (1964); 164, 1215 (1967); 177, 1435 (1963).

PARTICLES AS NORMAL MODES OF A GAUGE FIELD THEORY*

Fred Cooper
Yeshiva University
New York, New York

and

Alan Chodos
University of Pennsylvania
Philadelphia, Pennsylvania

(Presented by Fred Cooper)

I. INTRODUCTION

In this report we would like to show how one can describe the physical properties of the hadrons, such as the mass spectrum and electromagnetic form factors by considering the hadrons to be normal modes of an underlying field theory possessing gauge invariance of the second kind.[1,2,3] In this picture, there is at each point in space-time an infinite number of irreducible tensorial fields $\phi^{\mu_1\mu_2\cdots\mu_k}$ and Rarita-Schwinger fields, $\psi_\alpha^{\mu_1\mu_2\cdots\mu_k}$. These fields have bilinear interactions dictated by gauge invariance, and the system (at each point in space-time) is a one-dimensional lattice with interactions between nearest neighbors in the Lorentz index space k. The physical particles are the normal modes of this system and therefore have structure. We show in Section 3 how the masses and coupling constants of the underlying field theory determine the parameters of the physical particles. If we assume that the electromagnetic field couples to the underlying gauge fields via minimal coupling, then we find that the form factors F_1 and F_2 of the proton are given by a power series in the 3 + 1 dimensional Legendre

*Work supported in part by the National Science Foundation and by a Frederick Gardner Cottrell Grant in Aid.

polynomials $P_k(\frac{p\cdot p'}{m^2})$ and that an infinite number of narrow resonances appear in the amplitude for Compton scattering off hadrons. By further postulating that strong interaction scattering as well as Compton scattering can be considered as the excitation and deexcitation of the underlying gauge fields by external quanta coupling to appropriate currents, we are able to give a prescription for strong interaction scattering amplitudes in the Narrow Resonance tree approximation.

II. LAGRANGIAN FIELD THEORIES

In previous works[3] we have shown how the principle of gauge invariance of the second kind generates a particular class of Boson and Fermion field theories containing an infinite number of particles. Specifically we found that the field combinations

$$G^{\mu_1\mu_2\cdots\mu_k} \equiv \delta^{\mu_1\mu_2\cdots\mu_k}_{\nu_1\nu_2\cdots\nu_k}\partial_{\mu_k}\phi_{\mu_1\cdots\mu_{k-1}} + \alpha_k\phi^{\mu_1\mu_2\cdots\mu_k} \tag{1a}$$

(abbreviated by $G^k = \delta\ \partial\ \phi^{k-1} + \alpha_k\ \phi^k$)

and

$$G_\alpha^{\mu\mu_1\cdots\mu_k} \equiv \delta^{\mu\mu_1\cdots\mu_k}_{\nu_1\cdots\nu_{k+1}}\ \partial^{\nu_1}\psi_\alpha^{\nu_2\cdots\nu_{k+1}} + \bar{\alpha}_k\psi_\alpha^{\mu\mu_1\cdots\mu_k} \tag{1b}$$

$$(G_\alpha^{k+1} = \delta\ \partial\ \psi_\alpha^k + \bar{\alpha}_k\ \psi^{k+1})$$

are invariant under the gauge transformations of the second kind:

$$\phi_{\mu_1\cdots\mu_k} \rightarrow \phi_{\mu_1\cdots\mu_k} + \gamma_k\ \delta^{\nu_1\cdots\nu_k}_{\mu_1\cdots\mu_k}\ \partial_{\nu_1}\cdots\partial_{\nu_k}\Lambda(x) \tag{2a}$$

$$(\phi_k \rightarrow \phi_k + \gamma_k\ \delta\ \partial_k\Lambda(x))$$

$$\psi_{\mu_1\cdots\mu_k\alpha} \rightarrow \psi_{\mu_1\cdots\mu_k\ \alpha} + \lambda_k\ \delta^{\nu_1\cdots\nu_k}_{\mu_1\cdots\mu_k}\ \gamma_{\nu_1}\ \partial_{\nu_2}\cdots\partial_{\nu_k}U_\alpha(x) \tag{2b}$$

provided $\gamma_{k-1} + \alpha_k \gamma_k = 0$ and $\lambda_k + \bar{\alpha}_k \lambda_{k+1} = 0$. ($\delta^{\mu_1 \cdots \mu_k}_{\nu_1 \cdots \nu_k}$ is the projection operator onto the space of traceless symmetric tensor of rank k).

Thus the Lagrangians

$$L^{\text{boson}} = \sum_{k=0}^{\infty} \eta_k \, [G^k(\delta\partial\phi'_{k-1} + \alpha_k\phi_k - \frac{1}{2} G_k)]$$

$$(\eta_k = (-1)^k) \tag{3a}$$

and

$$L^{\text{fermion}} = \sum_{k=1}^{\infty} \eta_k \, [\delta(\bar{\psi}^k\gamma^\mu)(i\partial_\mu\psi_k + \bar{\alpha}_k\psi_{k\mu}) + \text{h.c.}] \tag{3b}$$

are invariant under the above gauge transformations of the second kind. ($\delta\bar{\psi}^k\gamma^\mu$ is automatically invariant under (2b) since for any tensor A, $\delta(\gamma\gamma A) = 0$). These Lagrangians lead to the field equations

$$\Box^2 \phi^k + (k+1)\alpha_k^2 \phi^k + \frac{k^2}{k+1} \delta\partial(\partial\cdot\phi^k)$$

$$= -(k+1)\alpha_{k+1} \partial\cdot\phi^{k+1} - \alpha_k \delta\partial\phi^{k-1} \tag{3a}$$

and

$$i(\gamma\cdot\delta\partial\psi^k + \partial\cdot\delta\gamma\psi^k) = \bar{\alpha}_{k-1}^* \delta\gamma\psi^{k-1} - \bar{\alpha}_k^* \gamma\cdot\psi^{k+1} . \tag{3b}$$

We see that α_k is related to the mass of the underlying gauge field $\phi_k(\psi_{k\alpha})$ and also gives the nearest neighbor coupling constants: The source of the field $\phi_k(\psi_{k\alpha})$ are the fields ϕ_{k+1} and ϕ_{k-1} ($\psi_{k+1\alpha}$ and $\psi_{k-1\alpha}$). The exact gauge invariance leads to the conditions that there are no spin zero ($\frac{1}{2}$) particles in these theories[2,3] so that they describe families of particles like the ρ and N^* trajectories. To discuss the π and N trajectories we broaden our class of Lagrangians to include those which are invariant under restricted gauge transformations

$$\Lambda(x): \quad (\Box^2 + m^2)\Lambda(x) = 0$$

$$U_\alpha(x): \qquad (i\,\gamma\cdot\partial - m)U_\alpha(x) = 0 \quad .$$

Thus we are led to consider the Boson Lagrangian

$$L = \sum_{k=0}^{\infty} \eta_k G^k (\partial\phi_{k-1} + \alpha_k\phi_k + \beta_k\partial\cdot\phi_{k+1} - \frac{1}{2}G_k) \qquad (4)$$

which is invariant under

$$\phi_k \to \phi_k + \gamma_k^j \;\; \delta\partial^{(k-j)} V_j(x)$$

if

$$(\Box^2 + m_j^{\,2})V_j(k) = 0 \qquad \partial\cdot V_j = 0$$

and

$$-\frac{m_j^{\,2}}{2}\;\frac{(k-j+1)(k+j+2)}{(k+1)^2}\;\beta_k\,\gamma_{k+1}^j + \alpha_k\,\gamma_k^j + \gamma_{k-1}^j = 0$$

and the Fermion Lagrangian

$$L = \frac{1}{2}\sum \eta_k [\bar\psi^k\gamma^\mu(i\;\delta\;\partial_\mu\psi_k + \bar\alpha_k\;\psi_{k\mu} + i\;\bar\beta_k\partial^\lambda\psi_{k\lambda\mu}) + \text{h.c.}]$$

which is invariant under

$$\psi_k \to \psi_k + \lambda_k^j\;\delta\gamma\partial^{(k-j-1)}\;u_j(x)$$

if

$$(i\gamma\cdot\partial - m_j)u_j = 0 \qquad\qquad \gamma\cdot u_j = \partial\cdot u_j = 0$$

and

$$-\,i\;m_j^{\,2}\;\bar\beta_k\;\lambda_{k+i}^j\;\frac{(k+1-j)(k+3)(k+j+3)}{(k+2)^3} + 2(i\;\lambda_k^j + \bar\alpha_k\lambda_{k+1}^j) = 0\;.$$

III. MASS SPECTRUM

In order to determine the mass spectrum we assume the existence of a set of normal mode free fields $\tilde\phi_{Nj}$ of mass m_{Nj} ($\tilde\psi_j^{(N)}$ for fermions) and spin j ($j + \frac{1}{2}$). Being free fields they satisfy

$$(\Box^2+m_{Nj}^2)\tilde{\phi}_j^{(N)} = 0 \qquad \partial\cdot\tilde{\phi}_j^{(N)} = 0$$
$$(i\gamma\cdot\partial-m_{Nj})\tilde{\psi}_j^N = 0 \qquad \partial\cdot\tilde{\psi}_j^{(N)} = \gamma\cdot\tilde{\psi}_j^{(N)} = 0 \ . \tag{6}$$

Since the underlying field theory has only bilinear interactions, each field ϕ_k, $\psi_{k\alpha}$ can be written as an infinite sum over those normal modes which do not exceed its maximum spin ($j \lesssim k$). Thus

$$\phi_k = \sum_{\substack{N_j \\ j\leq k}} a_k^{(Nj)} \ \delta\partial_{(k-j)} \ \tilde{\phi}_j^{(N)}$$
$$G_k = \Sigma \ b_k^{(Nj)} \ \delta\partial_{(k-j)} \ \tilde{\phi}_j^{(N)} \tag{7}$$
$$\psi^k = \Sigma \ \bar{a}_k^{(Nj)} \ \delta\partial^{(k-j)}\tilde{\psi}^{j(N)} + b_k^{(Nj)}\delta\gamma\partial^{(k-j-1)}\tilde{\psi}^{j(N)} \ .$$

By demanding that the wave equation (3a and its generalization for $\beta \neq 0$) be satisfied for each $\phi_j^{(N)}$, $\psi_j^{(N)}$ separately, we find that $a_k^{(Nj)}$ and $\bar{b}_k^{(Nj)}$ obey second order inhomogeneous difference equations, whereas $b_k^{(Nj)}$ and $\bar{a}_k^{(Nj)}$ obey homogeneous difference equations.

The coefficients $a_k^{(Nj)}$ and $b_k^{(Nj)}$ are related to the wave function of the Nth particle of spin j and m_{Nj} is the mass of that particle and its value is determined by the allowed eigenvalues of the difference equation for the coefficients $a_k^{(Nj)}$.

The "potential" of the difference equation is governed by α_k and β_k, the masses and couplings of the underlying field theory. The boundary conditions are that

$$a_k^{(Nj)} = 0 \quad k < j \qquad \text{origin condition}$$

and that the wave function of the physical state $|Njp\lambda\rangle$ is normalizable. This is implemented by the condition

$$\langle N'j'p'\lambda'|Q|Njp\lambda\rangle = \delta_{N'N}\delta_{j'j}\delta_{\lambda'\lambda}\delta^3(p'-p) \tag{8}$$

where Q is the electromagnetic charge (or T_3 for the neutron or T_{00} for particles without other quantum numbers). This condition gives us information about the allowed behavior of $a_k^{(Nj)}$ for large values. For example, for the bosons

$$1 = \sum_{k=j}^{\infty} (m^2)^{k-j} \frac{(-1)^{k+j+1}}{2^{k-j+1}} \frac{(j!)^2 (k+j+2)!(k+j+1)!}{(2j+1)!\ [(k+1)!]^2} (b_{k+1}^{(Nj)} a_k^{(Nj)} + \beta_k a_{k+1}^{(Nj)} b_k^{(Nj)}). \quad (9)$$

With a rescaling of b_k and $\bar{a}_k$, into $\tilde{b}_k$ and γ_k, we have the following second order difference equations:

$$-\frac{m_{Nj}^2}{2} \frac{(k-j+1)(k+j+2)}{(k+1)^2} \tilde{b}_{k+1}^{(Nj)} = \frac{k^2}{2} \tilde{b}_{k-1}^{(Nj)} + m_k \tilde{b}_k^{(Nj)} \quad \text{(bosons)} \quad (10a)$$

$$-\frac{m_{Nj}^2}{2} \frac{(k-j)(k+j+2)}{(k+1)^2} \gamma_{k+1}^{(Nj)} = \frac{k^2}{2} \gamma_{k-1}^{(Nj)} + m_k \gamma_k^{(Nj)} \quad \text{(fermions)} \quad (10b)$$

where

$$m_k m_{k-1} = \begin{cases} -\dfrac{k^2}{2} \dfrac{\alpha_k \alpha_{k-1}}{\beta_{k-1}} & \text{in (10a)} \\[2ex] \dfrac{k^3}{k+1} \dfrac{(\bar{\alpha}_{k-1}\bar{\alpha}_{k-2})^*}{2\beta_{k-2}} & \text{in (10b)} \end{cases} .$$

In reference (5) Chodos and Haymaker showed that if $m_k m_{k-1} \sim k^3$ for large k, then the spectrum is asymptotically linear. Specifically we can solve these equations for fixed j_0, arbitrary N by the choice

$$m_k\ m_{k-1} = \frac{1}{\gamma} (n+j_0+a)(n+j_0+a-1)(n+2j_0+1) \ . \quad (11)$$

We find that the difference equation is then analogous to the Laguerre polynomial differential equation and we obtain for the eigenvalue condition

$$m_{Nj_0}^2 = \frac{1}{\gamma} (N+j_0+a) \quad . \quad (12)$$

Therefore we find that the $j = j_0$ daughters are evenly spaced.

Since we know that asymptotically the trajectories are linear, we get families of linear trajectories (the parent and an infinite number of daughters).

It is interesting to notice that although the fermion wave equation is first order, the difference equation determining the mass spectrum for the fermions depends only on m^2. This dependence of the fermion trajectory on m^2 and not on m is a consequence of the gauge invariance of the Fermion Lagrangian, and is also an experimentally observable fact.

IV. ELECTROMAGNETIC FORM FACTORS

If we write the normal mode field $\tilde{\phi}^{(N)}_{j=0}$ in terms of the underlying gauge field, we would get an expansion of the form

$$\tilde{\phi}^{(N)}_{j=0}(x) = \sum_k C^{(N)}_k \, \partial^{(k)} \, \phi_k(x) \quad . \tag{13}$$

This contains an infinite number of derivatives, so we expect that the physical states have non-trivial form factors. If we assume that the photon couples directly to the underlying gauge fields with the usual minimal interaction then we get the following current for the proton trajectory. Using the usual Gell-Mann Levy equations, letting

$\psi_k(x) \to e^{-i\alpha(x)e} \psi_k(x)$, we have

$$j^\mu_{em} = \frac{1}{e}\frac{\delta L}{\delta\partial_\mu \alpha(x)} = \frac{1}{2}\sum_k \eta_k (\bar{\psi}_k\gamma)^{k+1}\delta^{k\mu}_{k+1}\psi_k + \bar{\beta}_k(\bar{\psi}_k\gamma)_{k+1}\psi^{k+1\mu} + h.c. \tag{14}$$

where $(A)^k B_{k\mu}$ means $A^{\nu_1 \ldots \nu_k} B_{\nu_1 \ldots \nu_{k\mu}}$. Similar expressions for the boson trajectories are found in reference 2.

To find the matrix element of this current between physical states we decompose the fields ψ_k into their normal modes as in (eq. 7). The normal mode fields have the following non-vanishing matrix elements between the vacuum and the proton trajectory

$$<0|\psi^{(N')}_{j'}(x)|Nj\lambda p> = \frac{e^{-ip\cdot x}}{(2\pi)^{3/2}}\left(\frac{m_{Nj}}{E}\right)^{\frac{1}{2}} \delta_{NN'}\,\delta_{jj'}\, u^j(p\lambda) \; , \tag{15}$$

where $u^j(p\lambda)$ is a spinor for spin $j + \frac{1}{2}$ and momentum p and helicity λ as given for example, by Scadron[5]. (λ goes from $- j - \frac{1}{2}$ to $j + \frac{1}{2}$). The current (eq. 14) has both diagonal and off diagonal matrix elements. These matrix elements are power series in the 3+1 dimensional Legendre polynomials

$$P_k(\gamma) = \frac{\sinh(k+1)\ \Theta}{(k+1)\sinh\ \Theta} \qquad \gamma = \frac{p\cdot p'}{mm'} = \cosh\ \Theta \quad . \tag{16}$$

For example the matrix element of j^μ between the $N = 0$ $j = 0$ state (the proton) is

$$\langle N = 0,\ j = 0\ \lambda' p' |\ j^\mu(x)\, | N = 0,\ j = 0\ \lambda p\rangle = \frac{m}{(EE')^{\frac{1}{2}}} \frac{e^{i(p'-p)\cdot x}}{(2\pi)^3}$$

$$\times\ [\bar{u}(p'\lambda')\gamma^\mu u(p\lambda)F_1(q^2) + (p+p')^\mu \bar{u}(p'\lambda')u(p\lambda)F_2(q^2)]\ , \tag{17}$$

where

$$F_1(q^2) = \Sigma\eta_k m^{2k}[2^k(k+1)]^{-1}(|a_k|^2(\frac{k+2}{k+1}) + \frac{m^2}{2}\mathrm{Re}\bar{\beta}_k a^*_{k+2} a_k)P'_{k+1}(\gamma)$$

$$F_2(q^2) = \Sigma\eta_k m^{2k-1}[2^k(k+1)]^{-1}[(|a_k|^2(\frac{k+2}{k+1})P''_{k+1} - (k+1)P'_k + (\gamma-2)P''_k)$$

$$-\ \mathrm{Re}(\beta_k a^*_{k+2} a_k)\ \frac{m^2}{2}\ \{(k+1)P''_k + (k+3)P''_{k+2} - 2(k+2)P''_{k+1}\}]\ .$$

F_1 and F_2 are thus a function of $\bar{\alpha}_k$ and $\bar{\beta}_k$, whereas the mass spectra is a function of the ratio $\frac{\bar{\alpha}_k \bar{\alpha}_{k-1}}{\bar{\beta}_{k-1}}$. Thus in principle we might be able to choose $\bar{\alpha}_k$ and $\bar{\beta}_k$ such that the trajectories are linear and the form factors behave as $\frac{1}{q4}$ asymptotically. At the present we have found no simple choices of $\bar{\alpha}_k$ and $\bar{\beta}_k$ that allows a simple summation of eqs. 17 and their off diagonal counterparts. Although we can write an analogous expression for the off diagonal elements (and thus for the structure functions νW_2 and W_1 of inelastic electron scattering) our inability to sum these series has prevented us from making any real progress.

V. SU(2) ⊗ SU(2) CURRENTS AND STRONG INTERACTION DYNAMICS

We noticed that the electromagnetic current $j^\mu(x)$ had sufficient structure to produce both form factors and transition between the states. In fact the Compton amplitude to order e^2

$$
{}_{out}\langle \gamma(k'\lambda')N'j's'p'|\gamma(k\lambda)Njsp\rangle_{in} = \delta_{fi}
$$

$$
+\frac{i^2}{2}e^2\varepsilon^{\mu}(k\lambda)\varepsilon^{\nu*}(k'\lambda')\int d^4x d^4y \langle N'j's'p'|T(j^{\mu}_{em}(x),j^{\nu}_{em}(y)|Njsp\rangle
$$

$$
\times\, e^{-i(kx-k'y)} \tag{18}
$$

contains an infinite number of "narrow resonances" in the s and u channel. Therefore if we are only interested in the "narrow resonance" approximation to scattering we can picture hadron-hadron scattering as the excitation of the underlying field theory by an external hadron coupling to appropriate currents inherent in the "free" boson and fermion Lagrangians. We will, therefore, use as our first approximation to a strong interaction theory this external field excitation picture and assume that external pions, rho mesons, protons, etc. couple respectively to the axial vector, vector and Baryonic currents inherent in the Lagrangians (eqs. 4,5).

To obtain N-point functions we treat any of the N legs as the underlying field in a normal mode, and absorb and emit external quanta consistent with the scattering process, with the underlying field being excited and de-excited, finally returning to another normal mode state. We then sum over treating each particle as the underlying field in a normal mode to obtain crossing symmetric amplitudes.

To get an idea of the structure of the currents to which the external quanta couple, we look at the trilinear interactions of ordinary field theory, treat each of the three fields as an external field, and write the interaction as $L^{int} = \phi^{ext}_{\mu(x)} j^{\mu}(x)$. Where ϕ^{ext}_{μ} is a 4-vector constructed from the particle considered as an external field (or vector spinor) and $j^{\mu}(x)$ is a bilinear current we try to generalize to the infinite component case. In the case that the external field is a π, or ρ meson, usual lore tells us that $j^{\mu}(x)$ should be the axial vector and vector currents obeying the SU(2) $\otimes$ SU(2) algebra. When the external field is a nucleon we use baryonic currents which are bilinear in the meson and baryon infinite component field generated directly from the free Lagrangian by gauge transformations as a generalization of the ordinary field theory coupling.

To generalize our Lagrangian to include the nucleon, π and ρ trajectories, we add the isospin index i=1,2 to the field $\psi_{k\alpha}$ and the indices a=1,2,3 (for isospin) and $\eta = 0,1$ for normality (parity = $(-1)^{j+\eta}$) to ϕ_k and get the Lagrangian

$$L = \sum_{k,a,\eta} \eta_k G^k_{a\eta} (\delta \partial \phi^{a\eta}_{k-1} + \alpha^\eta_k \phi^{a\eta}_k + \beta^\eta_k \partial - \phi^{a\eta}_{k+1} - \frac{1}{2} G^{a\eta}_k)$$
$$+ \frac{1}{2} \sum_{ki} \eta_k \delta (\bar{\psi}^k_i \gamma^\mu)(i \partial_\mu \psi^i_k + \bar{\alpha}_k \psi^i_{k_\mu} + i \bar{\beta}_k \partial^\lambda \psi^i_{k\lambda_\mu}) + \text{h.c.} \tag{19}$$

We obtain the vector and axial vector currents by considering the transformations

$$\psi \to e^{-i\alpha \cdot \tau/2} \psi \qquad\qquad \psi \to e^{-i\alpha \cdot \tau/2 \gamma_5} \psi$$
$$\phi_i \to \phi_i + \varepsilon_{ijk} \alpha_j \phi_k \qquad\qquad \phi_i \to \phi_i + \varepsilon_{ijk} \alpha_j \varepsilon \phi_k \tag{20}$$

where $\varepsilon_{\eta\eta'} = \begin{pmatrix} 0 & 1 \\ 1 & 0 \end{pmatrix}$ exchanges the π and ρ trajectories.

Using the usual Gell-Mann Levy equations we find for V^μ and A^μ,

$$\vec{V}^\mu = \frac{1}{2} \sum_k \eta_k [\delta (\psi^k \gamma^\mu) \frac{\vec{\tau}}{2} \psi_k + \beta_{k-2} \bar{\psi}_{k-2} \gamma_\lambda \frac{\vec{\tau}}{2} \psi^{k-2\lambda\mu} + \text{h.c.}]$$
$$- \sum_k \eta_k (\vec{G}^{k-1\mu} \times \vec{\phi}_{k-1} + \beta_k \vec{G}_k \times \vec{\phi}^{k\mu}) \tag{21a}$$

$$\vec{A}^\mu = \frac{1}{2} \sum \eta_k \{\delta (\bar{\psi}^k \gamma^\mu) \frac{\vec{\tau}}{2} \gamma_5 \psi_k + \bar{\psi}^k \frac{\vec{\tau}}{2} \delta (\gamma^\mu \gamma^5 \psi^k)$$
$$+ \bar{\beta}_{k-2} [\bar{\psi}_{k-2} \gamma_\lambda \gamma_5 \frac{\vec{\tau}}{2} \psi^{k-2\lambda\mu} + \bar{\psi}^{k-2\lambda\mu} \frac{\vec{\tau}}{2} \gamma_\lambda \gamma_5 \psi_{k-2}]\} \tag{21b}$$
$$- \sum \eta_k [\vec{G}^{k-1\mu} \varepsilon \times \vec{\phi}_{k-1} + \beta^\eta_k \vec{G}^\eta_k \varepsilon_{\eta\eta'} \times \vec{\phi}^{k\mu}_{\eta'}] \quad .$$

The vector current is conserved. One can verify using the canonical commutation relations for the fields ϕ_k (anti-commutators for ψ_k), that the time components A^o and V^o obey the chiral SU(2) ⊗ SU(2) algebra.

VI. π-N ELASTIC SCATTERING

We want to use our rules to calculate $\pi^- p$ elastic scattering in the narrow resonance approximation. If we restrict our world to the nucleon, π and ρ trajectories, the pole diagrams of ordinary field theory would be the neutron pole in the s channel and the ρ^o pole in the t channel. Thus the elementary vertices are

$$
\begin{aligned}
&g_{\rho\pi\pi}\ \vec{\rho}^{\mu}\cdot(\partial_\mu\vec{\pi}\times\vec{\pi})\\
&g_{\rho NN}\ \vec{\rho}^{\mu}\cdot(\bar{N}\ \gamma^\mu\ \frac{\vec{\tau}}{2}\ N) \qquad (22)\\
&g_{\pi NN}\ \partial_\mu\vec{\phi}\cdot(\bar{N}\ \gamma^\mu\ \frac{\vec{\tau}}{2}\ \gamma_5 N) \quad .
\end{aligned}
$$

For the process $\pi^- p \to \pi^- p$ the external particles are pions and nucleons and we have said, they couple to the axial current $\bar{N}\ \gamma_\mu\ \frac{\tau}{2}\ \gamma_5\ N + \rho_\mu \times \pi$ and the baryonic current $(\bar{N}\gamma^\mu)_\alpha\ j^{\mu\alpha}$

where
$$j^{\mu\alpha} = \frac{\tau}{2}\ \gamma_5\ N \cdot \partial^\mu\phi + c\ \frac{\tau}{2}\ N\ \rho^\mu \ . \qquad (23)$$

Clearly (eq. 21b) is the generalization to the infinite component case of $\bar{N}\ \gamma_\mu\ \frac{\tau}{2}\ \gamma_5\ N + \rho^\mu \times \pi$. To find the appropriate generalization of (eq. 23) we consider the following transformations:

$$
\begin{aligned}
&\phi^i_\eta(x) \to \phi^i_\eta(x) + g_o(\bar{f}(x)\lambda^*_\eta\ \frac{\tau^i}{2}\ \psi(x) + \bar{\psi}(x)\ \lambda_\eta\ \frac{\tau^i}{2}\ f(x)] \qquad (24)\\
&\psi(x) \to \psi(x) + i\ g_o\ \lambda_\eta\ \frac{\tau^i}{2}\ f(x)\phi_{i\eta}(x) \quad .
\end{aligned}
$$

Here f(x) is a spinor in isospin and space time, and λ_η is a Dirac matrix defined to be

$$\lambda_\eta = I \quad \text{if}\ \eta = 0$$

$$\lambda_\eta = i \frac{g_1}{g_0} \gamma_5 \quad \text{if} \quad \eta = 1 \quad .$$

These transformations lead to the following baryonic current

$$j^\mu_\alpha(x) = \frac{\delta L}{\delta g_0 \partial_\mu \bar{f}_\alpha(x)}$$

$$= \Sigma\, \eta_k [\vec{G}^{k-1\mu}_\lambda \cdot (\frac{\vec{\tau}}{2} \lambda^*_\eta \psi_{k-1})_\alpha + \beta_k\, G^\eta_k \cdot (\frac{\tau}{2} \lambda^*_k \psi^{k\mu})_\alpha \tag{25}$$

$$+ \vec{\phi}_k \cdot (\frac{\tau}{2} \lambda^*_\eta \delta(\gamma^\eta \psi^k)_\alpha + \bar{\beta}^\lambda_k \vec{\phi}^{k\mu\lambda} \cdot (\frac{\tau}{2} \lambda^*_\eta (f\gamma_\lambda \psi_k)_\alpha)] \quad .$$

This is the natural generalization of eq. (23). Using these currents and assuming that the external π, nucleons couple to these currents, we get the diagrams of fig. 1. We see we get the ρ trajectory exchanged in the t channel and the nucleon trajectory in the s channel. For example, figs.1(a) and 1(b) are given by

$$\frac{\bar{u}(p_2)}{(2\pi)^3} \gamma^\mu \int d^4x\, d^xy\, e^{-i(k_1 \cdot x - p_2 \cdot y)} \left(\frac{m}{2(p_2)_0 (k_1)_0}\right)^{\frac{1}{2}}$$

$$\times \langle \pi^-(k_2) N{=}0 j{=}0 \eta{=}1 | T(\partial_\lambda A^\lambda(x), j^\mu_\alpha(y)) | p(p_1) N{=}0 j{=}0\ s \rangle \quad . \tag{26}$$

When we expand the axial and baryonic currents into the normal modes, the above expression leads to both an s channel neutron trajectory and a t channel ρ trajectory. The expansions are given in references 2 and 3.

VII. CONCLUSIONS

We have found that using gauge invariant Lagrangians provides a framework for discussing the masses and form factors of the particles found in nature. At worst one can probably find some set of parameters α_k, β_k to agree with the observed world. We then venture that we have enough structure in the matrix elements of the currents to consider them as containing strong interaction physics information. Thus we could hope that the matrix elements of the free Lagrangian vector and axial vector currents are useful in describing vector meson and pseudoscalar meson induced reactions in the narrow resonance approximations. At present these matrix elements are too unwieldy to pursue such calculations.

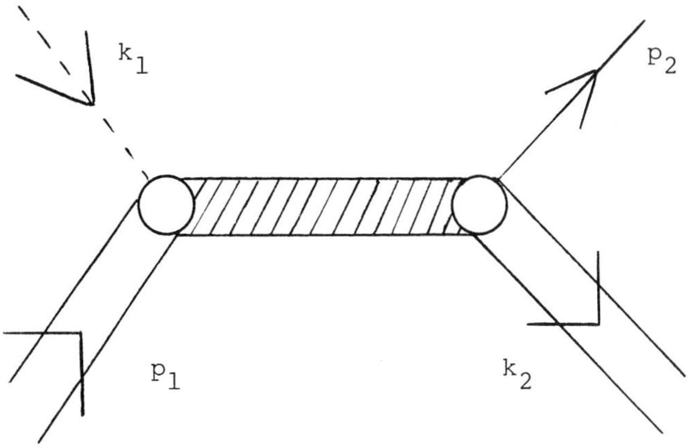

(a)

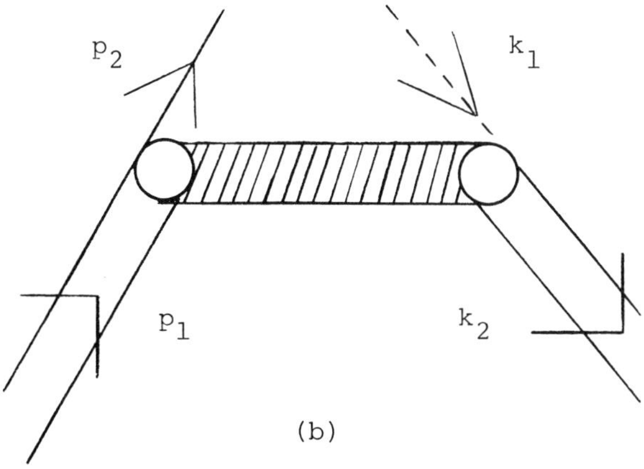

(b)

Figure 1: Diagrams contributing to $\pi^- p$ elastic scattering. Dashed line is external π^- coupling to $\partial_\mu A^\mu$; solid line is external proton coupling to the baryonic current $j^{\mu\alpha}$. The shaded line denotes a propagating sum of normal modes.

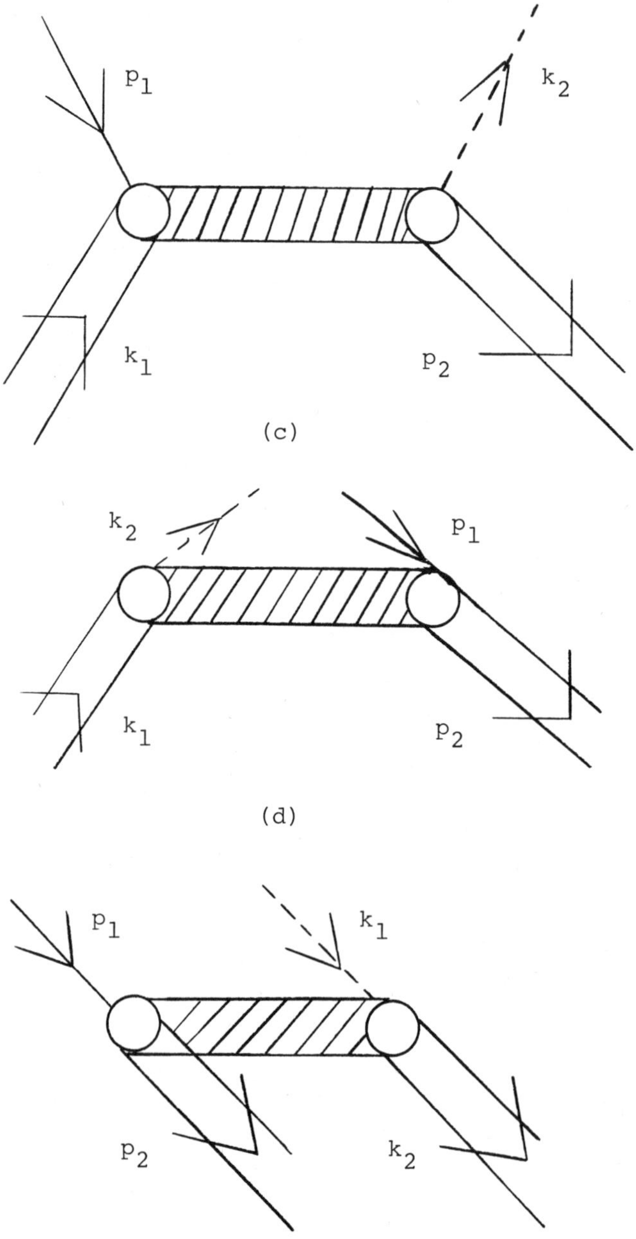

Figure 1 (contd)

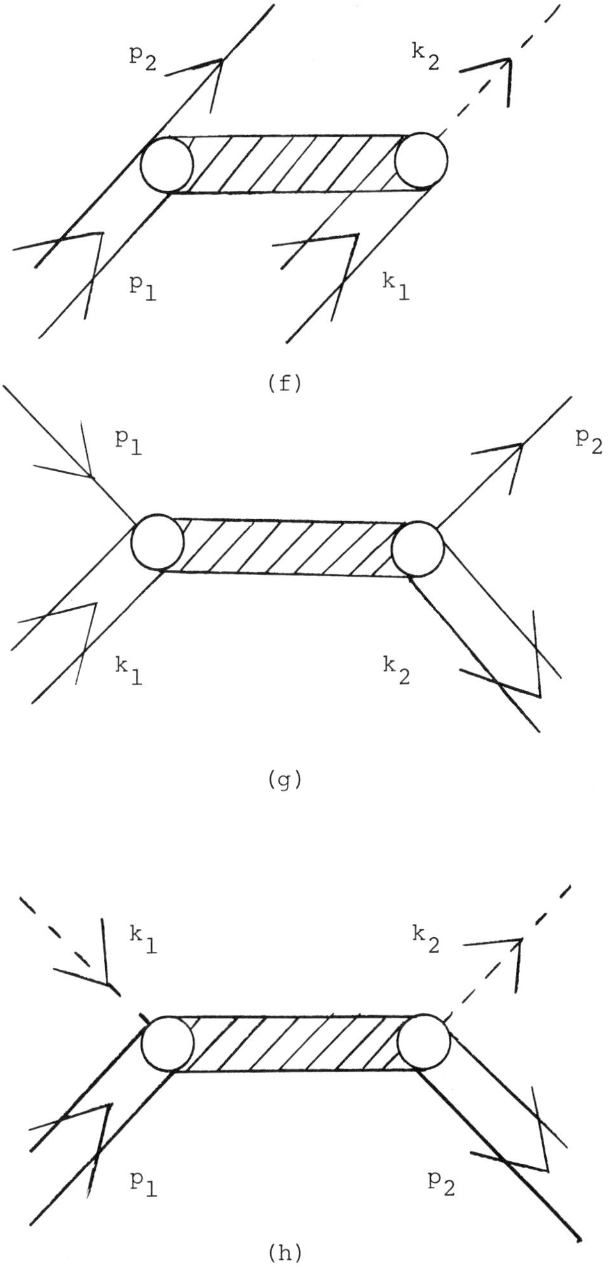

Figure 1 (contd)

REFERENCES

1. A. Chodos and F. Cooper, Phys. Rev. D3, 2461 (1971) (see this paper for earlier references).
2. F. Cooper and A. Chodos, Phys. Rev. D4, (to be published).
3. F. Cooper and A. Chodos, "Gauge Field Theory of Particles II - Fermions", (submitted to Phys. Rev.).
4. A. Chodos and R. Haymaker, Phys. Rev. D2, 793 (1970).
5. M. Scadron, Phys. Rev. 165, 1640 (1968).

LIST OF PARTICIPANTS

Paul R. Auvil
Department of Physics
Northwestern University

M. A. B. Beg
Department of Physics
Rockefeller University

Gregory Breit
Department of Physics
State University of New York
Buffalo

Laurie M. Brown
Department of Physics
Northwestern University

Arthur A. Broyles
Department of Physics and Astronomy
University of Florida

Richard H. Capps
Department of Physics
Purdue University

Anton Z. Capri
Department of Physics
University of Alberta, Canada

Peter Carruthers
Laboratory of Nuclear Studies
Cornell University

Ngee-Pong Chang
Department of Physics
City University of New York

Norman H. Christ
Department of Physics
Columbia University

David B. Cline
Department of Physics
University of Wisconsin

Fred Cooper
Belfer Graduate School of Science
Yeshiva University

Horace Crater
Department of Physics
Vanderbilt University

S. R. Deans
Department of Physics
University of South Florida

Robert S. Dickman
Nuclear Physics Division
Air Force Office of Scientific Research

P. A. M. Dirac
Department of Physics
Florida State University

Loyal Durand, III
Department of Physics
Florida State University

H. P. Durr
Max-Planck-Institut fur Physik und Astrophysik
Munchen

Karl-Erik Eriksson
Institute of Theoretical Physics
Goteborg, Sweden

Frank L. Feinberg
Laboratory for Nuclear Science
Massachusetts Institute of Technology

Bernard T. Feld
Laboratory for Nuclear Science
Massachusetts Institute of Technology

George B. Field
Department of Astronomy
University of California

Peter Freund
Enrico Fermi Institute
University of Chicago

Austin M. Gleeson
Department of Physics
University of Texas at Austin

Marvin L. Goldberger
Department of Physics
Princeton University

O. W. Greenberg
Department of Physics and Astronomy
University of Maryland

David Gross
Department of Physics
Princeton University

Franz L. Gross
Department of Physics
College of William and Mary
Virginia

Morton Hamermesh
School of Physics and
Astronomy
University of Minnesota

Kerson Huang
Department of Physics
Massachusetts Institute of
Technology

C. J. Isham
Department of Physics
Imperial College of Science
and Technology
London

Geoffrey Iverson
Center for Theoretical Studies
University of Miami

Arthur M. Jaffe
Department of Physics
Harvard University

Hans A. Kastrup
Theoretische Physik
University of Munich

Ralph E. Kelley
Nuclear Physics Division
Nuclear Physics Division
Air Force Office of
Scientific Research

Toichiro Kinoshita
Laboratory of Nuclear Studies
Cornell University

Hans Kleinpoppen
Department of Physics
University of Stirling
Scotland

T. K. Kuo
Department of Physics
Purdue University

Behram Kursunoglu
Center for Theoretical Studies
University of Miami

Chun-Chian Lu
Center for Theoretical Studies
University of Miami

Rabi Majumdar
School of Mathematics
University of Dublin

Alfred K. Mann
National Accelerator Laboratory
Batavia

V. S. Mathur
Department of Physics and
Astronomy
University of Rochester

Sydney Meshkov
U. S. Department of Commerce
National Bureau of Standards

Yuval Ne'eman
Department of Physics and
Astronomy
Tel-Aviv University

Jan S. Nilsson
Institute of Theoretical Physics
Goteborg, Sweden

Reinhard Oehme
Enrico Fermi Institute
University of Chicago

Susumu Okubo
Department of Physics and
Astronomy
University of Rochester

Sadao Oneda
Department of Physics and
Astronomy
University ot Maryland

Lars Onsager
Department of Chemistry
Yale University

Claudio A. Orzalesi
Department of Physics
New York University

Konrad Osterwalder
Department of Physics
Harvard University

Heinz R. Pagels
Department of Physics
Rockefeller University

Michael Parkinson
Department of Physics and Astronomy
University of Florida

P. J. Peebles
Department of Physics
Princeton University

Arnold Perlmutter
Center for Theoretical Studies
University of Miami

Giuliano Preparata
Istituto Di Fisica
Universita di Romi, Italy

Pierre Ramond
Department of Physics
Yale University

Rudolf Rodenberg
Physikalisches Institut
Technische Hochschule Aachen
Germany

Fritz Rohrlich
Department of Physics
Syracuse University

Gerald Rosen
Department of Physics
Drexel University

Lon Rosen
Department of Physics
Princeton University

Howard J. Schnitzer
Department of Physics
Brandeis University

Robert Schrader
Department of Physics
Harvard University

Barry Simon
Department of Mathematics
Princeton University

Charles Sommerfield
Department of Physics
Yale University

E. A. Spiegel
Department of Astronomy
Columbia University

E. C. G. Sudarshan
Department of Physics
University of Texas at Austin

Edward Teller
Lawrence Radiation Laboratory
University of California
Berkeley

Francis J. Testa
Department of Physics
Drexel University

Dietrick E. Thomsen
Science News
Washington, D. C.

Ivan Todorov
University of Sofia
Sofia, Bulgaria

Alar Toomre
Department of Mathematics
Massachusetts Institute of Technology

Hiroomi Umezawa
Department of Physics
University of Wisconsin

Kameshwar Wali
Department of Physics
Syracuse University

Steven Weinberg
Department of Physics
Massachusetts Institute of Technology

Arthur Wightman
Department of Physics
Princeton University

Gaurang B. Yodh
Department of Physics and Astronomy
University of Maryland

Fredrik Zachariasen
Charles C. Lauritsen Laboratory
California Institute of Technology

Harold I. Zimmerman
Center for Theoretical Studies
University of Miami

Daniel Zwanziger
Department of Physics
New York University